普通高等教育“十一五”国家级规划教
全国高等医药院校药学类实验双语教

物理学实验与指导

（第二版）

Experiment and Guide for General Physics

（Second Edition）

主　编　陈　曙
副主编　周春红
编　者（以姓氏笔画为序）
马　骄（沈阳药科大学）
王众虎（中国药科大学）
尹　媛（中国药科大学）
陈　曙（中国药科大学）
周春红（中国药科大学）
姚峥嵘（中国药科大学）
黄耀庭（广东药学院）
蒋岩玲（中国药科大学）
韩东强（中国药科大学）

中国医药科技出版社

内容提要

本书共收载20个基本物理实验，详细介绍每一个实验的实验目的、实验器材、实验原理、实验内容、数据处理等，附有思考题，还有实验指导。既注重实验本身的训练，同时注意对学生实验规范的训练。适合药学及相关专业本科一年级学生使用。

图书在版编目（CIP）数据

物理学实验与指导/陈曙主编．—2版．—北京：中国医药科技出版社，2009.8
普通高等教育“十一五”国家级规划教材．全国高等医药院校学类实验双语教材
ISBN 978－7－5067－4285－6

Ⅰ．物…　Ⅱ．陈…　Ⅲ．物理学－实验－双语教学－医学院校－教材　Ⅳ．04－33

中国版本图书馆CIP数据核字（2009）第103124号

美术编辑　陈君杞
版式设计　郭小平
出版　中国医药科技出版社
地址　北京市海淀区文慧园北路甲22号
邮编　100082
电话　发行：010－62227427　邮购：010－62236938
网址　www.cmstp.com
规格　787×1092mm $^1/_{16}$
印张　21
字数　429千字
初版　2005年3月第1版
版次　2009年8月第2版
印次　2018年7月第2版第11次印刷
印刷　三河市航远印刷有限公司
经销　全国各地新华书店
书号　ISBN 978－7－5067－4285－6
定价　38.00元

全国高等医药院校药学类规划教材常务编委会

出版说明

全国高等医药院校药学类专业规划教材是目前国内体系最完整、专业覆盖最全面、作者队伍最权威的药学类教材。随着我国药学教育事业的快速发展，药学及相关专业办学规模和水平的不断扩大和提高，课程设置的不断更新，对药学类教材的质量提出了更高的要求。

全国高等医药院校药学类规划教材编写委员会在调查和总结上轮药学类规划教材质量和使用情况的基础上，经过审议和规划，组织中国药科大学、沈阳药科大学、广东药学院、北京大学药学院、复旦大学药学院、四川大学华西药学院、北京中医药大学、西安交通大学药学院、山东大学药学院、山西医科大学药学院、第二军医大学药学院、山东中医药大学、上海中医药大学和江西中医学院等数十所院校的教师共同进行药学类第三轮规划教材的编写修订工作。

药学类第三轮规划教材的编写修订，坚持紧扣药学类专业本科教育培养目标，参考执业药师资格准入标准，强调药学特色鲜明，体现现代医药科技水平，进一步提高教材水平和质量。同时，针对学生自学、复习、考试等需要，紧扣主干教材内容，新编了相应的学习指导与习题集等配套教材。

本套教材由中国医药科技出版社出版，供全国高等医药院校药学类及相关专业使用。其中包括理论课教材 82 种，实验课教材 38 种，配套教材 10 种，其中有 45 种入选普通高等教育“十一五”国家级规划教材。

全国高等医药院校药学类规划教材

编写委员会

2009 年 8 月 1 日

序

实验教学是高等药学院校最基本的教学形式之一，对培养学生科学的思维与方法、创新意识与能力，全面推进素质教育有着重要的作用。飞速发展的科学技术，已成为主导社会进步的重要因素。高等药学院校必须不断更新教学内容，以学科发展的前沿知识充实实验课程内容。

近年来，中国药科大学坚持以研究促教改，通过承担教育部“世行贷款——21世纪初高等教育教学改革项目”及立项校内教改课题等多种方式，调动了广大教师投身教学改革的积极性，将转变教师的教育思想观念与教学内容、教学方法的改革紧密结合起来，取得了实效。此次推出的国家“十一五”规划教材——药学专业双语实验教学系列，是广大教师长期钻研实验课程教学体系，改革教学内容，实现教育创新的重要成果。他们站在21世纪教育、科技和社会发展趋势的高度，对药学专业实验课程的教学内容进行了“精选”、“整合”和“创新”，强调对学生的动手能力、创新思维、科学素养等综合素质的全面培养。这套教材具有以下的特点：

1．教材将各学科的实验内容进行了广泛的“精选”，既体现了高等药学教育“面向世界、面向未来、面向现代化”，也考虑到我国药学教育的现状与实际；既体现了各门实验课程自身的独立性、系统性和科学性，又充分考虑到各门实验课程之间的联系与衔接，有助于学生在教学大纲规定的实验教学学时内掌握基本操作技术，提高动手能力，养成严谨、求实、创新的科学态度。

2．教材中新增的综合性、设计性实验有利于学生全面了解和综合掌握本门实验课程的教学内容。这一举措既满足了学生个性发展的需要，更注重培养学生分析问题、解决问题的能力和创新意识。

3．教材中适当安排一些反映药学学科发展前沿的实验，有利于学生在掌握实验基本技术的同时，对药学学科的新进展、新技术有所了解，激发他们学习药学知识与相关学科的兴趣。

4．教材以实践教学为突破口，采用双语体系编写，为实验课程改革构建数字化、信息化和外语教学的平台，有利于提高学生的科技英语水平。通过我校多年的药学系列实验课程双语教学实践，证明学生完全能够接受此套教材的教学。

国家十一五规划教材——药学专业双语实验教学系列教材的陆续出版，必将对推动我国高等药学教育的健康发展，产生积极而深远的影响。由于采用双语体系编写药学教学实验丛书尚属首次，缺乏经验，在内容选择及编写方法上的不妥之处，在所难免。欢迎从事药学教育的同行们批评赐教。

吴晓明

（中国药科大学校长、博士、教授、博士生导师）

Preface

Experimental teaching is one of the most fundamental teaching means in pharmaceutical colleges, playing an important role in training scientific thoughts and methods, creative consciousness and ability of the students as well as in promoting quality-oriented education in all-round way. Fast-advancing science and technology has come to be an important factor in dominating social progress. Teaching materials must be updated continually in pharmaceutical colleges, especially enriching the materials of experimental courses with the most advanced knowledge in the subject.

In recent years, China Pharmaceutical University have been stressing the promotion of teaching reform on the basis of research, succeeding in stimulating teachers' enthusiasm for teaching reform by various means such as undertaking the project of teaching reform in higher education at the beginning of 21st century sponsored financially by World Bank and entrusted by the Ministry of Education as well as approving and ratifying internal programs on teaching reform. Meanwhile, it yields fruits to integrate the transforming of teachers' educational ideology into the reform of teaching materials and methods. This series of textbook of national "11th five" planning-bilingual pharmaceutical experimental teaching series, is an important achievement made through studying ueaching system of experimental courses for long, reforming teaching materials and carrying out educational innovation of all the teachers concerned.

Meeting the new demands for education, science and technology and social growth, they select, integrate and innovate the teaching materials of pharmaceutical experimental courses, stressing the overall cultivation of comprehensive qualities, including experimental ability, creative thought and scientific attainments. This set of textbook possesses the following features:

1. These textbooks make an extensive "selection" of the experimental materials of each subject, reflecting the goal of facing the world, facing the future and facing the modernization in higher pharmaceutical education, and taking into account the status quota and reality of our pharmaceutical education; meanwhile embodying the individuality, systematicness and scientificalness of each experimental courses, which helps the students to grasp basic techniques of operation within the class hours of experimental teaching pre-

scribed by teaching syllabus and to improve their experimental ability and finally to cultivate a scientific approach of precision, practicality and creation.

2. The comprehensive designing experiments newly supplemented in the textbooks help the students to learn totally and grasp comprehensively the teaching materials of the experimental courses, which not only meets the students' needs for individual development but also trains their ability to analyze and solve problems and cultivates their creative consciousness.

3. Some experiments representing the latest development in pharmacy are properly included in the textbooks, which helps the students to learn about new advance and technology in pharmacy and to further arouse their interests in studying pharmacy and relevant subjects while grasping some basic techniques of experiment.

4. The textbooks take experimental teaching as starting point and are compiled in a system of bilingualism and aim to set up a platform of digitalization, information and foreign language teaching for the purpose of reforming experimental courses, which serves to enhance the students' level of technological English. It has been proved that the students have no difficulty being adapted to the teaching of this set of textbook through many years of bilingual teaching practice carried out in a series of pharmaceutical experimental courses of our university.

The successive publishing of the series of textbooks used for bilingual pharmaceutical experimental teaching-the national "11th-five" planning textbooks, will surely produce good and far-reaching influence in promoting the sound development of higher pharmaceutical education of our country. Since it is the first time that we have compiled this series of textbook of pharmaceutical teaching experiment in a bilingual system, we lack experience and thus some defects in choice of materials and way of compilation are inevitable. Experts engaged in pharmaceutical education are welcome to give any criticisms and advice.

Wu Xiaoming

Ph. D, prof., and supervisor of doctoral candidates
President of China Pharmaceutical University

第二版前言

本书自2005年3月第一版出版以来已经4年多了。这些年以来编者一直比较注意本书在使用中的反响，也很关注国内同类高校的物理理论和实验教学的状况，第二版依然优先考虑实用性和适用性。这次编写工作主要着重于以下几个方面。

1. 整合了一些实验，删除了两个实验，新增了两个实验，从原来22个实验调整为20个实验。作为实验教材不是手册性的书籍，不能讲究大而全。教师好教学生好用应是教材的基本出发点。注重基本知识学习和基本技能训练才是本课程的主要任务。

2. 中英文完全做到了一一对应。学生完全可以按照英文部分完成实验。几年来的教学实践表明双语物理实验对大学生在用英语准确地获取科技信息方面、操作指导和书写表达方面都是有帮助的。双语实验教学对教和学两个方面都是有益的。

3. 每个实验前均做了一些简要的介绍，修改了一些错误和不恰当的地方。

参加本书编写的有陈曙、周春红、王众虎、蒋岩玲、姚峥嵘、尹媛、韩东强、马骄和黄耀庭。他们都是长期参加物理实验教学的老师，有着较为丰富的教学经验。马军老师为本书的用图做了许多工作，施雪娟老师对本书的编写也给予了许多帮助。

这些年来使用本书的学生提出了许多意见和建议，特此表示感谢。美国加州大学伯克利分校的刘冰教授、中国药科大学的国际留学生等都对本书的英文编写给予过帮助。还要感谢中国药科大学教务处的领导为本书做的组织工作，他们一直鼓励双语教材的编写。

本书一定还有不妥或错误的地方，有待将来在使用中改过和改进。

陈　曙

2009年4月

Preface to the Second Edition

It has been more than 4 years from the publication of the 1st edition of this text book. During these years we have been paying a lot of attention on the response of this book and the situation of teaching and learning of physics course in the similar colleges in China. A much more consideration on the rewriting work of this 2nd edition is paid on practicability and applicability. We emphasize main work as following.

1. Some of the experiments have been combined. Two experiments have been omitted and two new have been added in. Totally there are 20 experiments in this 2nd edition other than the 22 in the 1st edition. As a lab text book it is not a handbook or a dictionary. The essential basis for a text of fundamental course is its practical purpose for students as well as teachers. The main role of physics lab course is to learn fundamental knowledge and basic abilities.

2. The English version is completely corresponding to its Chinese text. Students can find its English version working as Chinese. These years the practice of bilingual teaching is showed helpful for students not only obtaining accurate scientific information but also correct operation as well as written expression.

3. A brief introduction is given before every experiment. Some of the errors and unsuitable expression have been modified.

The writers of this edition are Chen Shu, Zhou Chunhong, Wang Zhonghu, Jiang Yanling, Yao Zhengrong, Han Dongqiang, Ma Jiao and Huang Yaoting. They are all working as lab instructors for many years and have good understanding on the physics lab course. Ma Jun has done some of the drawing work and Shi Xuejuan has done a lot of assistant work for this edition.

I am delight to acknowledge the readers of the 1st edition for their practical and creative suggestions. Professor Liu Bing of University of California at Berkeley and the international students of China Pharmaceutical University are all helpful to the English version. I wish to express my appreciation to the officials of the Teaching Administration Bureau of China Pharmaceutical University, they are all enthusiasm for the publication of bilingual text books.

There could be incorrectness in this book which should be corrected in the further editions.

Chen Shu
April, 2009

目　　录

Contents

绪　　论

物理实验在物理学的发展过程中一直起着重要的和直接的推动作用。同时，物理实验也为其他学科、技术的发展提供了有力的方法和工具。作为一名理科学生，通过对物理实验课程的系统学习，必然会给后期其他课程的学习打下良好的基础。

一、教学要求

物理实验是一门独立设置的课程。开设这门课程的主要目的是培养学生具有良好的实验素养、基本实验技能、理论联系实际和独立工作的能力。以下是本课程的一些基本教学要求：

1. 在物理实验的基本知识、基本方法和实验技能方面受到较为系统的训练。

2. 学习如何认识实验思想、实验方案，正确使用基本仪器，掌握基本测量技术，正确分析实验数据，分析实验结果。

3. 培养严肃认真、实事求是的科学态度和工作作风。

二、教学环节

物理实验是学生在教师指导下独立进行的一种实践活动。学生是实验的主要完成者。一般说来，每个实验都有三个教学环节：预习、进行实验和完成实验报告。

1. 预习实验

仔细阅读实验教材，了解实验原理、方法和步骤。大致了解实验仪器的使用方法、实验中可能出现的一些现象。在此基础上，要写出实验预习报告。实验预习报告要包括实验目的、实验原理简介，实验步骤，实验用示意图，实验数据记录表格等项目。

2. 进行实验

实验时应遵守实验室规章制度。在教师指导下正确使用仪器。实验进行时，应合理操作，仔细观察，认真思考。在预习报告本上正确记录实验现象、实验数据。不要用铅笔记录数据。正确的删除或改正实验数据的方法是，可以用笔划去，再注明原因。在电磁学实验中，连接好实验线路后，应自己先检查一遍，再由指导老师检查，如正确无误，方可接通电源继续实验。

3. 实验报告

实验报告是实验者对实验情况的书面报告形式，一般应做到字迹清晰，文理通顺，逻辑合理。报告中的示意图要按照具体要求，用直尺或曲线板辅助画出。对实验现象、结果的分析讨论要有理并符合实际情况。

一般说来，实验报告应包括以下几个部分。

（1）实验名称、实验者姓名、实验日期。

（2）实验目的。

（3）实验原理。要求叙述简要。

（4）实验仪器器材。要记录名称、型号、规格、编号等。

（5）实验步骤及数据。记录数据一定要用表格形式，数据计算过程要简洁明了。

（6）实验结果及讨论。分析有关实验现象及实验结果，提出对实验中存在的一些问题的解决方案或建议，回答思考题，等。

对有特殊要求的实验报告，按具体要求完成。

最后，需要特别提醒的是安全。一旦发现异常现象，要立即切断电源、火源、气源，并报告指导老师，有必要时要拨打“119”火警电话。

Introduction

Physics experiment is an important part of physics science; it also plays a determined and propulsive role of the physics science. There are a lot of physics experimental methods and tools, which have been widely applied in the areas other than the physics science. As a science student, it is really helpful to be trained systematically in this lab course.

Learning Requirement

1. Learn to know the basic knowledge and methods of physics as well as experimental skills.

2. Learn to get a clear opinion of every experiment plan and technique. Operate the apparatus correctly. Learn to study and analyze the experimental data to get proper conclusion.

3. It is a right manner for a science student to be practical and realistic.

Lab Process

General speaking, there are three steps for one experiment:

1. Preparation

Read the text or other reference to understand the experiment principle and main method. Get a clear idea about the apparatus and experiment phenomenon. Write a preparation report which should include such items as objectives, principle, apparatus, steps, data tables, and so on.

2. Experiment

It is important to obey the lab bylaws. Your instructors may give you some necessary experimental instruction and explaining. You should try to be patient, careful and considerate. Record the lab phenomenon and data on you own book with a pen. Some electric circuits you joined should be checked by you instructor before switching on electric power.

3. Experiment Report

The paper report of your experiment should be clear and methodic. Usually an experiment report should include the following items:

(1) The experiment title, your name and date.

(2) Experiment objectives.

(3) Principle (briefly introduced) .

(4) Lab apparatus, with name, type, specification, serial number, etc.

(5) Experiment process and data. All the data should be recorded into suitable tables, and make the data processing being concise.

(6) Experiment results and discussing. The results and some phenomenon maybe discussed according to your experiment. If you have any suggestion you may put forward here. Usually there are some questions you should answer.

For any other special requirement, your report should be written in the form as your text claimed.

Please keep safety always in your mind. If there is anything abnormal, first you should cut off the electric power, gas or water supply, and then report you lab instructor. The emergency telephone number in our city is "119" .

实验一　关于测量的基本理论

物理实验的目的是研究观察各种物理规律，其中对物理量的定量测量是很重要的一个环节。通过对测量得到的数据进行处理运算，才能得到实验结果。但是由于实验条件、实验仪器、实验理论等因素限制，测量结果和实验结果都有一定的误差。这就需要了解有关测量误差、数据处理和误差分析的一些基本知识。

误差分析理论对实验工作在改进实验措施、仪器精度选择和确定最有利的实验条件等许多方面都具有指导作用。

【测量】

物理实验不仅要定性地观察各种物理现象，更重要的是要找出有关物理量之间的定量关系，因此就需要进行测量。测量在实验中占有重要地位。所谓测量一个量，就是用某种特定的方法将这个量和一个选为标准的、大小一定的同类量进行比较，确定所要测量的量是标准量的多少倍。选为标准的同类量就是一个单位，倍数就是测量数值。测量的结果一定是由测量数值和单位一起构成的。我国以国际单位制（SI）为国家法定计量单位，即以米（m）、千克（kg）、秒（s）、安培（A）、开尔文（K）、摩尔（mol）、坎德拉（cd）作为基本单位，其他量都由以上七个基本单位导出，称为国际单位的导出单位。

测量可分为直接测量和间接测量。在一般情况下，总是先测出与待测量有已知关系的诸量，然后利用该已知关系计算出待测量。例如，我们可以根据一个圆柱体的体积和质量来确定它的密度：

$$\rho = \frac{m}{V} = \frac{4m}{\pi D^2 L}$$

式中 m 为该物体的质量，D 为物体的直径，L 为物体的长度。m、D、L 都是可以直接测量的量，把测量的结果代入上式，便可以计算出物体的密度 ρ。这种测量就是间接测量。

【测量误差的基本知识】

物理量在客观上有着确定的数值，称为真值。然而在实际测量时，由于测量仪器、测量方法、实验条件以及种种有关因素的局限，测量是不可能无限精确的，测量结果与真值之间总有一定的差异，也就是说总存在着测量误差。测量结果的误差大小反映我们接近客观真值的程度。

测量误差与实验的各个方面都有着密切的关系，如测量方法、仪器结构和其准确度、实验外界环境条件、实验者本人等都会影响实验结果。每个实验对其测量结果的误差或精确度总有一定的要求。不能认为和要求仪器越高级越好，环境条件（如恒温、恒湿度、无噪声等）越稳定越好，测量次数越多越好，等等，否则，容易造成人力和物力的浪费。正确的做法是应该根据实验具体要求进行合理的设计，选择合适的实验仪器和合理的实验方案。重要的是，在给定的实验条件下，以最小的代价来取得最好的结果。

1. 测量的误差

一个待测量的测量值 x 与真实值 x_0 之差就称为误差 Δx：

$$\Delta x = x - x_0$$

误差存在于一切测量之中，而且贯穿测量过程的始终。因此，只有通过正确的误差分析，知道实验中哪些量对实验结果的影响大，哪些量对实验结果的影响小，从而努力测准那些对结果影响大的关键量，而不必化大力气在那些测不太准、且对实验结果影响很小的量上去。

2. 误差的性质和来源

既然误差总是存在，那么就应该对它的性质和来源有正确的认识，才能最大限度地减小测量误差。测量误差按其来源和性质，可分为系统误差、随机误差和过失误差三大类。

(1) 系统误差　系统误差的特点是有规律的：测量结果都大于真值，或者都小于真值，也就是说它总是使测量结果向一个方向偏离。在测量条件发生改变时，系统误差也按一定的规律在变化。

①由仪器引进的系统误差　如测量仪器的不完善、不够精密、安装调整不妥等。例如刻度不准、零点不对、砝码未经校准、天平臂不等长、应该水平放置的仪器没有放水平等都属于这一类误差。

②理论（方法）误差　由于某些理论公式本身的近似性，或实验条件不能达到理论公式所规定的要求，或测量方法本身所带来的误差。如测量电压时未考虑电压表内阻对电路的影响等。

③个人误差　由于实验者本人生理或心理特点造成的，使实验结果产生的偏向性误差。如有些人因习惯于侧坐斜视读数、眼睛辨色能力较差等，可使测量值偏大或偏小。

系统误差总是使测量结果偏向一边，或者偏大、或者偏小，因此多次测量求平均值并不能消除系统误差。而是要根据各种不同情况，找到产生系统误差的原因，采取一定的方法去尽量消除它的影响，或者对测量结果进行修正。

(2) 随机误差　在测量时，即使消除了系统误差（实际上不可能也不必要绝对排除），在相同条件下进行多次重复测量同一物理量时，发现各测量值之间也有差异，由此而产生的误差（即各测量结果与真值之间的差异）的绝对值和正负号是不确定的，这样的误差就是随机误差，也称为偶然误差。

随机误差的存在，表现为每次测量值的偏大或偏小是不一定的，但它服从一定的统计规律。统计理论表明：测量结果与真值偏差大的测量值出现的概率较小，偏差小的测量值出现的概率大；正方向误差和负方向误差出现的概率相等；并且绝对值很大的误差出现的概率趋近于零。这就是我们在实验中采取多次重复测量的依据。

随机误差主要来源于人们视觉、听觉和触觉等感觉能力的限制，以及实验环境偶然因素的干扰。例如温度、湿度、电源电压的起伏，气流波动以及振动等因素的影响。因此，用实验方法消除测量中的随机误差是不可能的。但是，用概率统计方法处理可以减小随机误差对最后结果的影响，并且可以估计误差的大小。

(3) 过失误差（错误） 在测量中还可能出现错误，如读数错误、记录错误、操作错误、估算错误等。过失误差应尽量注意避免。克服错误的方法，除端正工作态度，严格工作方法外，可用和另一次测量结果相比较的办法发现纠正，或者运用逻辑合理性判断的方法来发现并剔除异常数据。

3. 精密度、准确度和精确度

精密度、准确度和精确度可以用来评价测量结果的好坏。但这三个词的涵义是不同的，使用时要注意区别。

测量的精密度高，是指测量数据比较集中，随机误差较小。

测量的准确度高，是指数据的平均值偏离真值较小，测量结果的系统误差较小。

测量的精确度高，是指数据比较集中在真值附近，即测量的系统误差和随机误差都较小。

图 1－1 是对某量 x 的几次测量值在一维数轴上的图示，其中 x_0 是其真值。

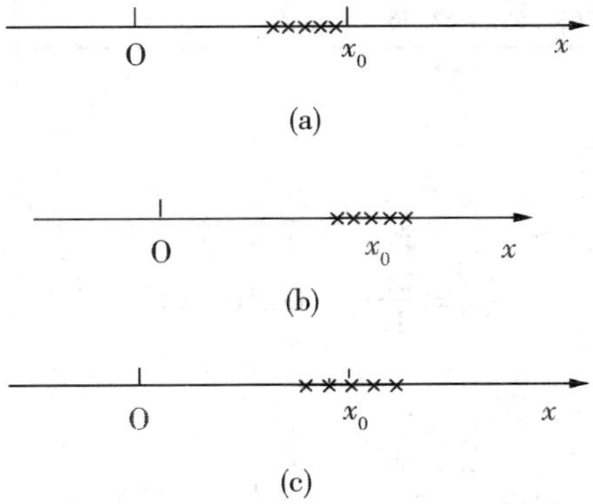

图 1－1 对 x 的几次测量值在一维数轴上的表示

(a) 测量值比较集中，但偏离真值，表示精密度高而准确度较差；

(b) 测量值比较分散，但平均值接近真值，表示准确度高而精密度较差；

(c) 精密度和准确度均较好，即精确度较高

值得提一下的是，某些仪器上注明有“精密度”是多少的字样，是指该仪器能准确分辨的最小的被测量值，不可与前述意义相混淆。

4. 直接测量时随机误差的估计

(1) 以算术平均值表示测量结果 设在相同条件下对一个物理量进行了 n 次重复测量，测量值分别为 x_1、x_2、$x_3 \cdots x_n$，如前所述，真值 x_0 是不能确切知道的，但作为对真值的最好近似，我们可以用这些测量值的算术平均值来表示：

$$x_0 = \bar{x} = \frac{1}{n}\ (x_1 + x_2 + \cdots + x_n) = \frac{\sum_{i=1}^{n} x_i}{n} \tag{1}$$

上式中 x_i 是第 i 次测量值。

(2) 平均绝对误差 算术平均值是接近真值的最佳值，但它仍非真值。对它与真值之

间的差别的一种估计方法是平均绝对误差$\overline{\Delta x}$。平均绝对误差$\overline{\Delta x}$的定义为：

$$\overline{\Delta x}=\frac{(|x_1-\bar{x}|+|x_2-\bar{x}|+\cdots+|x_n-\bar{x}|)}{n}=\frac{\sum_{i=1}^{n}|\Delta x_i|}{n} \qquad (2)$$

上式中 $\Delta x_i=x_i-\bar{x}$ 称为第 i 次测量的绝对误差。

统计理论可以证明，多次重复测量中的任一测量值的绝对误差 Δx_i 落在（$-\overline{\Delta x}$，$\overline{\Delta x}$）区间内的概率为57.5%，平均值与真值之差落在此区间内的概率就更大了。所以可以认为真值就在（$\bar{x}-\overline{\Delta x}$，$\bar{x}+\overline{\Delta x}$）区间内，但并不排除多次测量中有部分测量值落在这一区间之外。最后，按以上方法处理数据，得到相同条件下多次测量的算术平均值 $\bar{x}$ 和平均绝对误差$\overline{\Delta x}$以后，应将结果表示成 $x=(\bar{x}\pm\overline{\Delta x})$（单位）的形式。

例1　对某一长度在相同条件下作了10次测量，结果见表1－1。

表1－1　对某一长度的10次测量数据记录表

i	1	2	3	4	5	6	7	8	9	10
x_i（cm）	63.57	63.58	63.55	63.56	63.56	63.59	63.55	63.54	65.57	63.57

求测量结果。

解：先求平均值：

$$\bar{x}=\frac{\sum_{i=1}^{10}x_i}{10}=63.564\text{cm}$$

再计算平均绝对误差：

$$\overline{\Delta x}=\frac{\sum_{i=1}^{10}|\Delta x_i|}{10}=0.012\approx0.01\text{cm}$$

由于随机误差本身是一个估计值，所以结果一般只取一位有效数字。

处理结果表示为：

$$x=(63.56\pm0.01)\ \text{cm}$$

以上结果表示该长度 x 的真值在63.55～63.57cm之间。

（3）标准偏差和极限误差　对随机误差的另一种常见的估算方法是用标准偏差。对于在相同条件下重复 n 次测量中的某一次测量结果的标准偏差 σ_x 的定义为：

$$\sigma_x=\sqrt{\frac{\sum_{i=1}^{n}(x_i-\bar{x})^2}{n-1}}=\sqrt{\frac{\sum_{i=1}^{n}(\Delta x_i)^2}{n-1}} \qquad (3)$$

即 σ_x 是对各次测量误差平方求和后，除以（$n-1$）再计算平方根的，因此标准偏差又称为方均根误差。根据概率统计理论，标准偏差 σ_x 的物理意义是：在相同条件下重复 n 次测量中的任一次测量值的误差落在（$-\sigma_x$，σ_x）区间的概率为68.3%。标准偏差 σ_x 是科学文献中广泛应用的评价数据的一种依据。

同样，按统计理论有，对于在相同条件下重复 n 次测量结果的平均值 $\bar{x}$ 的标准偏差 $\sigma_{\bar{x}}$

$$\sigma_{\bar{x}}=\frac{\sigma}{\sqrt{n}}=\sqrt{\frac{\sum_{i=1}^{n}(x_i-\bar{x})^2}{n(n-1)}}=\sqrt{\frac{\sum_{i=1}^{n}(\Delta x_i)^2}{n(n-1)}} \tag{4}$$

上式表明，平均值的标准偏差 $\sigma_{\bar{x}}$ 是 n 次重复测量中任意一次测量值标准偏差的 $\frac{1}{\sqrt{n}}$ 倍。$\sigma_{\bar{x}}$ 的值要小于 σ_x，这一结论是显而易见的。因为算术平均值是测量结果的最佳值，它比任何一次测量值 x_i 更接近真值。

式（3）和式（4）都是用来估算随机误差的，理论上都要求测量次数相当多。但在物理教学实验中，往往重复测量的次数不可能很多，所以用这两个式子估算出来的随机误差带有相当程度的近似性。另外，在测量次数较少（$n<10$）时，$\sigma_{\bar{x}}$ 随测量次数 n 的增加而明显的减小，以后，随着测量次数 n 的继续增加，$\sigma_{\bar{x}}$ 的减小愈来愈不明显而逐渐趋近于恒定值。由此可见，一味地增加测量次数，其价值并不大。根据实际情况，一般取重复测量次数为 5 ~ 10 次即可。

有时会遇到测量对象本身不均匀的情况。例如，测量一个小钢珠的直径，钢珠各处的直径略有微小差异，即直径各处的真值并不完全一致，因此对所测得的各处测量值取平均值只是反映了钢珠直径的平均大小。多次测量不可能减小钢珠直径的不均匀性，所以计算平均值的标准偏差实属没有必要。而标准偏差则反映了钢珠直径的不均匀程度。

例 2　根据例 1 的数据，计算 σ_x。

解：由式（3）可得

$$\sigma_x=\sqrt{\frac{\sum_{i=1}^{n}(x_i-\bar{x})^2}{n-1}}=\sqrt{\frac{\sum_{i=1}^{10}(x_i-63.564)^2}{10-1}}=\sqrt{\frac{2.040\times10^{-6}}{9}}=0.015\approx0.02\text{cm}$$

所以结果可以表示为：

$$x=(63.56\pm0.02)\ \text{cm}$$

比较例 1 处理的结果 $\overline{\Delta x}$ 和本例的处理结果 σ_x，可见 $\overline{\Delta x}<\sigma_x$；这一关系表示任一次测量值落在（$\bar{x}-\overline{\Delta x}$，$\bar{x}+\overline{\Delta x}$）区间的概率，比落在（$\bar{x}-\sigma_x$，$\bar{x}+\sigma_x$）区间的概率要小。实际上两者的关系为 $\sigma_x=1.25\overline{\Delta x}$。

统计理论还证明，相同条件下多次测量中的任一次测量值的误差落在（$-3\sigma_x$，$3\sigma_x$）区间内的概率为 99.7%。因此，一般将 $3\sigma_x$ 称为极限误差。如果发现某次测量值的误差超过了 $\pm3\sigma_x$，可以认为该次测量失败，应该将此数据剔除。通常，将 $3\sigma_x$ 看作正常的随机误差的极限。

在用计算器处理具体实验数据时，可以利用计算器上的统计功能计算标准偏差 σ_x。本课程中，对相同条件下所作的重复多次测量，一般只要求计算平均绝对误差 $\overline{\Delta x}$，在测量次数 $n>5$ 次时可以计算标准偏差 σ_x。通常没有必要计算平均值的标准偏差 $\sigma_{\bar{x}}$。

（4）仪器误差　上述估算随机误差的方法是在测量次数较多的情况下使用的。有些实验中对某些物理量只进行一次测量，这时估计误差要根据仪器分度值大小、测量的环境条件等具体情况考虑，要尽量符合实际。一次测量的误差，可以包括随机误差和系统误差。在本课程中，我们根据所用测量仪器上注明的仪器误差来表示单次测量的误差。如果没有注明，可取仪器最小分度值（有时以精密度表示）的一半作为仪器误差；有时也可以根据实际情况，采用仪器最小分度甚至更大的合理数值。

（5）相对误差　平均绝对误差$\overline{\Delta x}$、标准偏差 σ_x 等都是有单位的，其单位和被测量的单位相同，都是绝对误差，用 Δx 表示。Δx 的大小表示了真值有一定的概率出现在（$x-\Delta x$，$x+\Delta x$）区间内。误差还有一种表示方法，即相对误差 E，也叫百分误差，它是绝对误差 Δx 与平均值（或标准值）x 的比值，结果常用百分数表示，即

$$E=\frac{\Delta x}{x}\times 100\% \tag{5}$$

相对误差表示了绝对误差在测量中所占的比重，它也可以用来比较测量对象不同时测量的好坏程度。例如，有两个测量结果：$x_A=(100\pm1)$ cm，$x_B=(1000\pm1)$ cm，两个结果的绝对误差都一样，但由于相对误差 $E_A=\frac{1}{100}=1\%$，$E_B=\frac{1}{1000}=0.1\%$，可知第二个测量结果较好。

5. 间接测量结果的误差

直接测量值不可避免地有误差存在，因此将直接测量值代入公式经过运算得到的间接测量结果也必然有误差存在。计算间接测量结果误差的公式称为误差的传递公式。

设 x_A、x_B 为两个直接测量值，其绝对误差分别为 Δx_A 和 Δx_B。N 是由 x_A、x_B 求出的间接测量值，ΔN 为其误差。

（1）加、减运算

先考虑加法运算：$N=x_A+x_B$，因为误差是来源于参与运算的量，所以：

$$\begin{aligned}N\pm\Delta N&=(x_A\pm\Delta x_A)+(x_B\pm\Delta x_B)\\&=(x_A+x_B)\pm(\Delta x_A+\Delta x_B)\end{aligned}$$

最大可能的误差为：$\Delta x_A+\Delta x_B$。

所以对加法运算的绝对误差为：$\Delta N=\Delta x_A+\Delta x_B$。

其相对误差为：$E=\frac{\Delta N}{N}=\frac{\Delta x_A+\Delta x_B}{x_A+x_B}$。

同样，可以推导得减法运算 $N=x_A-x_B$，有：

$$\Delta N=\Delta x_A+\Delta x_B$$

$$E=\frac{\Delta N}{N}=\frac{\Delta x_A+\Delta x_B}{x_A-x_B}$$

因此，几个直接测量值相加、减时，其结果的绝对误差为各直接测量值的绝对误差之和。

（2）乘、除法运算

先考虑乘法运算：$N = x_A \cdot x_B$，

$$N \pm \Delta N = (x_A \pm \Delta x_A) \cdot (x_B \pm \Delta x_B)$$
$$= x_A \cdot x_B \pm x_A \cdot \Delta x_B \pm \Delta x_A \cdot x_B \pm \Delta x_A \cdot \Delta x_B$$

由于 Δx_A、Δx_B 与 x_A、x_B 相比，可视为小量，故 $\Delta x_A \cdot \Delta x_B$ 为二级小量，可以忽略不计，又考虑到最大可能的误差，因此有：

$$N \pm \Delta N = x_A \cdot x_B \pm x_A \cdot \Delta x_B \pm \Delta x_A \cdot x_B$$
$$= x_A \cdot x_B \pm (x_A \cdot \Delta x_B + x_B \cdot \Delta x_A)$$

所以对乘法运算的绝对误差可为：

$$\Delta N = x_A \cdot \Delta x_B + x_B \cdot \Delta x_A$$

其相对误差为：

$$E = \frac{\Delta N}{N} = \frac{(x_A \cdot \Delta x_B + x_B \cdot \Delta x_A)}{x_A \cdot x_B}$$
$$= \frac{\Delta x_A}{x_A} + \frac{\Delta x_B}{x_B} = E_A + E_B$$

同样可以对除法 $N = \frac{x_A}{x_B}$ 推导得：

$$\Delta N = \frac{x_A \cdot \Delta x_B + x_B \cdot \Delta x_A}{{x_B}^2}$$

$$E = \frac{\Delta N}{N} = \frac{x_A \cdot \Delta x_B + x_B \cdot \Delta x_A}{{x_B}^2} \Big/ \frac{x_A}{x_B}$$
$$= \frac{\Delta x_A}{x_A} + \frac{\Delta x_B}{x_B} = E_A + E_B$$

因此，在乘除运算时，结果的相对误差为各直接测量值的相对误差之和。

（3）任意函数 $N = f(x_A, x_B\cdots)$ 的误差传递公式（误差的算术合成公式）。

对函数 $N = f(x_A, x_B\cdots)$ 求全微分

$$\mathrm{d}N = \frac{\partial f}{\partial x_A} \cdot \mathrm{d}x_A + \frac{\partial f}{\partial x_B} \cdot \mathrm{d}x_B + \cdots \qquad (6)$$

上式表示，当 x_A、$x_B\cdots$有微小变化 $\mathrm{d}x_A$、$\mathrm{d}x_B\cdots$时，N 变化了 $\mathrm{d}N$。通常误差远小于测量值，把 $\mathrm{d}x_A$、$\mathrm{d}x_B\cdots$看成误差 Δx_A、$\Delta x_B\cdots$，则

$$\Delta N = \frac{\partial f}{\partial x_A} \cdot \Delta x_A + \frac{\partial f}{\partial x_B} \cdot \Delta x_B + \cdots$$

注意到上式中右侧各分项的误差正负不定，在最不利情况下，各分项误差将累积起来。因此，将上式右侧各分项分别取绝对值相加，即：

$$\Delta N = \left|\frac{\partial f}{\partial x_A}\right| \cdot \Delta x_A + \left|\frac{\partial f}{\partial x_B}\right| \cdot \Delta x_B + \cdots \qquad (7)$$

上式右侧各直接测量值的误差前的非负系数叫做误差传递系数。

相对误差为：

$$E = \frac{\Delta N}{N} = \frac{1}{f}\left(\left|\frac{\partial f}{\partial x_A}\right| \cdot \Delta x_A + \left|\frac{\partial f}{\partial x_B}\right| \cdot \Delta x_B + \cdots \right)$$

$$= \left|\frac{\partial \ln f}{\partial x_A}\right| \cdot \Delta x_A + \left|\frac{\partial \ln f}{\partial x_B}\right| \cdot \Delta x_B + \cdots \qquad (8)$$

由式（7）、（8）可知，间接测量值的误差，不仅与各个直接测量值有关，而且还取决于各个误差传递系数。

实际应用过程中，对于和差形式的误差，用式（7）先求绝对误差较为方便；对于积商形式的误差，用式（8）先求相对误差较为方便。表 1－2 是一些常用函数的误差传递公式。

求间接测量结果的误差可按如下步骤进行。

①对函数求全微分；

②合并同一变量的系数；

③各传递系数取绝对值后，将微分号变为误差号 Δ。

例 3　通过测定圆柱体的直径 D 和高 h 求其体积。在等精度测量时，试讨论 D 和 h 的误差 ΔD 和 Δh 对结果 V 的误差 ΔV 的影响。

解：根据 $V = \frac{1}{4}\pi D^2 h$，可得：

$$E = \frac{\Delta V}{V} = \left|\frac{\partial \ln V}{\partial D}\right| \cdot \Delta D + \left|\frac{\partial \ln V}{\partial h}\right| \cdot \Delta h$$

$$= 2\frac{\Delta D}{D} + \frac{\Delta h}{h}$$

在等精度测量时，ΔD 和 Δh 相近，这两者对 ΔV 的影响还与圆柱体本身的尺寸 D 和 h 有关。由体积的相对误差 E 的表达式可以看出，$\frac{\Delta D}{D}$ 和 $\frac{\Delta h}{h}$ 相近时，ΔD 项对 $\frac{\Delta V}{V}$ 的影响大。在处理实际问题时，要根据具体情况分析误差，并不能一概而论。

表 1－2　一些常用函数的误差传递公式

函数 $N = f(x_A, x_B \cdots)$	绝对误差 ΔN	相对误差 $E = \frac{\Delta N}{N}$
$N = x_A + x_B$	$\Delta x_A + \Delta x_B$	$\frac{\Delta x_A + \Delta x_B}{x_A + x_B}$
$N = x_A - x_B$	$\Delta x_A + \Delta x_B$	$\frac{\Delta x_A + \Delta x_B}{x_A - x_B}$
$N = x_A \cdot x_B$	$x_A . \Delta x_B + x_B \cdot \Delta x_A$	$\frac{\Delta x_A}{x_A} + \frac{\Delta x_B}{x_B}$
$N = \frac{x_A}{x_B}$	$\frac{x_A . \Delta x_B + x_B \cdot \Delta x_A}{x_B{}^2}$	$\frac{\Delta x_A}{x_A} + \frac{\Delta x_B}{x_B}$
$N = k \cdot x$	$k\Delta x$	$\frac{\Delta x}{x}$
$N = \sqrt{x}$	$\frac{\Delta x}{2\sqrt{x}}$	$\frac{1}{2}\frac{\Delta x}{x}$
$N = \sin x$	$\lvert \cos x \rvert \Delta x$	$\mathrm{ctg} x \Delta x$
$N = \ln x$	$\frac{\Delta x}{x}$	$\frac{\Delta x}{x \ln x}$

【测量结果的有效数字及数据处理】

如前所述，任何测量都有误差，那么在记录测量数据时，如何才能有效地表示测量结果，而又能正确地反映测量的准确程度呢？还有，在处理数据过程中，不允许因为计算而带来误差，那么又如何才能既简明又正确地进行数据处理呢？

1. 有效数字的一般概念

如一个测量结果为 $L=1.352\text{cm}$，其绝对误差为 $\pm 0.01\text{cm}$，就应记录为 $L=1.35\text{cm}$。因为由其误差可知，L 以 cm 为单位时，它的第二位小数“5”就已经有误差而不可靠了，也就是可疑的了，以下位数上的小数写出来便没有多大意义了。

通常我们把测量结果表示成由几位可靠的数字加上最后一位可疑数字组成的数据，这样的数据就是有效数字。有效数字的最后一位虽然是可疑的，即有误差，但它还是在一定程度上反映了客观实际，因此它也是有效的。例如，用最小刻度为 1 毫米的米尺去测量一块铜板的宽度，得 15.3mm，有三位有效数字，最后一位数字“3”是估读的，另外的人去测量时，可能会估读为“2”或“4”。但这一位数字的存在，就表示了这块铜板实际长度为 15mm（准确的）多 0.3mm 左右，同时也反映了测量工具是最小刻度为 1 毫米的、或相近的最小刻度。如果用五十分游标尺（游标精度为 0.02mm）去测量这块铜板，得 15.30mm，它表示测量可进行到 10^{-2}mm，只不过它的读数刚好为零，因此它有四位有效数字。所以，物理量的数值和数学上的数有着不同的意义，在数学上 $15.3=15.30=15.300$，在物理量上就没有这种等号关系了。由此可见，有效数字最后的“0”不能随便去掉或添上。

有效数字的位数与十进制的单位的变换无关，即与小数点的位置无关，因此用来表示小数点的位置的“0”不是有效数字。如 $15.3\text{mm}=1.53\text{cm}=0.0153\text{m}=0.0000153\text{km}$，这四种表示方法完全等价，均为三位有效数字。

通常我们可以将测量结果的有效数字写成 $a\times 10^n$ 的形式，其中 $1\leqslant a<10$。则测量结果的有效数字由 a 来确定，十进制单位的变化也只是其中 n（整数）的变化而已。如地球的半径可以表示为 $r=6.371\times 10^3\text{km}=6.371\times 10^6\text{m}$。这样表示，清楚明了，又可避免加零可能造成的错误。

2. 如何确定测量结果的有效数字

根据有效数字的定义可知，有效数字的最后一位有误差，是不可靠的。因此我们写下的有效数字的最后一位应使之与误差发生的一位取齐。

误差本身也是有效数字表示的数据，由于它只是一个估计范围，通常只写一位。如 $L=(15.3\pm 0.5)\text{ mm}$ 是正确的，$R=(31.0\pm 2)\ \Omega$ 是错误的，$h=(83.5\pm 0.17)\text{ m}$ 是错误的。

从以上可知，绝对误差决定有效数字。因此在对仪器进行读数时，必须读到最后要估读的一位，不能随便增减。但有时当仪器误差较大或测量对象、测量方法比较粗糙时，应根据实际情况决定是否要估读到最小刻度一位。总之，所记录的有效数据，应使最后一位数字恰是误差所在的一位。

3. 有效数字的运算

对有效数字运算如同对间接测量结果的计算一样，也存在误差传递问题。严格地讲，应根据误差计算来确定有效数字的位数。在没有给出各量的误差时，可以根据以下三点有效数字运算规则，粗略地得到运算结果。

（1）准确数字与准确数字之间的运算结果仍为准确数字。

（2）可疑数字与准确数字（或可疑数字）之间的运算结果为可疑数字，但运算的进位可以是准确数字。

（3）运算结果只保留最后一位为可疑数字，去掉第二位可疑数字时用四舍五入法。

例 4　有效数字的运算，其中数字下面加横线的是可疑数字。

(a)

$$\begin{array}{r} 1.38\underline{9} \\ 17.\underline{2} \\ +)\ 8.6\underline{4} \\ \hline 27.\underline{229} \end{array}$$

结果为 $1.38\underline{9}+17.\underline{2}+8.6\underline{4}=27.\underline{2}$

(b)

$$\begin{array}{r} 12.\underline{2} \\ -)\ 1.38\underline{9} \\ -)\ 8.6\underline{4} \\ \hline 7.\underline{171} \end{array}$$

结果为 $17.\underline{2}-1.38\underline{9}-8.6\underline{4}=7.\underline{2}$

(c)

$$\begin{array}{r} 4.3\underline{8} \\ \times)\ 1.\underline{1} \\ \hline 0.\underline{438} \\ 4.3\underline{8} \\ \hline 4.\underline{818} \end{array}$$

结果为 $4.3\underline{8}\times 1.\underline{1}=4.\underline{8}$

(d)

$$\begin{array}{r} 4.3\underline{63} \\ 12\underline{1}\ )\ 52\underline{8} \\ 48\underline{4} \\ \hline 4\underline{40} \\ 36\underline{3} \\ \hline \underline{770} \\ \underline{726} \\ \hline \underline{43} \end{array}$$

结果为 $52\underline{8}\div 12\underline{1}=4.3\underline{6}$

结合上例，可以给出有效数字运算三个基本规则。

（1）有效数字相加减时，结果的最后一位保留到参与运算的各数都有的那一位为止。

（2）有效数字相乘除时，结果的有效数字位数和参与运算各数中位数最少的一个相同（或最多再保留一位）。

（3）其他如乘方、开方、三角函数、对数等运算中，通常取结果有效数字的位数与参与运算各数的有效数字位数相同。如 $(4.256)^2=18.11$，$\sqrt{4.256}=2.063$。

对于在公式中的一些常数，如 π、e、$\sqrt{2}$、$\frac{1}{2}$ 等，它们是准确数，可以认为它们有无穷多位有效数字，在运算过程中，一般应取比参与运算各数中位数最多的有效数字还要多一位。如求 $S=\pi R^2$，$R=5.082$，则 π 可取 3.1416。

在实验中正确运用有效数字，不仅能如实反映测量结果，而且可以简化运算，提高效率，避免浪费。因为一个间接测量量是由几个直接测量值通过运算得来的，所以在进行测量时应考虑各个直接测量值的有效位数要基本相仿，或者说它们的相对误差要比较接近。如果相差悬殊，那么精度过高的测量就失去意义。现在，我们经常使用计算器作一些数字运算，同样要养成合理取有效数字的习惯。特别要注意，将计算器运算的结果全部记录下来而不进行取舍，是对有效数字不理解、不习惯、缺乏意识的表现。

关于测量误差处理的两点说明：

（1）上述关于随机误差处理的方法的基础是概率统计理论，估算出的随机误差只能在数量级上对测量结果的可靠程度作出一个恰当的评价。因此，误差结果没有必要计算得非常精确。本课程中规定误差（包括绝对误差和相对误差）只要取一位有效数字即可。

（2）有时候会出现多次测量值都相等的情况，这并不表示误差不存在，只是说明仪器精度不够，不足以反映测量值的起伏；或者会有多次测量的绝对误差比仪器误差还要小。这时可以将仪器误差作为测量值的误差。

【物理实验结果的图示法】

物理量之间的关系，经常用图线来表示，这就是图示法。图示法具有简明形象并直观地显示物理量之间关系的特点，使人一目了然。即使在物理量之间的关系尚未被确认时，也可以通过大量实验画出图线，再对图线进行分析，用解析公式进行拟合。还可以在图线上求出变量的极值、微分、积分和其他运算。

1. 作图规则

（1）作图一定要用坐标纸。当决定了作图参量后，根据情况选用直角坐标纸、双对数坐标纸、单对数坐标纸或其他坐标纸。在物理实验中比较常用的是毫米方格纸。

（2）坐标纸大小及坐标轴的比例应根据所测得的数据的有效数字和结果的需要来定。原则上，以不损失实验数据的有效数字位数和能包括所有实验点作为选取坐标纸大小的最低限度，可以适当放大些。坐标轴要用铅笔标出，端头可以不画箭头。习惯上，横坐标代表自变量，纵坐标代表因变量。在坐标轴外侧要标明物理量和单位。并在轴上等间距标明物理量的数值，不要标得太密或太疏。以下是需要注意的几点。

①任一坐标点的读数的有效数字位数不能小于实验数据的有效数字位数。

②坐标轴比例应取 1：2、1：5 或它们的倍数的比例，不能取 1：3、1：7 这样的比例。合适的坐标比例可以让实验者准确、方便、迅速地找到坐标点或进行读数。

③在上述两点要求下，应尽量使图线比较对称地充满整个图纸，而不使图线偏于一角或一边，因此，坐标起点不一定是从零开始。

（3）根据实验数据标定数据的坐标点。坐标点可用符号“+”表示。符号“+”要用直尺及笔尖画出，“+”要求能体现误差。一张图纸上要画几条曲线或有不同类型的点时，应该用不同的标记，如“×”、“⊙”等，以示区别。

（4）画线。用直尺、曲线板（尺）等，把点连成光滑、清晰的直线或曲线。注意，所画的直线、曲线不一定通过所有的点，而是要求线两侧的点均匀分布且离曲线较近。如发现有个别点偏离过大，应重新审核，然后决定取舍。

图线要求用铅笔绘制，并要求清晰、光滑，墨色要求深浅适宜。在图纸空白处应写上实验名称、图名、特征数据（如果有）、实验者姓名、合作者姓名、日期等。

2. 直线及曲线改直

（1）当所作图线是直线时，可以通过图解法求出直线方程：$y = kx + b$。其中 k 为直线的斜率，b 为直线在纵坐标轴上的截距。

如图 1－2，首先在直线上靠近直线的两端选取两点 P_1（x_1，y_1）、P_2（x_2，y_2），其

坐标尽量选取整数值，即尽量选直线与坐标纸的大格线的交点，并用不同于实验点的符号将它们表示出来，在旁边要注明其坐标读数。此两点不能用实验的原始数据，并尽量离远一点，但应在实验范围内。

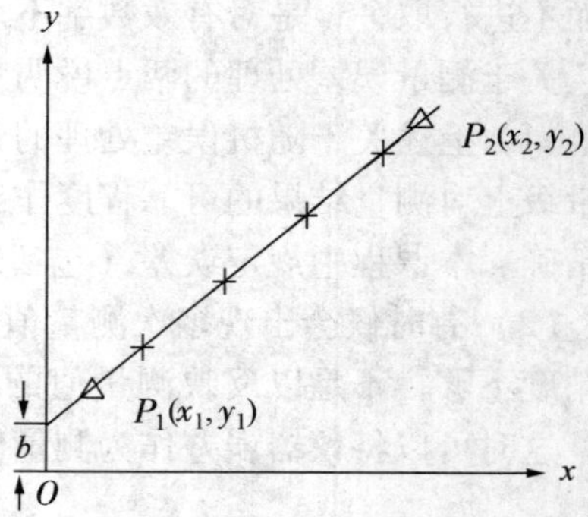

图1－2　直线图例

由此两点的坐标值，可求得斜率：

$$k=\frac{y_2-y_1}{x_2-x_1}$$

注意在具体情况下斜率 k 的单位。

直线在纵坐标轴上的截距 b，可直接从图上 $x=0$ 处读出。如果坐标起点不为零，可用下式计算得到截距：

$$b=\frac{x_2y_1-x_1y_2}{x_2-x_1}$$

（2）若两个物理量 x、y 满足线性关系，并由实验等精度地测得一组数据（x_i，y_i），$i=1$，2，…n，在理论上可以作出一条最佳拟合直线 $y=kx+b$，使所有的实验点到该直线的距离平方和为最小，这就是最小二乘法。依据最小二乘法原理可求得：

$$k=\frac{\sum_{i=1}^{n}x_i\cdot\sum_{i=1}^{n}y_i-n\sum_{i=1}^{n}(x_i\cdot y_i)}{(\sum_{i=1}^{n}x_i)^2-n\sum_{i=1}^{n}x_i^{\ 2}}=\frac{\bar{x}\cdot\bar{y}-\overline{x\cdot y}}{\bar{x}^2-\overline{x^2}}\tag{9}$$

$$b=\frac{\sum_{i=1}^{n}y_i}{n}-k\frac{\sum_{i=1}^{n}x_i}{n}=\bar{y}-k\bar{x}\tag{10}$$

上两式中的各平均值分别为：$\bar{x}=\frac{\sum_{i=1}^{n}x_i}{n}$，为 x_i 的平均值；$\bar{y}=\frac{\sum_{i=1}^{n}y_i}{n}$为 y_i 的平均值；$\overline{xy}=\frac{\sum_{i=1}^{n}(x_i\cdot y_i)}{n}$为 $x_i\cdot y_i$ 的平均值；$\overline{x^2}=\frac{\sum_{i=1}^{n}x_i^{\ 2}}{n}$为$x_i^{\ 2}$ 的平均值。

由式（10）可得 $y=kx+b=kx+\bar{y}-k\bar{x}$，即（$y-\bar{y}$）$=k$（$x-\bar{x}$），所以点（$\bar{x}$，$\bar{y}$）为该直线必定经过的点，因此作直线时可以先标出点（$\bar{x}$，$\bar{y}$），然后再画线。

必须指出的是，只有当 x 和 y 之间存在线性关系时，拟合的直线才有意义。为了检验拟合直线有无意义，数学上引进一个相关系数 r 的量，其定义是：

$$r=\frac{n\sum_{i=1}^{n}(x_i\cdot y_i)-\sum_{i=1}^{n}x_i\cdot\sum_{i=1}^{n}y_i}{\sqrt{[n\sum_{i=1}^{n}x_i^{\ 2}-(\sum_{i=1}^{n}x_i)^2]\cdot[n\sum_{i=1}^{n}y_i^{\ 2}-(\sum_{i=1}^{n}y_i)^2]}}$$

$$=\frac{\overline{x\cdot y}-\bar{x}\cdot\bar{y}}{\sqrt{(\overline{x^2}-\bar{x}^2)\cdot(\overline{y^2}-\bar{y}^2)}}\tag{11}$$

r 表示两变量的函数关系与线性关系的符合程度。r 值越接近 1，x 和 y 的线性关系越好；如果它接近于零，可以认为 x 和 y 之间不存在线性关系。物理实验中通常要求 r 值在

0.9 以上，如0.9、0.99、0.999 等，因此习惯上常称 r 值有几个“9”。一般的科学计算器都有用最小二乘法做线性拟合的功能，因此在实际应用中会很方便的。

（3）实际工作中，有许多函数的形式可以经过适当的变换，使变量的适当形式之间有线性关系，即把曲线改直。举例如下。

①$y = ax^b$（a、b 为常量）

$\ln y = b\ln x + \ln a$，$\ln y$ 与 $\ln x$ 成线性关系，斜率为 b，截距为 $\ln a$。

②$y = ae^{-bx}$（a、b 为常量）

$\ln y = -bx + \ln a$，$\ln y$ 与 x 成线性关系，斜率为 $-b$，截距为 $\ln a$。

③$y = ab^x$（a、b 为常量）

$\ln y = (\ln b)x + \ln a$，$\ln y$ 与 x 成线性关系，斜率为 $\ln b$，截距为 $\ln a$。

④$xy = c$（c 为常量）

$y = \frac{c}{x}$，y 与$\frac{1}{x}$成线性关系，斜率为 c。

⑤$y^2 = 2px$（p 为常量）

y^2 与 x 成线性关系，斜率为 $2p$。

⑥$y = ax + bx^2$（a、b 为常量）

$\frac{y}{x} = bx + a$，$\frac{y}{x}$与 x 成线性关系，斜率为 b，截距为 a。

例5　用直角坐标作图举例。

用惠斯登电桥测定铜丝在不同温度下的电阻值，数据见表1－3。试用作图法求铜丝的电阻与温度的关系。

表1－3　铜丝的电阻与温度的测量数据

t（℃）	R（Ω）	t（℃）	R（Ω）
15.5	2.807	40.3	3.059
24.4	2.897	45.0	3.107
26.5	2.919	49.7	3.155
31.1	2.969	54.9	3.207
35.0	3.003	60.0	3.261

解：以电阻 R 为纵坐标，温度 t 为横坐标。纵坐标选取2mm 代表0.010Ω，横坐标选取2mm 代表1.0℃，绘制铜丝的电阻与温度曲线，见图1－3。由图中数据点分布可知，铜丝的电阻与温度为线性关系，可满足如下方程：

$$R = kt + a$$

在直线的两端分别取两个代表点（20.0，2.853）和（60.0，3.255），应用前述公式可求出截距和斜率如下：

$$k = \frac{y_2 - y_1}{x_2 - x_1} = \frac{3.255 - 2.853}{60.0 - 20.0} = \frac{0.402}{40.0} = 0.0101\Omega \cdot ℃^{-1}$$

$$a = \frac{x_2y_1 - x_1y_2}{x_2 - x_1} = \frac{60.0 \times 2.853 - 20.0 \times 3.255}{60.0 - 20.0} = 2.652\Omega$$

所以，铜丝电阻与温度的关系为

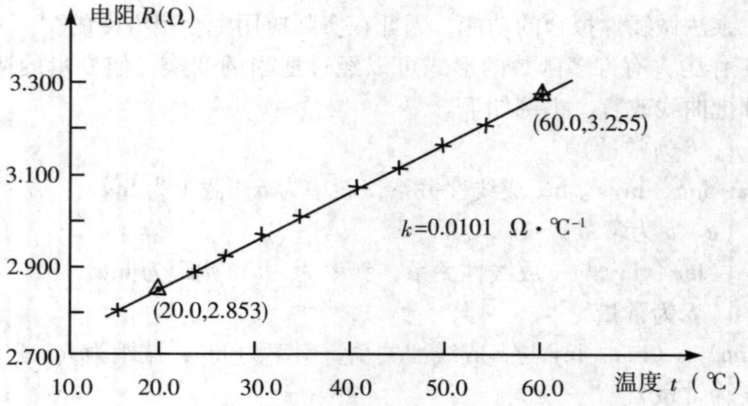

图 1－3 铜丝电阻与温度曲线

$$R = 0.0101t + 2.652\ \Omega$$

如果两个物理两成线性关系，在实验中常作多次测量，用图解法求比例系数。这样处理得到的结果比单次测量准确得多。

【思考题】

1. 简述系统误差、随机误差的区别。指出下列情况是属于系统误差，还是随机误差。

A. 读数误差

B. 天平零点漂移

C. 游标尺零点不准

D. 电表的接入误差

E. 水银温度计毛细管不均匀

2. 比较下列三个测量结果中哪个精密度高。

A. $l_1 = (54.98 \pm 0.02)$ cm

B. $l_2 = (0.498 \pm 0.002)$ cm

C. $l_3 = (0.0098 \pm 0.0002)$ cm

3. 写出下列函数的误差表达式（等式右边各量均为直接测量值）。

A. $N = x + y - 2z$

B. $N = \dfrac{x - y}{y - z}$

C. $Q = \dfrac{k}{2}(A^2 + b^2)$（$k$ 为常量）

D. $I_2 = I_1 \dfrac{r_2^2}{r_1^2}$

E. $n = \dfrac{\sin i}{\sin r}$

4. 用米尺（分度值为 1mm）测某一长度 l（17cm 左右），测量的极限误差为 3mm，求测量结果 $l \pm \sigma_l$ 的表达式。

5. 用米尺（分度值为1mm）测某一长度 l，其起点在10cm刻线上，终点在20cm刻线上。试以正确的有效数字来表达 l 的结果。

6. 对某一物理量 x 重复测量10次，结果如表1－4。

表1－4　思考题6的数据

i	1	2	3	4	5	6	7	8	9	10
x_i（cm）	20.2	20.0	20.4	20.3	20.2	20.4	20.1	20.3	20.4	20.5

请分别对全部数据和前5个数据求 $\bar{x}\pm\sigma_x$，并对两个结果进行比较。

7. 下面各题所用数据均为有效数字，试按有效数字运算规则进行计算。

A. $2.00\times4.00+50.0\times1.00+20.0\times0.1$

B. $\dfrac{1}{(0.1000)^2}-\dfrac{1}{(0.5000)^2}$（其中分子“1”是准确数）

C. $6.80\times10^3-20$

D. $\dfrac{2.480-2.0\times5.989}{2.00}$

E. $2.500\times3.000\times\left(1+\dfrac{4}{500}\right)$（其中分子“4”和数“1”是准确数）

8. 由单摆实验得到如表1－5的测量数据，其中 T 为摆长，L 为单摆周期。请按作图规则作一直线图，并计算直线斜率和重力加速度。

表1－5　思考题8的数据

L（cm）	61.5	71.2	81.0	89.5	95.5
T^2（s^2）	2.468	2.877	3.262	3.618	3.861

9. 试用最小二乘法对上题中的数据进行直线拟合。求直线的斜率、截距，并由斜率求重力加速度。

实　验　指　导

本实验介绍了测量误差的一些基本知识。这里大致介绍一下概率统计理论中有关正态分布规律的一些基本知识及其在处理误差方面的一些应用。同学们初次接触这些知识，可能不容易一下子完全接受，但在本课程的整个学习过程中可以经常翻阅，对于理解和掌握测量误差的基本知识是有帮助的。

【正态分布】

大量实验证明，在相同条件下对某一待测量 x 进行多次重复测量，其结果服从正态分布（亦称高斯分布）规律：

$$f(x)=\frac{1}{\sigma\sqrt{2\pi}}e^{-\frac{(x-\mu)^2}{2\sigma^2}},\quad |x|<\infty \tag{12}$$

上式中 x 为测量结果，$x=u$ 处对应正态分布 $f(x)$ 曲线的峰值，称为正态分布的均值，σ 是曲线拐点处的横坐标与 u 值之差的绝对值，称为标准偏差。

$f(x)$ 是概率密度函数，它的意义是测量值 x 落在 $x \to x+\mathrm{d}x$ 区间内的概率是 $f(x) \cdot \mathrm{d}x$。$x=\mu$ 的分布曲线如图 1－4。

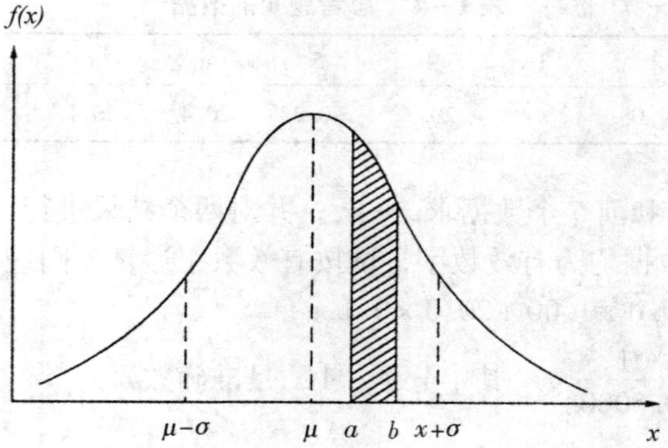

图 1－4　正态分布曲线

在相同条件下对某一待测量进行多次重复测量，所得测量值统称为测量列。对于一个确定的测量列，如果已知其 μ、σ 的值，则由 $f(x)$ 可以计算出实验测量值 x 落在 (a, b) 内的概率为：

$$P(a < x < b) = \int_a^b f(x) \cdot \mathrm{d}x = \int_a^b \frac{1}{\sigma\sqrt{2\pi}} e^{-\frac{(x-\mu)^2}{2\sigma^2}} \cdot \mathrm{d}x \tag{13}$$

上式的积分值等于 $x=a$ 和 $x=b$ 之间与曲线 $f(x)$ 及横坐标所包围的面积（图 1－4 中的阴影区）。取不同的 (a, b) 区间，计算式（13）中的定积分可得：

x 落在 $(\mu-\sigma, \mu+\sigma)$ 内的概率为 68.3%；

x 落在 $(\mu-2\sigma, \mu+2\sigma)$ 内的概率为 95.4%；

x 落在 $(\mu-3\sigma, \mu+3\sigma)$ 内的概率为 99.7%。

极限误差 3σ 即来源于此。当然，如取 $a=-\infty$，$b=+\infty$，则：

$$P(-\infty < x < +\infty) = \int_{-\infty}^{+\infty} f(x) \cdot \mathrm{d}x = 1$$

概率为 1，是必然事件，即任意的 x 总是落在 $(-\infty, +\infty)$ 区间内的。

将图 1－4 中的纵坐标轴 $f(x)$ 移到峰值 $x=\mu$ 处，亦即令 $\delta=(x-\mu)$，可得如下形式的正态分布曲线：

$$\Psi(\delta) = \frac{1}{\sigma\sqrt{2\pi}} e^{-\frac{\delta^2}{2\sigma^2}}, \quad |\delta| < \infty \tag{14}$$

该曲线如图 1－5，其中横坐标代表随机误差或偶然误差 $\delta=(x-\mu)$，图中三条曲线分别代表三个不同的 σ 值。

由图 1－5 可以看出曲线的几个特性。

①单峰性：绝对值小的误差出现的概率比绝对值大的误差出现的概率大；

②对称性：绝对值相等的正负误差出现的概率相等；

③有界性：绝对值很大的误差出现的概率接近于零，亦即误差有一定的实际限度。

④抵偿性：在实际测量中，对同一个量的测量，其误差的算术平均值随测量次数的增加而趋近于零。

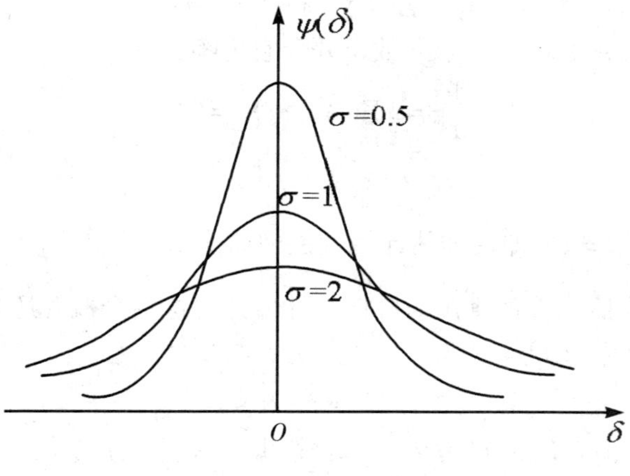

图 1－5　不同 σ 值的三条正态分布曲线

由于曲线 Ψ（δ）与横坐标围成区域的面积为 1，是定值，所以曲线拐点对应的横坐标值 σ 越大，曲线峰值越小，曲线越平坦；反之，σ 越小，曲线峰值越大，曲线越陡峭。σ 大表示变量 δ（即随机误差）的分散程度高，σ 小相对来说分散程度低。

【算术平均值和标准偏差】

设实验已消除了系统误差，在相同条件下对某量 x 进行 n 次等精度的重复测量，得到结果 x_1、x_2、$\cdots x_n$，它们满足正态分布 $f(x)=\dfrac{1}{\sigma\sqrt{2\pi}}e^{-\frac{(x-\mu)^2}{2\sigma^2}}$。概率统计理论证明（此处从略），其中 μ 即为待测量 x_1、x_2、$\cdots x_n$ 的真值。x_1、x_2、$\cdots x_n$ 的算术平均值（简称平均值）为：

$$\bar{x}=\frac{x_1+x_2+\cdots+x_n}{n}=\frac{\sum x_i}{n}$$

（上式中 $\sum x_i$ 是 $\sum\limits_{i=1}^{n}x_i$ 的简写。下同。）以下我们从三个方面论述取平均值 $\bar{x}$ 作为测量结果的理由，并且它是对真值 μ 的最佳估计值。

（1）由于每一测量值 x_1、x_2、$\cdots x_n$ 是等精度的，因此对它们的信赖程度应该是一样的，利用它们来求解测量的最后结果时，各个测量值应占有相同的比重，而算术平均值就具有这一特点。

（2）因为测量误差中已经消除了系统误差，故每一测量值的随机误差为

$$\delta_i=x_i-\mu$$

注意上式中的 μ 为真值。将各测量值的随机误差相加后除以 n，得到

$$\frac{\sum\delta_i}{n}=\frac{\sum x_i}{n}-\mu=\bar{x}-\mu$$

根据随机误差的抵偿性，当 $n\to\infty$ 时，$\dfrac{\sum\delta_i}{n}\to 0$，则有 $\bar{x}\to\mu$。

（3）最小二乘法告诉我们，对一组测量值 x_1、x_2、$\cdots x_n$ 所取的最佳估计值 X，与各测量值之间的偏差的平方和应为最小，即

$$Q = \sum (X - x_i)^2 \rightarrow min$$

为求最小值，我们将上式对 X 求导数，并令它为零：

$$\frac{dQ}{dX} = 2\sum(X - x_i) = 0$$

$$nX - \sum x_i = 0$$

最后可得：$X = \frac{\sum x_i}{n} = \bar{x}$，即最佳估计值就是平均值 $\bar{x}$。

现在我们知道平均值 $\bar{x}$ 是真值 μ 的最佳估计值，但这两个概念还是有区别的。例如各测量值 x_i 与平均值 $\bar{x}$ 的偏差为

$$\Delta x_i = x - \bar{x}$$

Δx_i 并不是 x_i 与真值 μ 的偏差 $\delta_i = x_i - \mu$。因此，Δx_i 称为残差。注意，本实验中直接称之为误差了，没有区分这其间的微小差异。

任意一次测量的标准偏差 σ_i 的定义是 δ_i 的方均根（平方和的平均值的平方根），即 $\sigma_i = \sqrt{\frac{\sum (\delta_i)^2}{n}} = \sqrt{\frac{\sum (x_i - \mu)^2}{n}}$。误差理论可以证明（可参阅方醉敏主编《物理实验与指导》，中国医药科技出版社 1992 年出版），当 n 足够大时，用残差 Δx_i 来估算标准偏差 σ_i 的计算公式为：

$$\sigma_i = \sqrt{\frac{\sum (\Delta x_i)^2}{n-1}} = \sqrt{\frac{\sum (x_i - x)^2}{n-1}}$$

这就是本实验中的式（3）。

【误差的传递及平均值的标准偏差】

对于独立的直接测量值 x_A、x_B…，经过函数 $N = f(x_A, x_B, \cdots)$ 运算后得间接测量值 N。本实验中已经给出 N 的绝对误差和相对误差为误差传递公式（7）和（8）。如果各个直接测量值 x_A、x_B…的绝对误差分别为 σ_A、σ_B…，则间接测量值 N 的标准偏差 σ_N 可用误差的方和根合成公式（公式的推导可参阅方醉敏主编《物理实验与指导》，中国医药科技出版社 1992 年出版），即：

$$\sigma_N = \sqrt{\left(\frac{\partial f}{\partial x_A}\sigma_A\right)^2 + \left(\frac{\partial f}{\partial x_B}\sigma_B\right)^2 + \cdots}$$

相对误差为：

$$E = \frac{\sigma_N}{N} = \frac{1}{f}\sqrt{\left(\frac{\partial f}{\partial x_A}\sigma_A\right)^2 + \left(\frac{\partial f}{\partial x_B}\sigma_B\right)^2 + \cdots}$$

现在，我们应用上述公式来求平均值的标准偏差。

首先，平均值是由各个直接测量值运算得到：$\bar{x} = \frac{(x_1 + x_2 + \cdots + x_n)}{n}$，则：

$$\sigma_{\bar{x}} = \sqrt{\left(\frac{\partial \bar{x}}{\partial x_1}\sigma_{x_1}\right)^2 + \left(\frac{\partial \bar{x}}{\partial x_2}\sigma_{x_2}\right)^2 + \cdots + \left(\frac{\partial \bar{x}}{\partial x_n}\sigma_{x_n}\right)^2}$$

$$= \sqrt{\frac{1}{n^2}\sigma_{x_1}{}^2 + \frac{1}{n^2}\sigma_{x_2}{}^2 + \cdots + \frac{1}{n^2}\sigma_{x_n}{}^2} = \sqrt{\frac{1}{n^2}\sum \sigma_{x_i}{}^2}$$

$$= \sqrt{\frac{1}{n^2} \cdot n\sigma_{x_1}{}^2} = \sqrt{\frac{\sigma_x{}^2}{n}} = \frac{\sigma_x}{\sqrt{n}} = \sqrt{\frac{\sum (x_i - x)^2}{n(n-1)}}$$

即为本实验中的式（4）。它告诉我们，平均值的标准偏差等于任一次测量值的标准偏差除以测量次数的平方根。也就是说，测量结果的可靠性随着测量次数的增加得到改善。但是以 100 次测量和 10 次测量相比较，结果的可靠性增加为 $\sqrt{100/10} \approx 3$ 倍。所以，这一做法通常不是很值得的。

1 Basic Knowledge about Measurement

Physics lab work is to study and observe physics rules. It is one of the most important tasks that measuring physical quantities. After processing data one can get rational experiment results. At the same time all the measurements have errors, as well as the final results, because of the factors such as lab conditions, instruments and theories. That is reason one has to learn the knowledge about measurement, data processing and error analysis.

The theory about error analysis plays a guide role for the lab work such as improving experiment methods, instrument precision selection and determining favorable experiment conditions.

Measurement

The goal of physics experiment is not only observing the physics phenomena, but the most important is finding out the quantitative relationships among the physics quantities. So the measurement has to be made properly. To measure a quantity is to compare it with a standard, uniformity quantity by using some special means and determine the multiple of it to the standard quantity. The uniformity quantity which is selected as a standard quantity is the unit of the measurement and the multiple is the magnitude of the measurement. The result of measurement is always constituted by magnitude and unit. In China the International System of units (SI) is the legal national units of measurement, of which meter, kilogram, second, ampere, kelvin, mole and candela are the basic units. All the other units are the products of the 7 units, which are called derived units.

There are two kinds of measurements: direct and indirect measurement. Generally, these quantities which have known relations with the objective quantities are measured first, and then, the objective quantities can be calculated out with the relations or equations. For instance, the density of a cylinder can be determined by its volume and mass:

$$\rho = \frac{m}{V} = \frac{4m}{\pi D^2 L}$$

m is the mass of the cylinder, D is its diameter and L its length. m, D, L are all quantities which can be measured directly. By calculating with these directly measured quantities we can get the density ρ, it is called indirect measurement.

Error of measurement

There are actual values of physical quantities, which are called true values. Because of the limits of measuring instruments, methods, lab conditions and other interrelated facts, measurements cannot be done exact unlimited. There is always a difference between the measurement result and the true value. It means that there is always a measurement error. The magnitude of the error shows the close extent of the measurement to the true value.

There are a lot of facts that will work on the error of measurements, such as measuring method, precision of the instruments, lab environment, and even the personality of experimenter. Usually there are proper requirements on the errors or accuracy of the measurement results. The ideas that the more precise of the instruments or the more stable of the environment (temperature, humidity, noise etc.) or more times of measurement, the better of the experiment results is unsuitable. Or it will be wasteful of resource. So, according to practical requirement we should make reasonable plan and select suitable instruments of the experiment. The most important point is how to get the best result with smallest cost under certain lab condition.

1. Error of measurement

The difference between the result of a measurement x and its true value x_0 is called the error Δx:

$$\Delta x = x - x_0$$

Error occurs to every measurement and during the total measuring process. The right way to handle error is analyzing error correctly, and we can get proper information about what kind of quantities has more influence on results than the others, and then take greater effort on the measuring of these quantities, not on the others.

2. The properties and sources of errors

Since the errors always exist, only if know their properties and sources can we reduce it to the most extant. According to their properties and sources the measurement errors can be classified into three kinds: systematic error, random error and mistake.

(1) Systematic error

Systematic error has the characteristic of making the result of measurement being bigger or smaller than the true value, it is an error in one direction. When the measurement condition changed, systematic error will also change in some certain way.

① Systematic error caused by lab apparatus The imperfection, inaccuracy and unsuitable adjustment of lab apparatus will cause systematic errors. For example: improperly scaled pan, inaccurate zero point, inaccuracy of weights, unequal hands of a balance, not horizontal placement of an instrument which should be placed horizontally, etc. All these situations may cause systematic errors.

② Systematic error caused by theories or methods Roughness of theories, imperfect lab condition, or the rough measurement method itself will cause systematic errors. For example: the internal resistance of a voltage meter has not been taken into account when measuring voltage of a certain electric circuit.

③ Systematic error caused by individual The physiological and mental characteristic of experimenter will be able to cause errors in a certain direction. For example: habit of improperly reading, weak-eyed, etc, will cause the measurement result bigger or smaller.

Because system errors always make the measurement results incorrect in one direction, being bigger or smaller, averaging many measurements cannot eliminate systematic errors. We should try to find out the source of systematic error basing on various situations, and try to eliminate its affection or modify the measurement results.

(2) Random error

Even if the systematic error has been eliminated (actually is impossible and unnecessary), there will also be differences among the values of the repeated measurements under the same measuring condition. This kind of errors, the differences between measurement results and the true value, are found uncertain with their magnitude and sign. It is called random error.

Random error makes every measurement bigger or smaller than the true value. But its behavior obeys some statistical rules: The possibility of big difference between measurement and true value is smaller than the small difference; the possibility of positive error equals to possibility of negative; the possibility of very big error almost equals zero. All these are the reasons we adopt method of repeated measurement.

The source of random error is the limit ability of human sense, such as seeing, hearing, feeling, and also the influence of some occasional facts of the experiment surrounding, such as the fluctuation of temperature, humidity, voltage of electrical source, the wind stream in the lab, unexpected vibration in the surrounding and so on. It's impossible to eliminate the random error with certain measurement method. But, by using statistical method we can reduce the influence of random error on the final measurement result, and manage to calculate the magnitude of the error.

(3) Mistake

There maybe a lot of measurement mistake such as reading mistake, recording mistake, operation mistake, calculation mistake and so on. We should pay more attention to avoid mistake. Correct method and working manner may be helpful to avoid mistake. We also can find out mistake by comparing measurements each other or judging by logic reason, and then delete it.

3. Precision, accuracy and exactness

Words of precision, accuracy and exactness are used to evaluate measurement. But there are differences among the three words.

For measurements, the precision is good, means the data are concentrated, the random error is small; the accuracy is good, means the difference between the mean value and the true value is small, the systematic error is small; the exactness is good, means the data are concentrated near the true value, the systematic error and random error are both very small.

Figure 1 – 1 shows the data of measurements of quantity x on the number axis, where x_0 is the true value.

Be careful with the word "precision" marked on some instruments, which indicates the smallest quantity this instrument can tell.

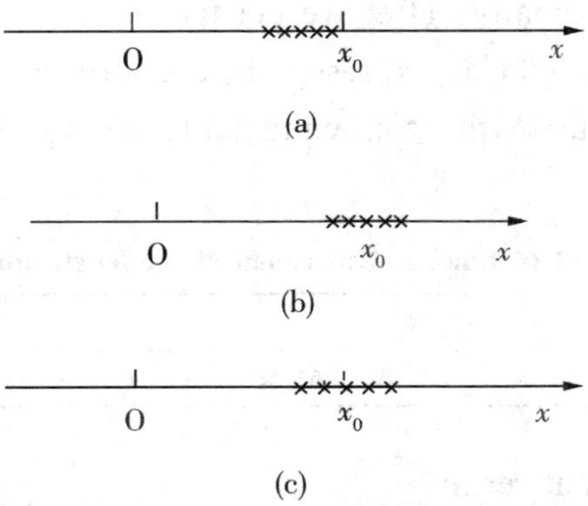

Figure 1 – 1 measurements of quantity x on the number axis

(a) the data are concentrated, but deviating the true value, precision is good.

(b) the data are decentralized, but the mean value is close to true value, accuracy is good.

(c) both precision and accuracy are good, exactness is good.

4. Estimate of random errors of direct measurement data

(1) Express the result of measurement with mean value

Say, n times of measurement of a physical quantity has been made under the same condition, the data are denoted as x_1、x_2、$x_3 \cdots x_n$. It is pointed out ahead that the true value x_0 is unknown or cannot be known exactly, but we can express it with the mean value of the data, as its best approximate value:

$$x_0 = \bar{x} = \frac{1}{n}(x_1 + x_2 + \cdots + x_n) = \frac{\sum_{i=1}^{n} x_i}{n} \tag{1}$$

Where, is x_i the i time's measurement value.

(2) Average absolute uncertainty

The mean value is the best approximate of the true value, but it is not the true value at last. The difference between it and the true value can be estimated as the average absolute uncertainty $\overline{\Delta x}$. The average absolute uncertainty can be defined as:

$$\overline{\Delta x}=\frac{(|x_1-\bar{x}|+|x_2-\bar{x}|+\cdots+|x_n-\bar{x}|)}{n}=\frac{\sum_{i=1}^{n}|\Delta x_i|}{n} \qquad (2)$$

Where, $\Delta x_i = x_i - \bar{x}$, is called the absolute uncertainty of the i time's measurement.

It can be proved that there is a 57.5% probability that the range $(-\overline{\Delta x},\ \overline{\Delta x})$ contains absolute uncertainty Δx_i of any one value of the repeated measurement, therefore the probability of the range containing the difference between mean value and the true value is much more great. So it is reasonable that the range $(\bar{x}-\overline{\Delta x},\ \bar{x}+\overline{\Delta x})$ contains the true value. But it is also possible that some measurement value are beyond the range $(\bar{x}-\overline{\Delta x},\ \bar{x}+\overline{\Delta x})$. Finally, after we get the mean value $\bar{x}$ and average absolute uncertainty $\overline{\Delta x}$, the result should be expressed in the form of $x=(\bar{x}\pm\overline{\Delta x})$ (unit).

EXAMPLE 1 The data in the following table are measurement 10 times of the length of a certain object:

Table 1-1 Data of 10 times measurement of the length for a certain object

i	1	2	3	4	5	6	7	8	9	10
x_i (cm)	63.57	63.58	63.55	63.56	63.56	63.59	63.55	63.54	65.57	63.57

What is the measurement result?

SOLUTION: First, the mean value is:

$$\bar{x}=\frac{\sum_{i=1}^{10}x_i}{10}=63.564\text{cm}$$

Then, the average absolute uncertainty:

$$\overline{\Delta x}=\frac{\sum_{i=1}^{10}|\Delta x_i|}{10}=0.012\approx 0.01\text{cm}$$

Because the average uncertainty is an approximate figure, usually it is rounded to one digit of significant figure.

Finally, the result can be kept as:

$$x=(63.56\pm 0.01)\ \text{cm}$$

The above expression means that the true value of the length x is located between the range of 63.55 ~ 63.57cm.

(3) Standard deviation and limited error

Another important way to evaluate the random error is standard deviation. For the measurements of n times, the standard deviation σ_x of any measurement is defined as:

$$\sigma_x = \sqrt{\frac{\sum_{i=1}^{n}(x_i - \bar{x})^2}{n-1}} = \sqrt{\frac{\sum_{i=1}^{n}(\Delta x_i)^2}{n-1}} \qquad (3)$$

σ_x is the square root of the quotient of the sum of all the square of the absolute uncertainties, also it is called root-mean-square error. The physical meaning of σ_x is: for the measurements of n times, there is a 68.3% probability that the range of value $(-\sigma_x, \sigma_x)$ contains the uncertainty or error of any one measurement. Nowadays, the standard deviation σ_x is commonly used by science literature for evaluation of measurement data.

The standard deviation of the mean value, $\sigma_{\bar{x}}$, is defined as:

$$\sigma_{\bar{x}} = \frac{\sigma}{\sqrt{n}} = \sqrt{\frac{\sum_{i=1}^{n}(x_i - \bar{x})^2}{n(n-1)}} = \sqrt{\frac{\sum_{i=1}^{n}(\Delta x_i)^2}{n(n-1)}} \qquad (4)$$

This expression shows that $\sigma_{\bar{x}}$ is $\frac{1}{\sqrt{n}}$ times of σ_x, the standard deviation. Obviously, $\sigma_{\bar{x}}$ is smaller than σ_x, because the mean value is the best result comparing with all the measurements, it's more closer to the true value than any measurement .

Theoretically, the number of measurement times should be very large when using formula (3) and (4) . But practically, most situations cannot meet this requirement, so the results of the two formulae are approximate. In addition, when the number of measurement is small, for instance $n<10$, $\sigma_{\bar{x}}$ becomes small evidently with the increasing of n, the measurement times; but when n gets larger and larger, $\sigma_{\bar{x}}$ becomes smaller not so evidently and tends to a fixed value. So it's unsuitable that repeating too many more times of measurements. In most of the case, 5 ~ 10 times is enough.

Sometime the object being measured maybe uneven or unstable, for example, the diameter of a small steel bead may have slight difference in different places. So the mean value of diameter is just the average size of the steel dead. Measurement cannot make it even, so it's unnecessary to calculate the standard deviation of the mean value. And the standard deviation can express the uneven extent of the steel bead.

EXAMPLE 2 For the data in example 1, calculate $\sigma_{\bar{x}}$.

SOLUTION: According to equation (3):

$$\sigma_x = \sqrt{\frac{\sum_{i=1}^{n}(x_i - \bar{x})^2}{n-1}} = \sqrt{\frac{\sum_{i=1}^{10}(x_i - 63.564)^2}{10-1}} = \sqrt{\frac{2.040\times10^{-6}}{9}} = 0.015 \approx 0.02\text{cm}$$

This result can be expressed as:

$$x = (63.56 \pm 0.02)\ \text{cm}$$

Compare σ_x with $\overline{\Delta x}$ of example 1: $\overline{\Delta x} < \sigma_x$; and also the probability that the range $(\bar{x} - \overline{\Delta x},\ \bar{x} + \overline{\Delta x})$ contains a measurement is smaller than the range $(\bar{x} - \sigma_x,\ \bar{x} + \sigma_x)$. Actually, the relation is $\sigma_x = 1.25\overline{x}$.

The statistical theory shows that there is a 99.7% probability that the range $(-3\sigma_x,\ 3\sigma_x)$ contains the error of a measurement. Therefore is $3\sigma_x$ called limit error. If an error of a measurement is found beyond $\pm 3\sigma_x$, this measurement can be suggested as an unsuccessful measurement, it should be deleted. So, $3\sigma_x$ is the limitation of normal random error.

Usually there are special function keys on a scientific calculator to get standard deviation σ_x. In this course, for many measurements it is required to calculate $\overline{\Delta x}$; for $n > 5$, σ_x is required. It's not always necessary to calculate $\sigma_{\bar{x}}$.

(4) Error of instrument

The way to handle random error given above is useful for repeated measurements. But how about the measurement of just one time? For this kind of situation, the uncertainty is determined by the scale of instrument, the surrounding, etc; it should be make out practically. Uncertainty of a measurement may include both random error and system error. In this course, we take the error marked on the instrument as the uncertainty of result. If it is not marked, half of one scale can be taken as the uncertainty. Sometimes, some rational datum also works other than the scale.

(5) Relative uncertainty

The average absolute uncertainty $\overline{\Delta x}$ and standard deviation σ_x both have their own units; their units are the same with the physical quantities. They are all absolute uncertainty, expressing them as Δx. Δxmeans that the true value is located on the range $(x - \Delta x,\ x + \Delta x)$ by some certain probability. There is another way to express uncertainty, that is relative uncertainty E, it is also called uncertainty in percentage. It is the ratio of absolute uncertainty Δx with mean or standard value x, usually the result is expressed in percentage:

$$E = \frac{\Delta x}{x} \times 100\% \qquad (5)$$

Relative uncertainty shows the proportion of absolute uncertainty in the measurement. It can be used for comparing different kind of measurement. For example, there are two expressions of: $x_A = (100 \pm 1)$ cm, $x_B = (1000 \pm 1)$ cm, the two absolute uncertainties are same, but the two relative uncertainties are $E_A = \frac{1}{100} = 1\%$ and

$E_A = \frac{1}{1000} = 0.1\%$, so the second measurement is better.

5. Uncertainty of indirect measurement

The uncertainty is unavoidable in the direct measurement, thus, there will be uncertainty in the indirect measurement getting from the direct measurement. Expression to calculate uncertainty of indirect measurement is called formula of propagation of error.

Suppose x_A、x_B are values of two direct measurement, their absolute uncertainties are Δx_A and Δx_B. N is a quantity given by x_A、x_B, ΔN is its uncertainty.

(1) Sum and difference

First, the sum: $N = x_A + x_B$

It can be expressed as:

$$N \pm \Delta N = (x_A \pm \Delta x_A) + (x_B \pm \Delta x_B)$$
$$= (x_A + x_B) \pm (\Delta x_A \pm \Delta x_B)$$

The possible greatest uncertainty: $\Delta x_A \pm \Delta x_B$

So, for N the sum we get: $\Delta N = \Delta x_A \pm \Delta x_B$

The relative uncertainty is:

$$E = \frac{\Delta N}{N} = \frac{\Delta x_A + \Delta x_B}{x_A + x_B}$$

To the same reason, for difference: $N = x_A - x_B$

$$\Delta N = \Delta x_A + \Delta x_B$$

$$E = \frac{\Delta N}{N} = \frac{\Delta x_A + \Delta x_B}{x_A - x_B}$$

In the sum and difference, the absolute uncertainty is determined by the sum of absolute uncertainties of the measurements.

(2) Multiplication and division

First, multiplication: $N = x_A \cdot x_B$

$$N \pm \Delta N = (x_A \pm \Delta x_A) \cdot (x_B \pm \Delta x_B)$$
$$= x_A \cdot x_B \pm x_A \cdot \Delta x_B \pm \Delta x_A \cdot x_B \pm \Delta x_A \cdot \Delta x_B$$

Because Δx_A、Δx_B are much more smaller than x_A、x_B, $\Delta x_A \cdot \Delta x_B$ should be infinitesimal, which can be neglected. For the possible greatest uncertainty, we can get:

$$N \pm \Delta N = x_A \cdot x_B \pm x_A \cdot \Delta x_B \pm \Delta x_A \cdot x_B$$
$$= x_A \cdot x_B \pm (x_A \cdot \Delta x_B + x_B \cdot \Delta x_A)$$

So, for multiplication:

$$\Delta N = x_A \cdot \Delta x_B + x_B \cdot \Delta x_A$$

The relative uncertainty is:

$$E = \frac{\Delta N}{N} = \frac{(x_A \cdot \Delta x_B + x_B \cdot \Delta x_A)}{x_A \cdot x_B}$$

$$=\frac{\Delta x_A}{x_A}+\frac{\Delta x_B}{x_B}=E_A+E_B$$

To the same reason, for division $N=\frac{x_A}{x_B}$, we can get:

$$\Delta N=\frac{x_A\cdot\Delta x_B+x_B\cdot\Delta x_A}{x_B^2}$$

$$E=\frac{\Delta N}{N}=\frac{x_A\cdot\Delta x_B+x_B\cdot\Delta x_A}{x_B^2}\Big/\frac{x_A}{x_B}$$

$$=\frac{\Delta x_A}{x_A}+\frac{\Delta x_B}{x_B}=E_A+E_B$$

In multiplication and division, the relative uncertainty of result is determined by the sum of the relative uncertainties of the direct measurements.

(3) Formula of propagation of uncertainty in function $N=f(x_A, x_B\cdots)$

Consider a mathematical function $N=f(x_A, x_B\cdots)$, its total differential is:

$$dN=\frac{\partial f}{\partial x_A}\cdot dx_A+\frac{\partial f}{\partial x_B}\cdot dx_B+\cdots \quad (6)$$

This expression shows there is a change of dN in N when x_A、$x_B\cdots$ changes slightly as dx_A、$dx_B\cdots$. Usually, uncertainty is much more smaller than its measurement value, therefore, it is rational to take dx_A、$dx_B\cdots$ as $d\Delta x_A$、$d\Delta x_B\cdots$ respectively:

$$\Delta N=\frac{\partial f}{\partial x_A}\cdot\Delta x_A+\frac{\partial f}{\partial x_B}\cdot\Delta x_B+\cdots$$

The signs of the items in the right side of the above expression are not decided. In the worst situation, all the uncertainties of the items will be added up to the final result. Thus, let these absolute values of all the items be added up:

$$\Delta N=\left|\frac{\partial f}{\partial x_A}\right|\cdot\Delta x_A+\left|\frac{\partial f}{\partial x_B}\right|\cdot\Delta x_B+\cdots \quad (7)$$

The positive coefficients of every item are called uncertainty propagation coefficients.

The relative uncertainty is:

$$E=\frac{\Delta N}{N}=\frac{1}{f}\left(\left|\frac{\partial f}{\partial x_A}\right|\cdot\Delta x_A+\left|\frac{\partial f}{\partial x_B}\right|\cdot\Delta x_B+\cdots\right)$$

$$=\left|\frac{\partial \ln f}{\partial x_A}\right|\cdot\Delta x_A+\left|\frac{\partial \ln f}{\partial x_B}\right|\cdot\Delta x_B+\cdots \quad (8)$$

For the formula (7) and (8), the uncertainty of indirect measurement is not only determined by the direct measurements, but also the uncertainty propagation coefficients.

Practically, it may be convenient to use formula (7) for sums and differences, and formula (8) for multiplication and division. Table 1-2 is a list of uncertainty propagation formula of common functions.

Here is a summary of three steps to get uncertainty of indirect measurement:

① Get the total differential of the function;

② Put the coefficients of the same variable together;

③ Be sure to make all the propagation coefficients positive, change all the signs of differential into the sings of uncertaintyΔ.

EXAMPLE 3 Get the volume V of a cylinder by measuring its diameter D and height h. Under the condition of equal observations, discuss the influence of the uncertainties of ΔD and Δh on the result uncertainty ΔV.

SOLUTION: $V=\frac{1}{4}\pi D^2 h$, then:

$$E=\frac{\Delta V}{V}=\left|\frac{\partial \ln V}{\partial D}\right|\cdot \Delta D+\left|\frac{\partial \ln V}{\partial h}\right|\cdot \Delta h$$

$$=2\frac{\Delta D}{D}+\frac{\Delta h}{h}$$

By equal observations, ΔD and Δh are nearly the same magnitude, the influence of the two items on the ΔV is related with the sizes of cylinder, D and h. For the expression of E, when $\frac{\Delta D}{D}$ is near to $\frac{\Delta h}{h}$, the item of ΔD has more influence on $\frac{\Delta V}{V}$. When solving actual problems, we should pay more attention to error analysis.

Table 1 – 2 A list of uncertainty propagation formula of common function

Function $N=f(x_A, x_B \cdots)$	Absolute uncertainty ΔN	Relative error $E=\frac{\Delta N}{N}$
$N=x_A+x_B$	$\Delta x_A+\Delta x_B$	$\frac{\Delta x_A+\Delta x_B}{x_A+x_B}$
$N=x_A-x_B$	$\Delta x_A+\Delta x_B$	$\frac{\Delta x_A+\Delta x_B}{x_A-x_B}$
$N=x_A\cdot x_B$	$x_A\cdot\Delta x_B+x_B\cdot\Delta x_A$	$\frac{\Delta x_A}{x_A}+\frac{\Delta x_B}{x_B}$
$N=\frac{x_A}{x_B}$	$\frac{x_A\cdot\Delta x_B+x_B\cdot\Delta x_A}{x_B^2}$	$\frac{\Delta x_A}{x_A}+\frac{\Delta x_B}{x_B}$
$N=k\cdot x$	$k\Delta x$	$\frac{\Delta x}{x}$
$N=\sqrt{x}$	$\frac{\Delta x}{2\sqrt{x}}$	$\frac{1}{2}\frac{\Delta x}{x}$
$N=\sin x$	$\lvert\cos x\rvert\Delta x$	$\text{ctg}x\Delta x$
$N=\ln x$	$\frac{\Delta x}{x}$	$\frac{\Delta x}{x\ln x}$

Significant figure and data processing

As it is discussed above, there are errors of all the measurements. How can we do to write down the data of measurement effectively or well and truly? In addition, it is not allowed to make mistake when handling data. So, how can we do to handle data effectively and correctly?

1. Concept of significant figure

Suppose a measurement is $L = 1.352$, its absolute uncertainty is $L = 1.35$, the right expression of the measurement is L. We know that when L is in the unit of cm its second decimal figure "5" is no longer reliable because of its uncertainty, this figure is doubtful, it makes no sense to write down figures after it.

Usually, we express a measurement as some reliable numbers and one doubtful number in the last digit. This kind of data is called significant figure. Although the number of the last digit is doubtful, it reflects the actual value to some extent, and is rational. For example, the result of measuring the width of a copper board with a ruler of scale 1mm is 15.3mm, a significant figure of 3 digit numbers, the last number "3" is a result of estimating. The others may read it by "2" or "4". The existence of this digit means that the actual size of the board is 15mm (accurate) and about 0.3mm. It also reflects the scale of the ruler is 1mm or about 1mm. If measure the board with a vernier caliper of 50 scale markers (1 scale is 0.02mm), and get 15.30mm, it means a length of 10^{-2} can be measured, and the last digit happens to be zero. So there are 4 significant digits in this figure. Therefore, the value of physical quantity doesn't have the same meaning with the figure of mathematics. Mathematically, $15.3 = 15.30 = 15.300\cdots$, but it is not correct for physics measurement. Adding or deleting "0" at the end of a significant figure casually is not suitable.

The number of digits keeps unchanged when the unit of decimal system changes or the position of decimal point changes. Thus the "0" which indicates the position of decimal points cannot be included in significant figure. For example, 15.3mm = 1.53cm = 0.0153m = 0.0000153km, all the four expressions are equivalent, they all have significant figure of three digits.

Usually, the measurement result is expressed in the form of $a \times 10^n$, for $1 \leqslant a < 10$. Its number of digits is determined by a, change n (integral) will make its unit changed. For example, the radius of Earth can be expressed as $r = 6.371 \times 10^3$ km $= 6.371 \times 10^6$ m. This way of expression is clear and cannot be confused with "0" s.

2. Significant figure of measurement

The last digit of a significant figure is uncertain or doubtful. So we should write down a significant figure to the last digit on which the error occurs.

Uncertainty is a datum of significant figure itself. It is a rough range; normally it will have only one digit. For example, $L=(15.3\pm0.5)$ mm is a correct expression, $R=(31.0\pm2)\ \Omega$ is not correct, $h=(83.5\pm0.17)$ m is not correct too.

For the explaining above, we know that the number of digits of significant figure is determined by absolute uncertainty. When keep data from an instrument, be sure write it to the last digit which have to make estimate, not more or less. Sometimes, it maybe unnecessary to keep it to that digit, because the error of instrument is bigger, or the measuring method is not very accurate. In a word, be sure to keep data to the last digit of which the uncertainty occurs.

3. Calculation of significant figures

It is a like way of calculation of significant figures and getting the result of indirect measurement. There is a problem of propagation of uncertainty. Strictly, the significant figure of the result of calculation should be determined by it uncertainty. When the uncertainties of the data are unknown, we can get the result by following the rules:

(1) The result of calculation of exact (reliable) numbers is reliable number.

(2) The result of calculation of doubtful number and exact (reliable) or doubtful number is doubtful.

(3) Be sure to keep just one digit of doubtful number at the end of result, round the number after it.

EXAMPLE 4 Calculation of significant figures, where the number with a bar under it is doubtful.

(a)

$$\begin{array}{r} 1.38\underline{9} \\ 17.\underline{2} \\ +)\ 8.6\underline{4} \\ \hline 27.\underline{229} \end{array}$$

The result is $1.38\underline{9}+17.\underline{2}+8.6\underline{4}=27.\underline{2}$

(b)

$$\begin{array}{r} 12.\underline{2} \\ -)\ 1.38\underline{9} \\ -)\ 8.6\underline{4} \\ \hline 7.\underline{171} \end{array}$$

The result is $17.\underline{2}-1.38\underline{9}-8.6\underline{4}=7.\underline{2}$

(c)

$$\begin{array}{r} 4.3\underline{8} \\ \times)\ 1.\underline{1} \\ \hline 0.\underline{438} \\ 4.3\underline{8} \\ \hline 4.\underline{818} \end{array}$$

(d)

$$\begin{array}{r} 4.3\underline{63} \\ 12\underline{1}\ \overline{)\ 52\underline{8}} \\ 48\underline{4} \\ \hline 4\underline{40} \\ 36\underline{3} \\ \hline \underline{770} \\ \underline{726} \\ \hline \underline{43} \end{array}$$

The result is $4.3\underline{8} \times 1.\underline{1} = 4.\underline{8}$ The result is $52\underline{8} \div 12\underline{1} = 4.3\underline{6}$

Summery:

(1) In sums and differences, keep the significant figure of result to the digit which all the data being calculated have.

(2) In multiplication and division, make the number of digits of result same as the datum of shortest digits, sometime it can be kept an additional digit.

(3) The other calculation such as power, exacting root, trigonometric function, logarithm, usually keep the digits of result as the same of variable. For example: $(4.256)^2 = 18.11$, $\sqrt{4.256} = 2.063$.

For the constants such as π, e, $\sqrt{2}$, $\frac{1}{2}$, etc, are exact figures and have digits as many as you need. When calculating, they should have one more digit than the datum which has the most digits among the data. For example: to calculate $S = \pi R^2$, for $R = 5.082$, then the π can be 3.1416.

By handling significant figure properly, not only the measurement can be written suitably, but also can make the process concise, and make the process be more efficient. By the calculating of direct measurements we can get an indirect measurement result or value, so when we design our experiment, we should make the relative uncertainties of all the direct measurements are similar in magnitude. Or some of measurements may be too precise to have practical value, it also a kind of waste of resource. When we get our result from a scientific calculator, it a good habit to keep significant figures suitably. Specially, it not a comprehensive way, or bad habit, to write down all the numbers of the calculation result and make no round off. It also makes no sense.

Some additional comments on data handling:

(1) The method of data handling is based on the theory of probability statistics; it is a suitable way of estimating the accuracy of the indirect measurement result. It is not necessary to make the uncertainty precisely. In this course, it is enough to keep the uncertainties (including absolute and relative uncertainties) as a significant figure of one digit.

(2) It doesn't show that there is no error if the values of measurements are the same, the reason may be the instrument is not precision enough, it cannot detect the small difference of the measuring object; or the average absolute uncertainty of the measurements is smaller than the error of instrument. Here, the error of instrument should be taken as the uncertainty of measurement result.

Graphical presentation of experiment

The relations between physics quantities are often expressed by graphs, which is called the graphical presentation. It will make the relations between physics quantities clear, concise and visualization. Even the math relations are not clear, we also can make graphs based on the measurements, analyze it and try to get a fitted equation. We also can try to get the of variables, differential coefficient, integral, or other calculation.

1. Rules of graphical presentation

(1) A graph should be made on a plotting paper. First, determine the parameters, then select a proper plotting paper: rectangular coordinate plotting paper, logarithm plotting paper, or other plotting paper. The plotting paper in millimeter scale is commonly used in physics experiment.

(2) The size of plotting paper and the proportion of its axes should be determined by the significant figures of measurements. In principle, the minimal size of the paper should be able to contain all the data and also can express the uncertainties of data. Mark the coordinate axes with a pencil, and it's unnecessary to mark out the arrows of axis. Usually the horizontal axis denotes the independent variable and the vertical axis denotes the dependent variable. The names of physics quantities and their units should be written along axis in the out side, and mark values in equidistance.

Pay attention when draw a graph:

① The digits of the read of any point cannot be shorter than the significant figures of measurements.

② The proportion of the axis should be 1 : 2, 1 : 5, or the multiple of them. It's not suitable to be 1 : 3, 1 : 7. We can find out a point or read out a point quickly and accurately with proper proportion.

③ Be sure to make the graph in the center of the plotting paper and take the most space of the paper, so it's not necessary to start the coordinate from zero.

(3) Mark the points on the plotting paper according to measurements. Mark the point with the sign " + " . The sign " + " should be marked with a ruler, and its size should be about the uncertainty of measurement. If there are several graphs in one paper, different signs such as " × ", " ⊙" should be used to make them different.

(4) Draw line. Draw a smooth and clear straight – line or curve linking all the points by the help of ruler or French curve. It is not necessary to make the straight – line or the curve passes through all the points, but make the points distributing equally on the two sides of the curve and as possible as near the curve. Pay attention to the exceptional points which have a big difference with the curve, and then decide accepting it or not.

The graph should be smooth, clear and with suitable darkness, drew with a pencil. Names of experiment, graph, experimenter, cooperator, and some specific data and experimental date should be written down on the blank space of the paper.

2. Straight – line and making straight of a curve

(1) When the relation between x and y is linear, we can get the equation from the graph: $y = kx + b$, where k is the slope and b is the intercept on the vertical coordinate.

For the straight – line showing in Figure 1 – 2, we choose two points P_1 (x_1, y_1), P_2 (x_2, y_2) near the two ends of the straight – line. It's easy to get the reads if we choose the cross points of the line with even number line of the paper. The two points should be marked with different sign and their coordinate value should be written down nearby (See figure 1 – 2). It is suggested that the two points which used to calculate slope not to be the original data and be separated enough but in the range of experiment.

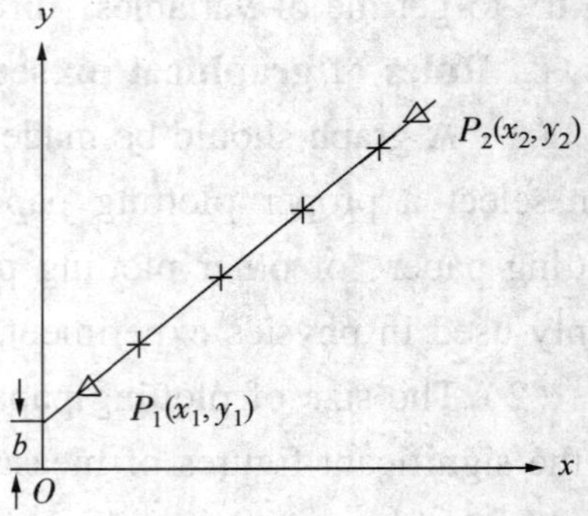

Figure 1 – 2 Straight – line

According to the coordinate values of the two points, we can get the slope:

$$k = \frac{y_2 - y_1}{x_2 - x_1}$$

Pay attention to the unit of slope k.

The intercept b of a straight – line can be read out directly vertical axis. It also can be calculated by the formula:

$$b = \frac{x_2 y_1 - x_1 y_2}{x_2 - x_1}$$

(2) If the two physics quantities x, y have a leaner relationship, and a series of data has gotten under equal observations: (x_i, y_i), $i = 1, 2, \cdots, n$, we can get a best straight – line $y = kx + b$ theoretically, which make the sum of all the square of the distances between experiment points and the line be least. This method is called least – square procedure:

$$k = \frac{\sum_{i=1}^{n} x_i \cdot \sum_{i=1}^{n} y_i - n\sum_{i=1}^{n} (x_i \cdot y_i)}{(\sum_{i=1}^{n} x_i)^2 - n\sum_{i=1}^{n} x_i^{\ 2}} = \frac{\bar{x} \cdot \bar{y} - \overline{x \cdot y}}{\bar{x}^2 \cdot \overline{x^2}} \qquad (9)$$

$$b = \frac{\sum_{i=1}^{n} y_i}{n} - k\frac{\sum_{i=1}^{n} x_i}{n} = \bar{y} - k\bar{x} \qquad (10)$$

where $\bar{x} = \frac{\sum_{i=1}^{n} x_i}{n}$ is the mean value of x_i; $\bar{y} = \frac{\sum_{i=1}^{n} y_i}{n}$ is the mean value of y_i;

$\overline{xy} = \frac{\sum_{i=1}^{n} (x_i \cdot y_i)}{n}$ is the mean value of $x_i \cdot y_i$; and $\overline{x^2} = \frac{\sum_{i=1}^{n} x_i^2}{n}$ of $x_i^{\ 2}$.

For $y = kx + b = kx + \overline{y} - k\overline{x}$, that is $(y - \overline{y}) = k(x - \overline{x})$, so the point $(\overline{x}, \overline{y})$ has to be passed through by the line. Thus we can mark the point $(\overline{x}, \overline{y})$ first, and then draw the straight – line.

Only if there is a linear relationship between $x$and $y$that can we fit them into a straight – line. There is a related coefficient r which can verify the significance of the fitting. Its definition is:

$$r = \frac{n\sum_{i=1}^{n}(x_i \cdot y_i) - \sum_{i=1}^{n}x_i \cdot \sum_{i=1}^{n}y_i}{\sqrt{(n\sum_{i=1}^{n}x_i^2 - (\sum_{i=1}^{n}x_i)^2) \cdot (n\sum_{i=1}^{n}y_i^2 - (\sum_{i=1}^{n}y_i)^2)}}$$

$$= \frac{\overline{x \cdot y} - \overline{x} \cdot \overline{y}}{\sqrt{(\overline{x^2} - \overline{x}^2) \cdot (\overline{y^2} - \overline{y}^2)}} \qquad (11)$$

r shows the difference of the relationship between the real relationship with linear function. The closer of the value r with 1 the better of the linear relationship between x and y; if r close to zero, it indicates that there is no linear relationship between x and y. It is required that r should be larger than 0.9, for instance, 0.99, 0.999. Usually we express r with the number of "9" .

(3) Practically, we can transfer some functions into a special linear relationship, this is called making straight of a curve. For example:

① $y = ax^b$ (a, b are constants)

$\ln y = b\ln x + \ln a$, $\ln y$ has a linear relationship with x, the slope is b and the intercept is $\ln a$.

② $y = ae^{-bx}$ (a, b are constants)

$\ln y = -bx + \ln a$, $\ln y$ has a linear relationship with x, the slope is $-b$ and the intercept is $\ln b$.

③ $y = ab^x$ (a, b are constants)

$\ln y = (\ln b)x + \ln a$, $\ln y$ has a linear relationship with x, the slope is b and the intercept is $\ln a$.

④ $xy = c$ (c is constants)

$y = \frac{c}{x}$, y has a linear relationship with $\frac{1}{x}$, the slope is c.

⑤ $y^2 = 2px$ (p is constants)

y^2 has a linear relationship with x, the slope is $2p$.

⑥ $y = ax + bx^2$ (a, b are constants)

$\frac{y}{x} = bx + a$, $\frac{y}{x}$ has a linear relationship with x, the slope is b, and the intercept is a.

Example 5 Draw a straight – line on plotting paper.

Table 1 – 3 shows the data of the resistance of brass wires at different temperatures.

Draw a straight – line to show the relationship between resistance and temperature.

Table 1 – 3 Data of the resistance of brass wires at different temperatures

t (℃)	R (Ω)	t (℃)	R (Ω)
15. 5	2. 807	40. 3	3. 059
24. 4	2. 897	45. 0	3. 107
26. 5	2. 919	49. 7	3. 155
31. 1	2. 969	54. 9	3. 207
35. 0	3. 003	60. 0	3. 261

SOLUTION: Let vertical and horizontal coordinate represent the resistance R and temperature t respectively. Length of 2mm on the vertical axis denotes 0. 010Ω, and 2mm on the horizontal axis denotes 1. 0℃, plot the points. It is showed on the Figure 3. We can find that the relation between resistance and temperature is linear, and can be expressed as the following straight – line equation:

$$R = kt + a$$

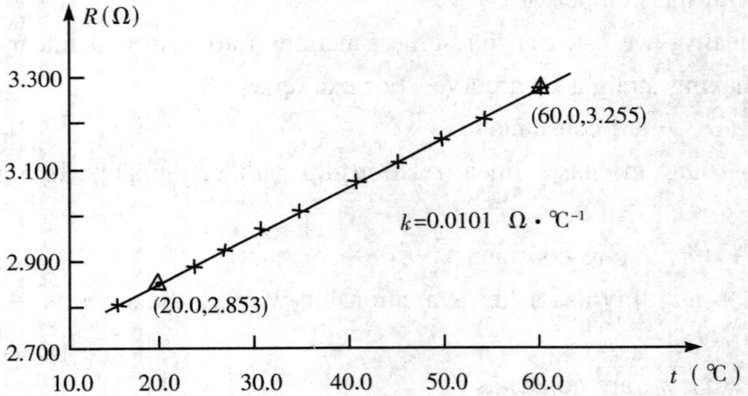

Figure 1 – 3 A graph of resistance and temperature of brass wire

Take two points (20. 0, 2. 853), (60. 0, 3. 255) near the two ends of the line, to calculate the slop and intercept:

$$k = \frac{y_2 - y_1}{x_2 - x_1} = \frac{3.255 - 2.853}{60.0 - 20.0} = \frac{0.402}{40.0} = 0.0101\Omega \cdot ℃^{-1}$$

$$a = \frac{x_2 y_1 - x_1 y_2}{x_2 - x_1} = \frac{60.0 \times 2.853 - 20.0 \times 3.255}{60.0 - 20.0} = 2.652\Omega$$

So we can get the relation between resistance and temperature:

$$R = 0.0101t + 2.652\ \Omega$$

When two physical quantities have a linear relationship, multiple times of measurements are often taken to get proportion coefficient from graph. It's a useful method and more accuracy than single measurement.

Problems

1. What's the difference between systematic error and random error? Which of the following error is systematic or random error?

A. Reading error

B. Wandering of zero point of a balance

C. Inaccuracy of a vernier

D. Connecting resistance of a ammeter

E. Non – uniform of capillary of a mercury thermometer

2. Which one of the following measurements is more accurate?

A. $l_1 = (54.98 \pm 0.02)$ cm

B. $l_2 = (0.498 \pm 0.002)$ cm

C. $l_3 = (0.0098 \pm 0.0002)$ cm

3. Get the uncertainty expressions of following functions (the quantities in the right hand are all direct measurement).

A. $N = x + y - 2z$

B. $N = \dfrac{x-y}{y-z}$

C. $Q = \dfrac{k}{2}(A^2 + b^2)$ (k is constant)

D. $I_2 = I_1 \dfrac{r_2^2}{r_1^2}$

E. $n = \dfrac{\sin i}{\sin r}$

4. A length l (about 17cm) is measured by a meter – ruler (scale 1mm), the limitation of the measurements error is 3mm. Write down the expression of $l \pm \sigma_l$.

5. Measuring a length I by a meter – ruler (scale 1mm), it starts at the line of 10cm and ends at 20cm. Please write down the expression of I by correct significant digits.

6. Table 1 – 4 is the data of the measurement of ten times for a certain quantity x.

Table 1 – 4 Data for Problem 6

i	1	2	3	4	5	6	7	8	9	10
x_i (cm)	20.2	20.0	20.4	20.3	20.2	20.4	20.1	20.3	20.4	20.5

Please calculate $\bar{x} \pm \sigma_x$ for the first 5 data and all the 10 data respectively, compare the results.

7. Calculate the results of the following expressions according to rules of significant figure:

A. $2.00 \times 4.00 + 50.0 \times 1.00 + 20.0 \times 0.1$

B. $\frac{1}{(0.1000)^2} - \frac{1}{(0.5000)^2}$ (where numerator "1" is accuracy)

C. $6.80 \times 10^3 - 20$

D. $\frac{2.480 - 2.0 \times 5.989}{2.00}$

E. $2.500 \times 3.000 \times \left(1 + \frac{4}{500}\right)$ (where, the numerator "4" and number "1" is accuracy)

8. The data of single pendulum experiment are showed in Table 1 – 5, where L is the length of the pendulum and T is period. Please draw a straight – line, calculate its slope and the acceleration of gravity.

Table 1 – 5 Data of Problem 8

L (cm)	61.5	71.2	81.0	89.5	95.5
T^2 (s^2)	2.468	2.877	3.262	3.618	3.861

9. According to the data of Problem 8, try to fit a straight – line by the least – square procedure. And calculate the slope, intercept, and get the gravity acceleration finally.

Additional Instruction

Some basic knowledge about measurement has been presented in this chapter. Here is a brief introduction about normal distribution and its application in error processing. It's not very easy to learn it very well because there is a lot of new concept. But it will be helpful for you to understand the measurement uncertainty if you can read it once in a while during this course.

Normal Distribution

It is proofed that the result of many measurements under same condition will obey normal distribution (also called Gauss Distribution):

$$f(x) = \frac{1}{\sigma\sqrt{2\pi}} e^{-\frac{(x-\mu)^2}{2\sigma^2}}, \quad |x| < \infty \qquad (12)$$

where x is the measurement result.

$f(x)$ is a probability density function, $f(x) \cdot dx$ means the probability of measurement result x being in the range of $x \rightarrow x + dx$. $f(x)$ gets its peak value when $x = \mu$, which is called the mean value. σ is the absolute value of the difference on the horizontal axis between the point of inflection of the curve and μ, and it is called standard deviation. The graph of $f(x)$ is showed in figure 1 – 4.

The result of repeated measurements under same condition is called a measurement list. For a certain list, if the values of μ, σ are known, we can calculate the probability of x being in the range of (a, b):

$$P(a < x < b) = \int_a^b f(x) \cdot dx = \int_a^b \frac{1}{\sigma\sqrt{2\pi}} e^{-\frac{(x-\mu)^2}{2\sigma^2}} \cdot dx \qquad (13)$$

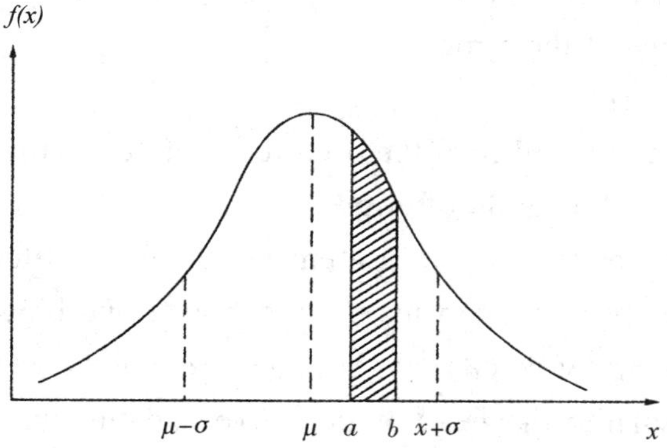

Figure 1 – 4 Normal distribution

This integral equals the shadow area surrounding by $x = a$, $x = b$, $f(x)$ and horizontal axis (showed in figure 1 – 4). Try different range of (a, b) and calculate the integral, we can get:

The probability of x being in the range of $(\mu - \sigma, \mu + \sigma)$ is 68.3%;

The probability of x being in the range of $(\mu - 2\sigma, \mu + 2\sigma)$ is 95.4%;

The probability of x being in the range of $(\mu - 3\sigma, \mu + 3\sigma)$ is 99.7%.

Here is the reason that 3σ is called limitation error. Take $a = -\infty$ and $b = +\infty$, of course we can get:

$$P(-\infty < x < +\infty) = \int_{-\infty}^{+\infty} f(x) \cdot dx = 1$$

The probability equals 1 means inevitable event, that is x always being in the range of $(-\infty, +\infty)$.

Shift the vertical axis $f(x)$ in the figure 1 – 4 to the peak $x = \mu$, that is let $\delta = (x - \mu)$, we can get the normal distribution as:

$$\Psi(\delta) = \frac{1}{\sigma\sqrt{2\pi}} e^{-\frac{\delta^2}{2\sigma^2}}, \quad |\delta| < \infty \qquad (14)$$

This curve is showed in figure 1-5, where the horizontal axis denotes the random error $\delta = (x-\mu)$. The three curve in Figure 1-5 represent three different value of σ. For the graph we can know three characteristics of it:

① Single peak There is a bigger probability of the occurrence of small errors than that of great errors;

② Symmetry There is a same probability of occurrence of the errors with same absolute value;

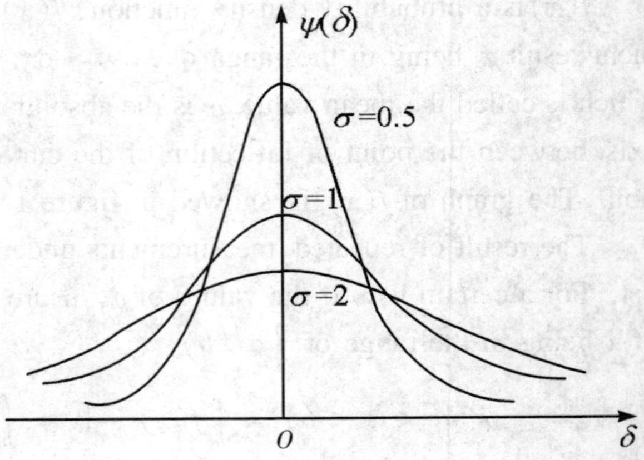

Figure 1-5 Normal distribution curves of 3 different σ values

③ Limitation The probability of the occurrence of very large error is near to be zero, or there is an actual limitation of error.

④ Compensation For the same measurement, the mean value of error will tend towards zero as the times of measurements become bigger and bigger.

The area surrounding by $\Psi(\delta)$ and horizontal axis is 1, which is a constant, thus the bigger σ is, the smaller the peak of the curve and the curve becomes more flat; contrarily, the smaller σ is, the higher the peak of the curve and the curve becomes sharper. The former means the variable δ (random error) is more dispersive, and the later means it is more concentrative.

Mean Value and Standard Deviation

Suppose systematic error is cleared up, we get the result $x_1, x_2, \cdots, x_n$ when measuring a certain quantity x under the same condition for n times. The data obey normal distribution $f(x) = \frac{1}{\sigma\sqrt{2\pi}} e^{-\frac{(x-\mu)^2}{2\sigma^2}}$. It is proved (omitted here) that μ is the true value of x. The mean value of $x_1, x_2, \cdots, x_n$ is:

$$\bar{x} = \frac{x_1 + x_2 + \cdots + x_n}{n} = \frac{\sum x_i}{n}$$

($\sum x_i$ is the shortening of $\sum_{i=1}^{n} x_i$) Now we discuss the reason choosing $\bar{x}$ as the measurement result and it is the best estimate value for the true value μ.

① The result should depend on every measurement $x_1, x_2, \cdots, x_n$ because they are equal observed, and every one of them should have the same weight. The mean value have this characteristic.

② Because the systematic error has been eliminated, the random error of every measurement is:

$$\delta_i = x_i - \mu$$

Here μ is the true value. Add up all the measurements and divide the sum by n:

$$\frac{\sum \delta_i}{n} = \frac{\sum x_i}{n} - \mu = \bar{x} - \mu$$

For the compensation of random error we know that for $n \to \infty$, get $\frac{\sum \delta_i}{n} \to 0$, so $\bar{x} \to \mu$.

③We know from the least – square method that the best estimate value X which gets from a set of measurement x_1, x_2, ⋯, x_n will be the smallest sum of the square of the difference between X and every measurement:

$$Q = \sum (X - x_i)^2 \to \min$$

To get the minimum, let the derivative of X be zero:

$$\frac{dQ}{dX} = 2\sum (X - x_i) = 0$$

$$nX - \sum x_i = 0$$

We get $X = \frac{\sum x_i}{n} = \bar{x}$, that is the best estimate value $\bar{x}$.

There is a subtle difference between the mean value $\bar{x}$ and the true value μ. For example the difference between x_i and $\bar{x}$ is:

$$\Delta x_i = x_i - \bar{x}$$

Δx_i is not the same as $\delta_i = x_i - \mu$. Thus Δx_i is called residual. In this course we pay no attention on the difference between them and call them error.

For one measurement the standard deviation σ_i is defined as the root – mean – square of $\delta_i = \sigma_i = \sqrt{\frac{\sum (\delta_i)^2}{n}} = \sqrt{\frac{\sum (x_i - \mu)^2}{n}}$

It can be proved theoretically ("Physics Experiments and Guidance", Edited by Zuimin Fang, China Medico – Pharmaceutical Science & Technology Press, 1992) that when n becomes large enough, the formula which is used to calculate σ_i from the residual error Δx_i will be:

$$\sigma_i = \sqrt{\frac{\sum (\Delta x_i)^2}{n-1}} = \sqrt{\frac{\sum (x_i - x)^2}{n-1}}$$

That is the equation (3) .

The propagation of uncertainty and standard deviation of mean value

The indirect measurement value N can be gotten by calculating from the function

$N=f(x_A, x_B, \cdots)$, in which x_A, x_B, $\cdots$ is independent direct measurement quantities. The uncertainty of N can be determined by the propagation formula (7) and (8). Suppose the absolute error of x_A, x_B, $\cdots$ are σ_A, $\sigma_B \cdots$, respectively. The standard deviation σ_N of N can also be determined by the following formula ("Physics Experiments and Guidance", Edited by Zuimin Fang, China Medico – Pharmacentical Science & Technology Press, 1992):

$$\sigma_N = \sqrt{\left(\frac{\partial f}{\partial x_A}\sigma_A\right)^2 + \left(\frac{\partial f}{\partial x_B}\sigma_B\right)^2 + \cdots}$$

Relative uncertainty is

$$E = \frac{\sigma_N}{N} = \frac{1}{f}\sqrt{\left(\frac{\partial f}{\partial x_A}\sigma_A\right)^2 + \left(\frac{\partial f}{\partial x_B}\sigma_B\right)^2 + \cdots}$$

Now we can get the standard deviation of mean value according to these formulas.

For $\bar{x} = \dfrac{(x_1 + x_2 + \cdots + x_n)}{n}$, we get

$$\sigma_{\bar{x}} = \sqrt{\left(\frac{\partial x}{\partial x_1}\sigma_{x_1}\right)^2 + \left(\frac{\partial x}{\partial x_2}\sigma_{x_2}\right)^2 + \cdots + \left(\frac{\partial \bar{x}}{\partial x_n}\sigma_{x_n}\right)^2}$$

$$= \sqrt{\frac{1}{n^2}\sigma_{x_1}{}^2 + \frac{1}{n^2}\sigma_{x_2}{}^2 + \cdots + \frac{1}{n^2}\sigma_{x_n}{}^2} = \sqrt{\frac{1}{n^2}\sum \sigma_{x_i}{}^2}$$

$$= \sqrt{\frac{1}{n^2} \cdot n\sigma_{x_1}{}^2} = \sqrt{\frac{\sigma_x{}^2}{n}} = = \frac{\sigma_x}{\sqrt{n}} = \sqrt{\frac{\sum (x_i - \bar{x})^2}{n(n-1)}}$$

That is the equation (4). The standard deviation of mean value equals the result of the standard deviation of one measurement divide by the square root of measure times. It means that the reliability of measurement result can be improved by increasing of measure times. But, comparing times of 100 times with 10 times, one can find that the reliability increases to $\sqrt{100/10} \approx 3$ times. So, this way of measuring is unworthy.

实验二　长度的测量

长度是最基本的物理量，是构成空间的最基本的要素。世界上任何物体都具有一定的空间形态，长度的测量对科学研究、生产活动和日常生活都有巨大的影响。所以长度的测量是最常用、最基本的测量之一。不仅限于对长度本身的测量，很多非长度的测量被归结为对长度的测量：例如医生通过体温计或血压计中水银柱的长度，读出患者的体温或血压。因此，长度的测量方法具有广泛的意义。随着科学技术的发展，长度测量仪器的测量范围是越来越广，测量精度是越来越高。长度的测量覆盖了从微观粒子到宇宙深处的整个物理学研究的尺度范围。根据被研究对象的尺度，应当采取不同的实验方法和实验仪器。通过本次实验，我们要学会利用游标卡尺和螺旋测微计进行长度的测量。

【实验目的】

1. 掌握游标卡尺及螺旋测微计的测量原理和使用方法。
2. 练习处理实验数据。

【实验器材】

游标卡尺、螺旋测微计、待测圆柱和小球。

【实验原理】

1. 游标卡尺

游标卡尺又称为游标测径器，是在米尺（主尺）上附加一个能够滑动的有刻度的副尺（游标），结构如图 2－1 所示。主尺 D 与量爪 A、A'相联，游标 E 与量爪 B、B'及深度尺 C 相联。推动小轮 G，游标即可紧贴着主尺滑动。F 为固定螺钉。量爪 A、B 用来测量外径，量爪 A'、B'用来测量内径，深度尺 C 用来测量深度。测量的读数值，都是由游标“0”刻度线和主尺的“0”刻度线之间的距离来表示。

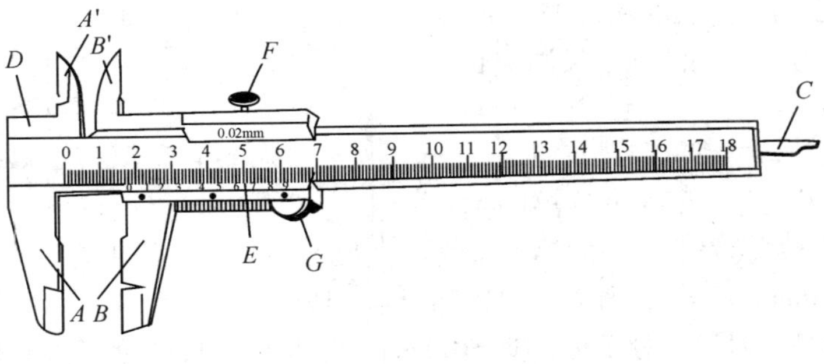

图 2－1　游标卡尺

游标卡尺利用游标原理进行长度测量：游标上 n 个等分格子的总长与主尺上（$n-1$）个等分格子的总长相等。设 x 代表游标上一个格子的长度，y 代表主尺上一个格子的长度，则有

$$nx=(n-1)y$$

那么，主尺与游标上每个格子的长度差值为

$$\delta x=y-x=\frac{1}{n}y$$

主尺的最小分度是1mm，即 $y=1\text{mm}$。当 n 取10时，称为十分游标，相应的 $\delta x=0.1$ mm。

下面以十分游标为例，说明游标卡尺的读数方法。如图2-2，当量爪 A、B 合拢时，游标上的“0”线与主尺的“0”线重合，游标上第一条刻线（除“0”刻线）在主尺第一条刻线（除“0”刻线）的左边0.1 mm处，第二条刻线在主尺第二条刻线左边0.2 mm处……，依次类推。

如果在量爪 A、B 间放进一张厚度为0.1 mm的纸片，这时游标就要向右移动0.1mm，如图2-3。游标上的第一条刻线与主尺上的第一条刻线重合，而游标上的其他任一刻线不与主尺的任一条刻线重合。如果纸片的厚度为0.2mm，那么游标上的第二条刻线与主尺上的第二条刻线重合……依次类推。

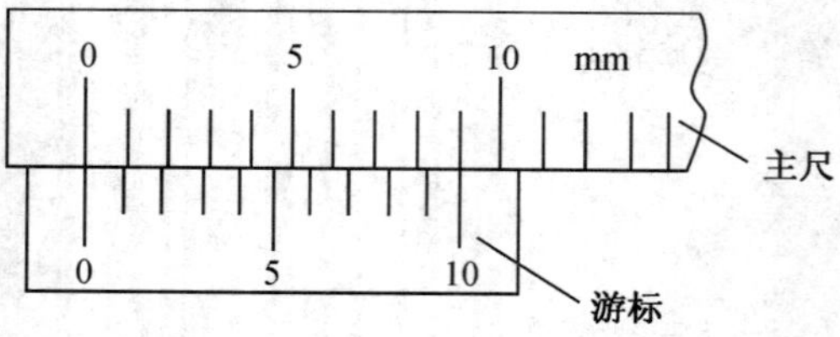

图2-2　十分游标

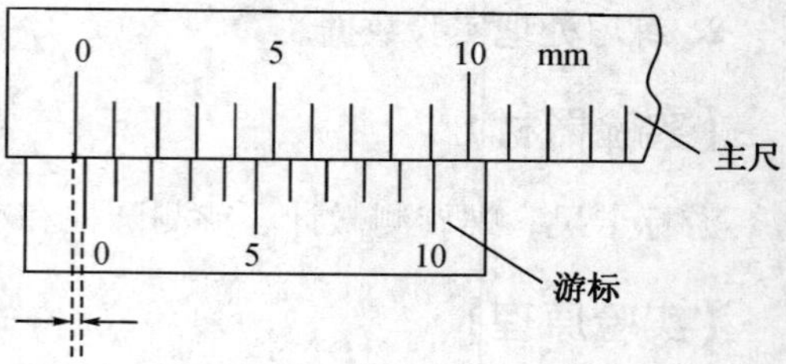

图2-3　十分游标测量0.1mm厚度

由上述可知，利用游标可以读出游标“0”线与主尺上前面最靠近的一条刻线之间的长度，即1 mm之内的长度。如图2-4，游标“0”线处在41 mm和42 mm刻线之间，并且游标上第五条线与主尺上的刻线对齐。因此，游标“0”线与41 mm刻线之间的长度是0.5 mm，那么物体的长度就是41.5mm。

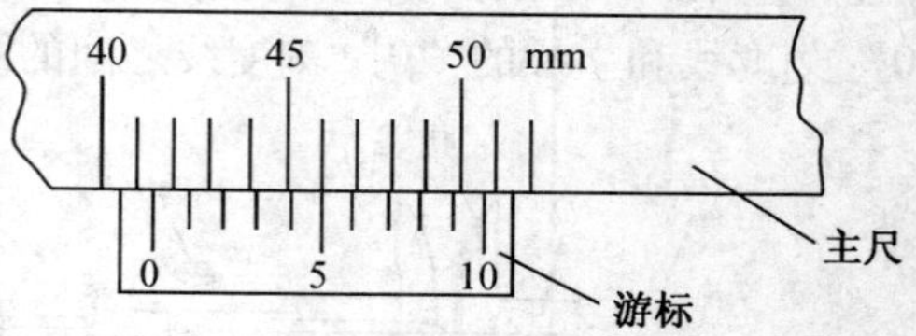

图2-4　十分游标卡尺的读数

至此，我们可以得到游标卡尺测量长度 l 的普遍表达式：

$$l=l'+k\delta x$$

l'是游标“0”线所在处可读出的主尺上的整毫米刻度，k'代表游标上的第 k 条线与主尺上的某一条线重合。图2-4所示的情况就是：$l'=41.0\text{mm}$，$k=5$，所以 $l=41.5\text{mm}=4.15\text{cm}$。

需要注意的是，对于十分游标，$\delta x=0.1\text{mm}$，这是由主尺的分度值 y 和游标尺的分度值 x 之差给出的，不是估读，它是十分游标能读出的最小数值，即十分游标的分度值。所

以在测量长度时，$k\delta x$ 是精确的。而根据有效数字的一般规则，读数的最后一位是读数的误差所在一位，仪器最大绝对误差 $\frac{1}{2}\delta x=0.05\text{mm}$，那么上面的长度 l 就应该改写为：$l=41.50\text{mm}=4.150\text{cm}$。最后一位加上“0”，表示读数误差数位。

由此可见，使用游标可以提高读数的准确程度。

另一种常见的游标是二十分游标（$n=20$），即将主尺上的 19mm 等分成游标上的20格（图 2－5），或者将主尺上的 39mm 等分成游标上的 20 格（图 2－6）。

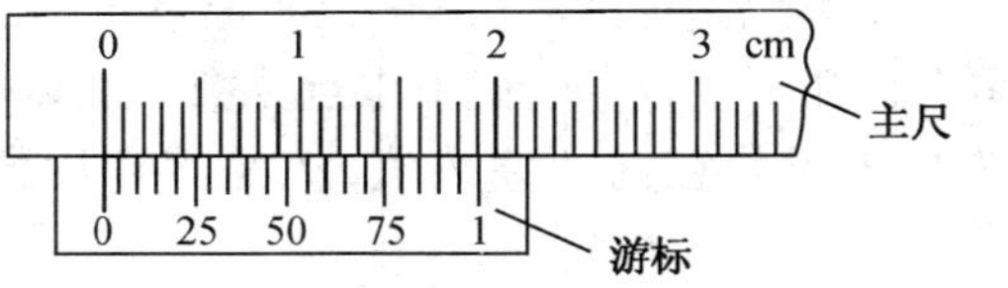

图 2－5　二十分游标

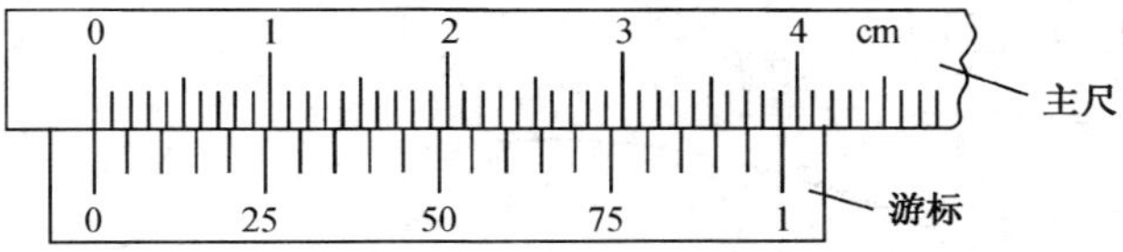

图 2－6　二十分游标

这样它们的最小分度 δx 分别为

$$\delta x=1.00-\frac{19}{20}=0.05\text{mm}$$

或

$$\delta x=2.00-\frac{39}{20}=0.05\text{mm}$$

在后一种情况下，δx 是主尺上两格（2mm）与游标上的一格之差。

二十分游标上常刻有 0、25、50、75 等标度，以便于直接读数。如游标上第 5 根线（标度 25）与主尺上某刻线对齐，则读数的尾数为 $5\times\delta x=0.25\text{mm}$，即可直接读出。其估读误差为 $\frac{1}{2}\delta x\approx0.03\text{mm}$，在百分之一毫米位数上。因此，读数的尾数之后不再加“0”。

还有一种常用的游标是五十分游标（$n=50$），即将主尺上 49mm 等分成游标上的 50 格，见图 2－7。五十分游标的分度值 $\delta x=1.00-\frac{49}{50}=0.02\text{mm}$，游标上刻有 0、1、2、……、9 标度，以便于读数，$\frac{1}{2}\delta x=0.01\text{mm}$，五十分游标的读数结果也写到百分之一毫米这一位上。

综上所述，游标卡尺的分度值是由主尺与游标上的一格的长度之差值决定的，各种常用的游标卡尺的读数都写到百分之一毫米这一位上。

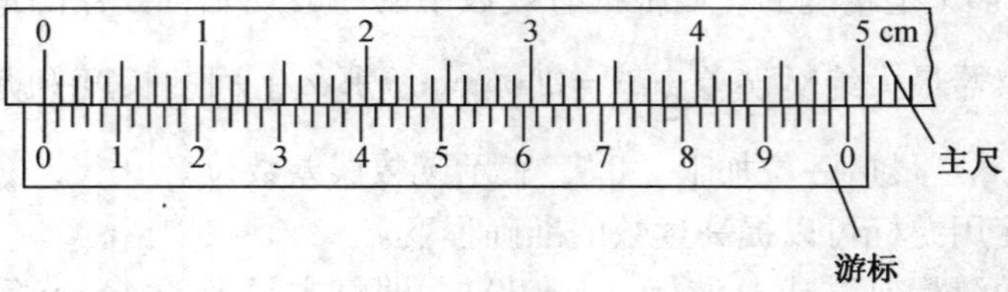

图 2－7　五十分游标

使用游标卡尺测量之前，应先进行零点修正，即把量爪 A、B 合拢，检查游标“0”线是否和主尺“0”线重合，如不重合，应记下零点读数，加以修正。即待测长度的结果应写为 $l=l_1-l_0$，l_1 是未作零点修正前的读数值，l_0 是零点读数（可以是正的，也可以负）。

使用游标卡尺时，可左手拿物体，右手持尺，如图 2－8 所示。要特别注意保护量爪不被磨损，使用时轻轻把物体卡住即可读数。不允许用游标卡尺来测量粗糙的物体，并切忌被夹紧的物体在卡口内挪动。

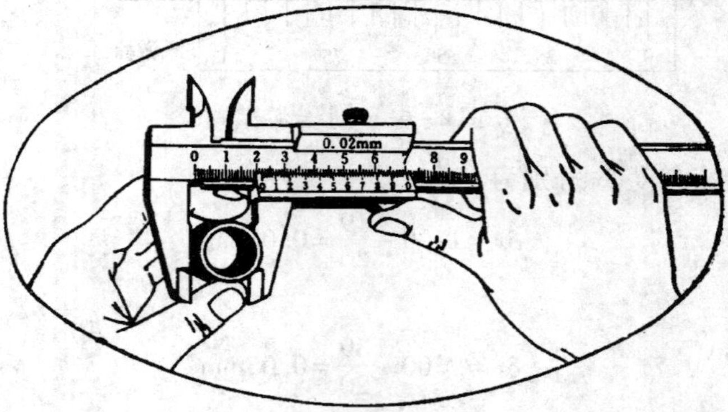

图 2－8　游标卡尺的使用

2. 螺旋测微计

螺旋测微计也叫千分尺，是一种比游标卡尺更精密的长度测量仪器。如图 2－9 所示，它的测量原理是将待测长度进行机械放大。其结构的主要部分是一个微动螺杆，螺距是 0.5mm。因此，当螺旋杆转动一周时，它沿轴线方向只前进 0.5mm。螺旋杆是和螺旋柄（微分筒）连着的，柄上附有沿圆周的刻度，共有 50 个等分格。当螺旋柄上的刻度转过一分格时，螺旋杆沿轴线方向前进 $\frac{0.5\text{mm}}{50}=0.01\text{mm}$，这就是螺旋测微计的分度值 δx，那么仪器误差为 $\frac{1}{2}\delta x=0.005\text{mm}$。

测量物体长度时，应轻轻转动螺旋柄后端的棘轮旋柄，推动螺杆，把待测物体刚好夹住。读数时，可以从固定标尺上读出整格数（每格 0.5mm）。0.5mm 以下的读数则由微分筒上的刻度读出，可估读到每格的1/10，即 0.001mm 位上。如图 2－10，（a）和（b）的读数分别为 5.650mm 和 5.150mm。

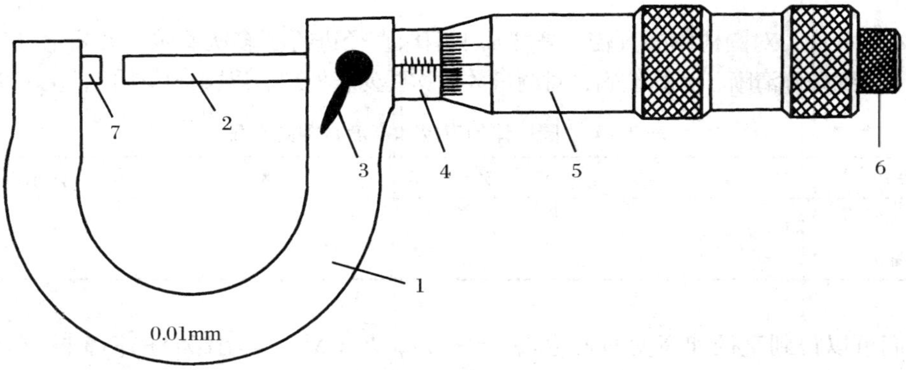

图 2－9　螺旋测微计

1．尺架；2．微动螺杆；3．锁紧装置；4．固定套管；
5．微分筒；6．棘轮旋柄；7．测砧

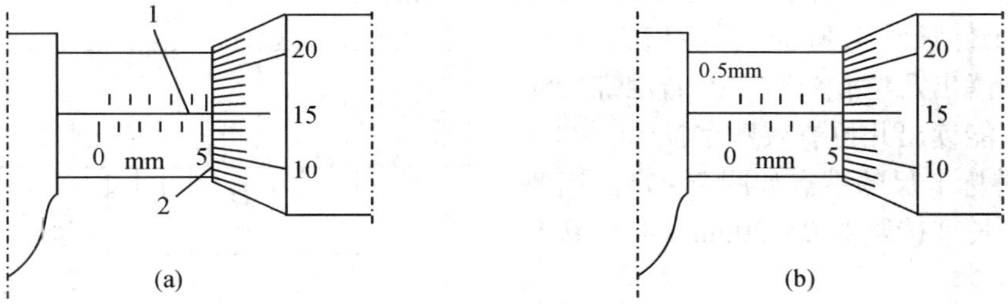

图 2－10　螺旋测微计的读数

1．圆周分度的读数准线；2．固定标尺的读数准线

使用时注意：

（1）进行零点修正。先记录下零点读数 l_0，并注意 l_0的正负。

（2）记录零点及将待测物夹紧测量时，必须缓慢转动棘轮旋柄推进螺杆，不要直接拧转螺杆，以免夹得太紧，影响测量结果及损坏仪器。当转动棘轮听到喀、喀的声音时，就不要再推进螺杆。这时就可以进行读数了。

（3）要保护仪器的测量面，不要用手去摸。使用完毕，放入仪器盒时应使两测量面之间留有一小空隙，以免仪器受热膨胀而损坏仪器。

【实验内容】

1. 用游标卡尺测金属圆柱体的高度和直径各 10 次，然后计算出圆柱体的体积和误差。

2. 用螺旋测微计测量小钢珠的直径 20 次，求出它的体积，并计算误差。

【数据处理】

在本实验中，对圆柱体的直径、高度及小钢珠直径进行了多次测量，并要计算体积和误差。为了做到条理清晰，计算准确，对测量数据均要列表处理。以圆柱体为例，见表2-1。

表2-1　圆柱体的直径和高度数据处理表

序号	1	2	3	4	5	6	7	8	9	10	平均值	标准差
d（cm）												
h（cm）												

我们可以得到直径和高度的表达式：$\bar{d} \pm \Delta d$，$\bar{h} \pm \Delta h$。算出圆柱体体积 $V = \frac{\pi}{4}\bar{d}^2\bar{h}$，再利用误差公式 $E = \frac{\Delta V}{V} = 2\frac{\Delta d}{d} + \frac{\Delta h}{h}$，最后得到 $\Delta V = E \cdot V$，并写出结果（$V \pm \Delta V$）。

【思考题】

1. 游标卡尺和螺旋测微计的仪器误差是怎样确定的？

2. 某工件长约1.8mm，若用五十分游标测量，能读出几位有效数字？若用螺旋测微计测量，能读出几位有效数字？

3. 一游标卡尺的零点如图2-11。当测量某物体的长度读数为25.20mm 时，其实际长度为多少？

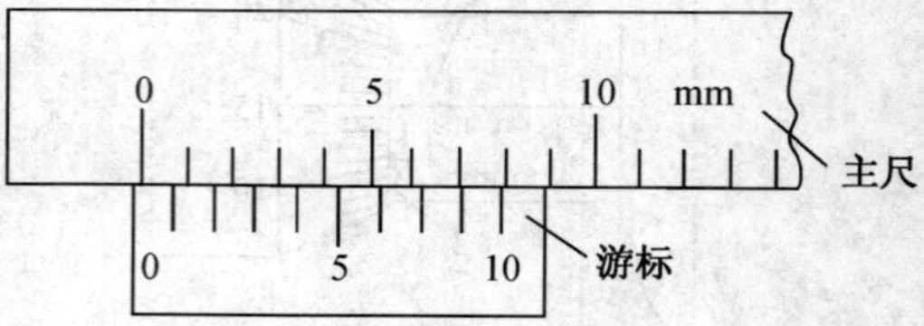

图2-11

4. 读出图2-12中螺旋测微计的零点读数。

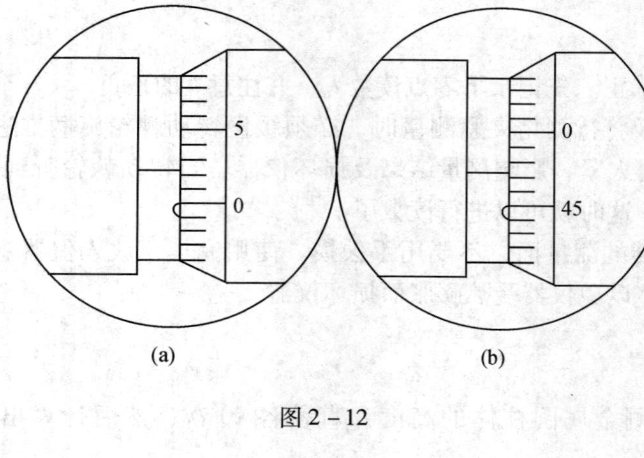

图2-12

实 验 指 导

【预习要求】

1. 首先要理解实验一中的测量误差的基本理论，特别是关于有效数字部分。
2. 为何利用游标可以提高测量的精确度?
3. 准确理解螺旋测微计的测量原理，即机械放大原理。

【操作要点】

1. 首先了解仪器使用的注意事项，切忌有违反操作规则的情况出现。

2. 在对圆柱体或钢珠进行多次重复测量时，注意数据点（测量点）要均匀分布，不能对某局部有过密或过疏的测量次数。

3. 通过本实验的操作，能熟悉各种游标的读数方法，熟悉螺旋测微计的读数方法，在以后遇到同类型的仪器，能正确操作使用。

【报告要求】

1. 认真按绪论中的实验报告要求，作出本实验报告。
2. 对各要求计算的项目，写出运算步骤，不能直接给出结果。

【实验后的思考】

1. 测量圆柱体体积时，d 和 h 二者中哪一个对结果 V 的误差影响大?
2. 如果要求结果的相对误差 $E<0.5\%$，如何安排测量的次数?

2 The Measurement of Length

Length is the most basic physical quantity, and the most fundamental element of space. Any object in the world has a certain degree of spatial patterns. The measurement of length has a great impact on scientific research, production activities and daily life. Therefore, the measurement of length is one of the most commonly used and the most basic measurement. Not confined to the length measurements, many non – length measurements are attributed to measurement of length. For example, doctors can find out their patients′ body temperature or blood pressure through the length of mercury column inside thermometer or sphygmomanometer. Therefore, the method of length measurement has a broad meaning. With the development of science and technology, the length measuring range of instruments is increasingly wide and the measurement accuracy is getting higher and higher. From the micro – particles to the depths of the universe, the measuring range covers the entire physics scales. According to the different object, different experimental methods and experimental apparatus should be taken. In this experiment, we learn to use vernier caliper and micrometer for length measurements.

Objectives

1. Learn the principle of measurement of vernier caliper and micrometer, and their application.
2. Practice handling experiment data.

Equipments

vernier caliper, micrometer, a steel cylinder and a steel bead

Principle

1. Vernier caliper

A short vernier scale, which can slide along the main scale, is attached to the main scale, and this combination is called vernier caliper, or caliper. As shown in figure 2 – 1, the jaws AA′ are fixed at main scale D; jaws BB, and the depth gauge C are fixed at vernier scale. Push the little wheel G, vernier can be moved along main scale. F is a fixing screw. The jaws AB are used to measure outside diameters and the jaws A′ B′ are used to measure inside diameters. The depth gauge C is used to measure depth. A measurement can be got by reading the distance between the zero mark of the vernier and the zero mark of the main scale.

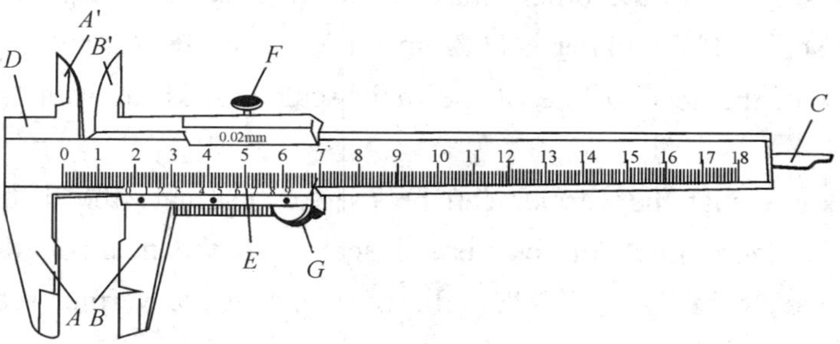

Figure 2 – 1 Vernier caliper

The vernier scale is graduated in this way: its certain number of divisions (say n) equals the total length of $(n-1)$ divisions of main scale divisions. Then if x is the length of a vernier division, and y the length of the corresponding division on main scale, we have

$$nx = (n-1)\ y$$

The difference between y and x is:

$$\delta x = y - x = \frac{1}{n} y$$

The graduation of the main scale is 1mm. When $n = 10$, that is a ten – graduation vernier, δx equals 0. 1mm.

Now, let's take a ten – graduation caliper for an example of reading direction. See figure 2 – 2, when the two jaws A and B are meeting together, the zero line of the main scale coincides with the zero line of the vernier scale. The fist line of the vernier scale (except the zero line) is at 0. 1mm left of the fist line (except the zero line) of the main scale, the second line of the vernier scale is at 0. 2mm left of the second line of the main scale, etc.

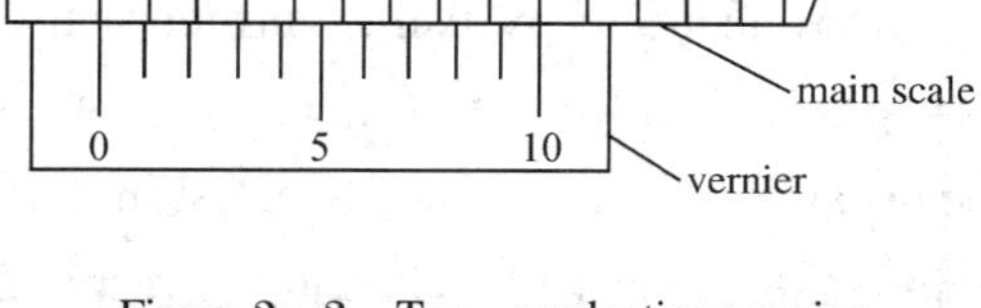

Figure 2 – 2 Ten – graduation vernier

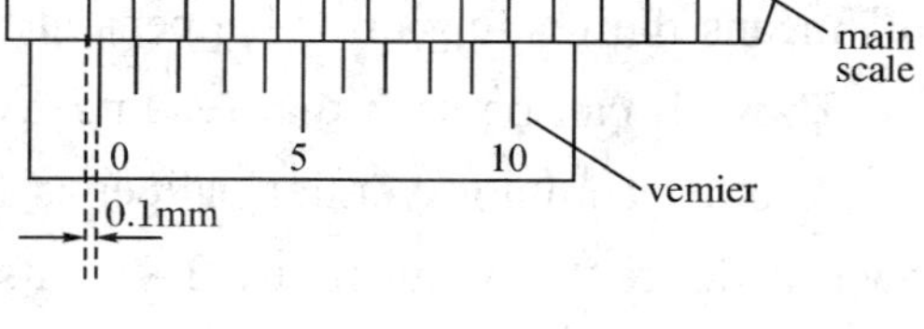

Figure 2 – 3 A thickness of 0. 1mm in a ten – graduation vernier

Place a paper of 0. 1mm thick between the jaws AB; the vernier will move 0. 1mm right, as shown in figure 2 – 3. The first line of the vernier scale coincides with the first line of the main scale, and the other lines of the main scale do not coincide with any line of the main scale. If the paper is 0. 2mm thick, then the second line of the vernier scale coincides with the second line of the main scale , it is the same reason for other situation.

Thus, we know that the vernier can be used to estimate the distance (less than 1mm) between the zero mark of the vernier scale and the nearest graduation line of main scale. As shown in figure 2 – 4, the zero line of the vernier scale lies between 41mm and 42mm, and the 5th line of the vernier scale is aligned with one of the marks on the main scale, so we can get the distance between 41mm and the zero mark of the vernier scale is 0. 5mm. The object length is 41. 5mm.

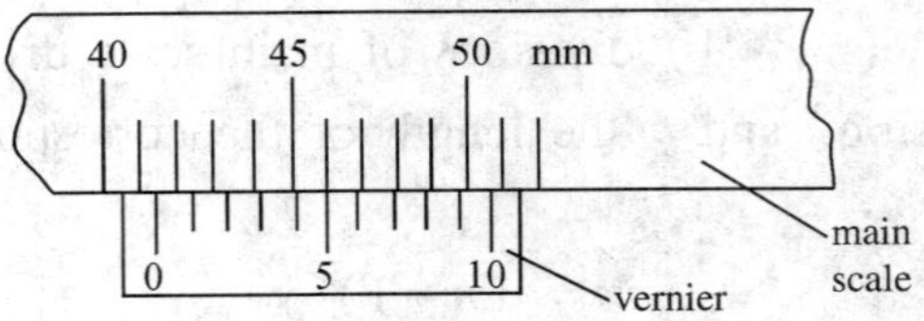

Figure 2 – 4 The reading of ten – graduation vernier

Now, we get a formula of measurement of vernier caliper:

$$l = l' + k\delta x$$

Where l' is the reading of integer millimeter on the main scale corresponding to the zero mark on the vernier scale, k represents the kth vernier scale mark coincides with a main scale mark. For figure 2 – 4, we get: $l' = 41.0\text{mm}$, $k = 5$, thus $l = 41.5\text{mm} = 4.15\text{cm}$.

For ten – graduation vernier caliper $\delta x = 0.1\text{mm}$ is the difference between y and x, not an estimated figure. So $k\delta x$ is accurate. By the significant figure rule, the last digit of reading is where error happens. The maximum error of the instrument is $\frac{1}{2}\delta x = 0.05\text{mm}$. So the above reading should be written as: $l = 41.50\text{mm} = 4.150\text{cm}$. The last "0" means the reading error happens at the last digit.

Thus, it can be seen that reading would be more accurate by using vernier scale.

Another common vernier caliper is twenty – graduation vernier ($n = 20$) . As shown in figure 2 – 5 and figure 2 – 6, its vernier scale is divided into 20 divisions, the total length corresponds to 19mm of the main scale, or 40 divisions corresponds to 39mm.

Therefore the graduation δx is:

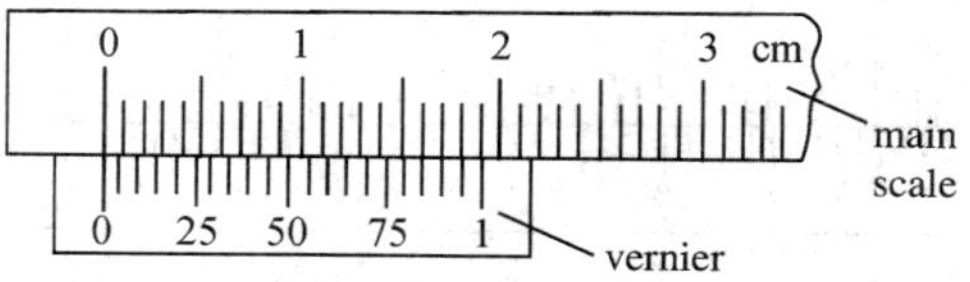

Figure 2 – 5 Twenty – graduation vernier

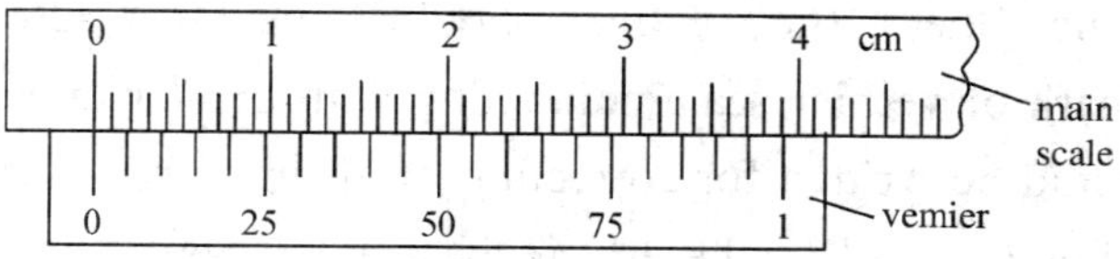

Figure 2 – 6 Twenty – graduation vernier

$$\delta x = 1.00 - \frac{19}{20} = 0.05\text{mm}$$

or

$$\delta x = 2.00 - \frac{39}{20} = 0.05\text{mm}$$

In the second case, δx is the difference between the length of two divisions of the main scale (2mm) and the length of one division of vernier scale.

It is convenience for reading when mark special figures of 0, 25, 50 and 75 on the vernier scale of twenty – graduation vernier caliper. For example, when the 5th line of the vernier scale is aligned exactly with one of the marks on the main scale, the mantissa of reading will be $5 \times \delta x = 0.25$mm, that exact the number marked at the right line. The reading error for a twenty – graduation vernier caliper is $\frac{1}{2}\delta x \approx 0.03$mm, and it happens at the digit hundredths of millimeters. So it is not necessary to write a "0" at the end of the reading.

Another commonly used vernier caliper is fifty – graduation ones ($n = 50$), it is a vernier caliper on which the length of 50 divisions on the vernier scale corresponds to 49mm of the main scale. See figure 2 – 7. Its graduation is $\delta x = 1.00 - \frac{49}{50} = 0.02$mm. The figure 0, 1, 2, ⋯, 9 are marked on the vernier scale for reading convenience. $\frac{1}{2}\delta x = 0.01$mm. The reading ends at the digit of hundredth of millimeters.

Now, we can see the graduation of vernier caliper is decided by the difference between a division on main scale and the one on vernier caliper, and the reading errors of commonly used vernier caliper are about several hundredths of millimeter.

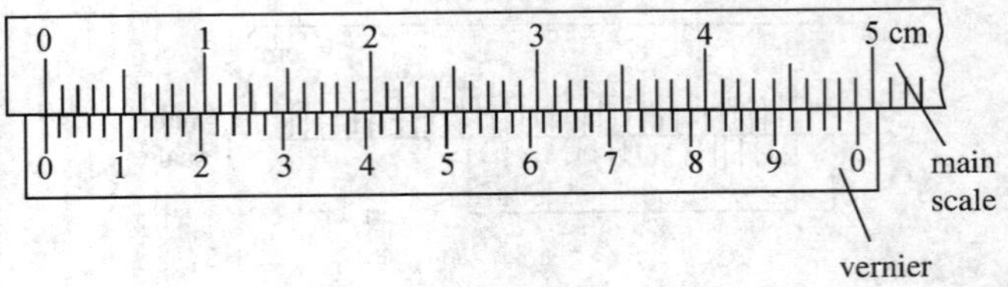

Figure 2 -7 Fifty - graduation vernier

Before measuring, the vernier caliper should have been zero corrected, that is checking the "0" mark of vernier scale coinciding with the "0" mark of main scale, if not, the reading should be written for correcting of all the other measurements. The result can be recorded as $l = l_1 - l_0$, where l_1 is the reading before correcting and l_0 is the reading of zero correcting. l_0 may be positive or negative either.

When measuring, it's better for holding object by left hand and holding vernier caliper by right hand, as figure 2 -8. All the jaws should be kept properly to avoid being worn.

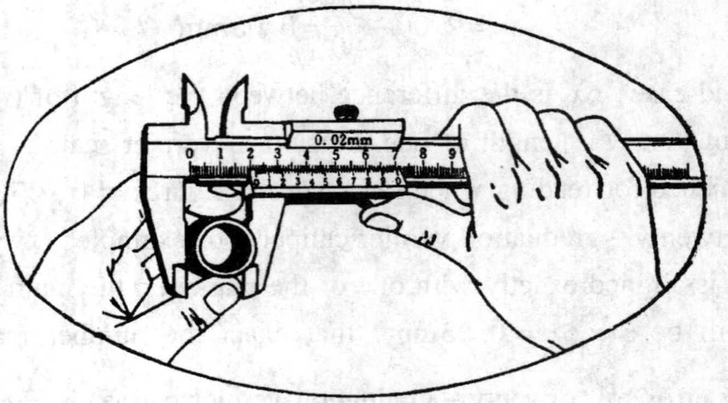

Figure 2 -8 The use of vernier caliper

2. Micrometer

The micrometer (also micrometer screw gauge) is a measurement instrument which is more accurate than the vernier calliper. The common one is shown in figure 2 -9. The measuring length has been mechanically magnified. Its essential part is a screw called spindle. If its pitch equals 0.5mm, the spindle will move forward by 0.5mm for turning one revolution of the graduated thimble. The graduated thimble is divided into 50 divisions, when the thimble turns for one division, the spindle will move $\frac{0.5\text{mm}}{50} = 0.01\text{mm}$, that is the smallest scale of the micrometer (δx). So the instrumental error is $\frac{1}{2}\delta x = 0.005\text{mm}$.

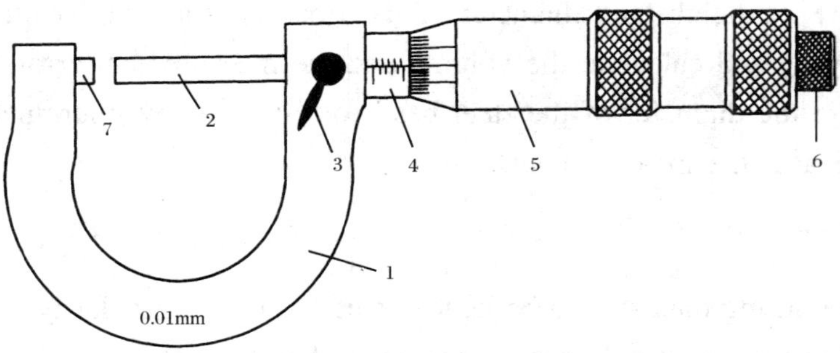

Figure 2 – 9　Micrometer

1. Frame　2. Screwed spindle　3. Fixing set　4. Measure scale

5. Graduated thimble　6. Handle of click pulley　7. Fixed spindle

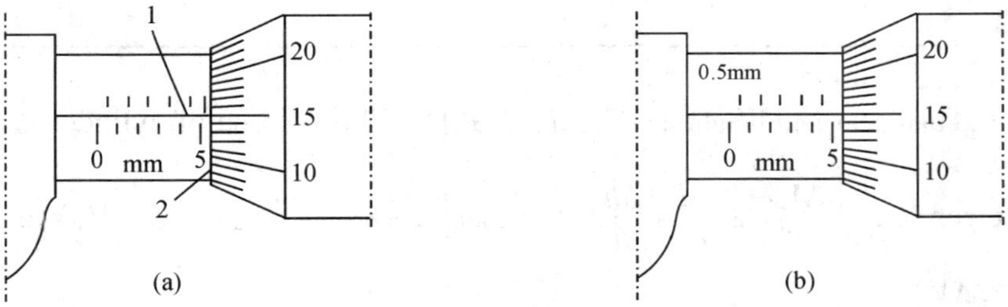

Figure 2 – 10　The reading of micrometer

1. Reading line for graduated thimble　2. Reading line for measure scale

When measuring, turn the handle of click pulley gently to make the screwed spindle just pressing the object. The integer divisions (0.5mm/div) can be read out on the measure scale, and length smaller than 0.5mm can be read by the graduated spindle, with an estimate of tenths of one division (0.001mm). For example, readings of figure 2 – 10 (a) and (b) are 5.650mm and 5.150mm, respectively.

When measuring, we must make sure that:

(1) Keep the zero error before measure. Note whether the zero error is positive or negative.

(2) The spindle is advanced by turning the handle, don't turn the spindle directly, in case that the pressure is too great to influence the data or damage the apparatus. When the ratchet head clicks, don't turn it any more and a reading can be taken.

(3) Don't touch the two measuring faces of spindles. When placed in the apparatus box after measurement, keep a distance between two faces in case that the instrument is damaged by a thermal expansion.

Experiment

1. Measure the height and diameter of the steel cylinder for 10 times respectively by vernier caliper, and calculate the volume and the measurement errors.

2. Measure the diameter of the steel bead for 20 times by micrometer, and calculate its volume and the measurement errors.

Data processing

All the measuring data should be kept in suitable tables for clarity and accurate calculation. We take the steel cylinder for example. See table 2 – 1.

Table 2 – 1 The measurement of diameter and height of the steel cylinder

Serial No.	1	2	3	4	5	6	7	8	9	10	mean value	standard deviation
d (cm)												
h (cm)												

We get the diameter and height as $\bar{d} \pm \Delta d$, $\bar{h} \pm \Delta h$. The volume of cylinder is $V = \frac{\pi}{4}\bar{d}^2\bar{h}$, its relative error is $E = \frac{\Delta V}{V} = 2\frac{\Delta d}{d} + \frac{\Delta h}{h}$. And then, according to $\Delta V = E \cdot V$, we get the last result: $(V \pm \Delta V)$.

Problems

1. What are the instrumental errors of vernier caliper and micrometer?

2. A certain object is about 1.8mm long. When it is measured by a fifty – graduation caliper, how many digits will it have for the reading? And what' s the answer for micrometer?

3. The zero point for a vernier caliper is shown in figure 2 – 11. What is the actual reading for a reading of 25.20mm?

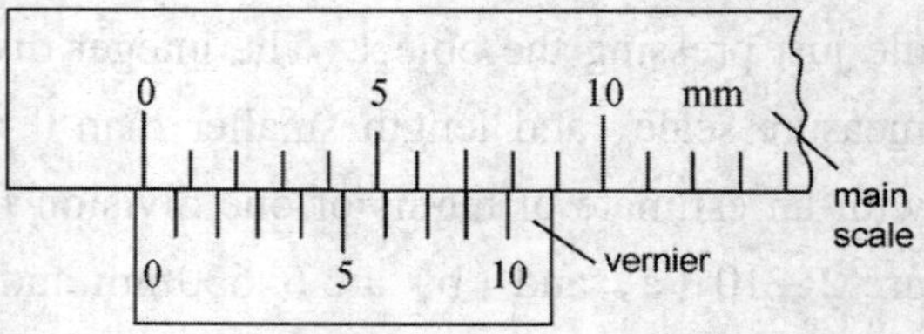

Figure 2 – 11

4. Read the zero points of micrometers in figure 2 – 12.

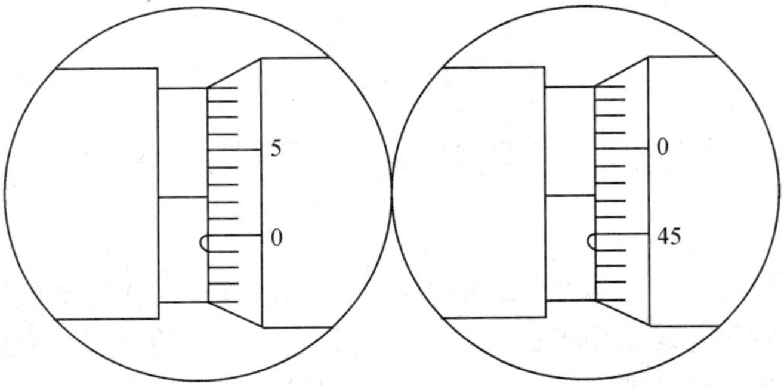

(a) (b)

Figure 2 – 12

Additional Instruction

Preparation requirement

1. Understand the basic knowledge of measurement of Exp. 1.
2. Why the measure precision could be increased for using a vernier caliper?
3. Understand the mechanical magnifying principle.

Operation key points

1. Observe operation rules.
2. Adequately distribute the measuring points on the surface of cylinder or bead.
3. The measure methods in this experiment are typical. Be sure to understand them.

Report requirement

1. Write this report according to the requirement in the preface.
2. It is necessary to write down every calculation step. It isn't a good habit to give out final result directly.

Problems after experiment

1. Which one of the two quantities, h and d, has a great effect on the error of the volume of cylinder?
2. How to arrange measuring times of different items for a requirement of relative error of final result, for instance, $E < 0.5\%$?

实验三　刚体转动惯量的测定

刚体的转动定律决定了刚体在外力矩作用下的运动规律。而转动惯量是表征刚体转动特性的物理量，是刚体在转动中惯性大小的量度，它与刚体的形状、刚体的质量、刚体的质量分布以及刚体转轴的位置有关。例如，舞蹈演员及花样滑冰运动员可以通过肢体的位置来改变自身的转动惯量。对于几何形状简单而有规则的刚体绕定轴转动的转动惯量，可以通过数学方法计算；但对于形状复杂或不规则的刚体，转动惯量就难以计算，所以常常通过实验的方法来测定。测量特定物体的转动惯量对炮弹飞行、飞轮设计、发动机叶片设计及卫星外形设计等生产研究工作具有十分重要的意义。本实验利用刚体转动实验仪测定刚体转动惯量。

【实验目的】

1. 验证刚体的转动定律和平行轴定理。
2. 学习用作图法处理数据——曲线改直。
3. 学习用最小二乘法进行数据处理。

【实验器材】

刚体转动实验仪、停表、米尺。

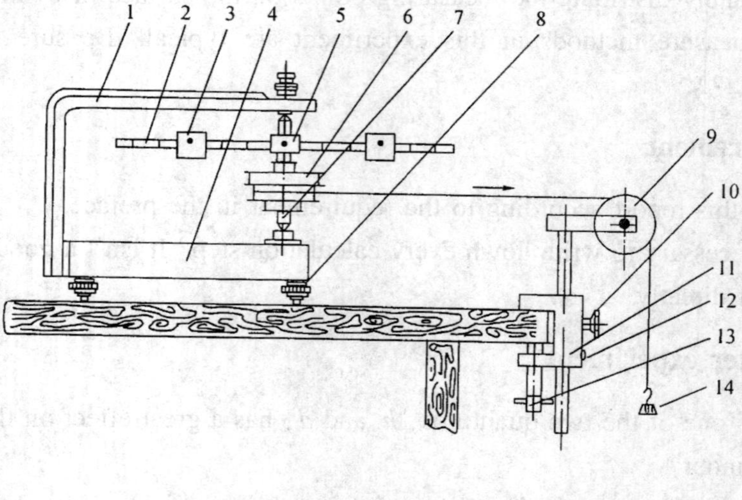

图 3－1　刚体转动实验仪

1. 臂；2. 杆；3. 重锤；4. 底座；5. 中心套；6. 塔轮；7. 心轴；8. 调平螺钉；9. 滑轮；10. 固定螺钉；11. 固定架；12. 指针；13. 夹紧螺杆；14. 砝码

刚体转动实验仪的结构如图 3－1 所示。塔轮（6）具有不同的半径，两边对称地伸出两根有等分刻度的均匀细柱，细柱上各有一个可移动的圆柱形重锤 m_0（3），它们一起组成一个可以绕固定轴转动的刚体。塔轮上绕一细线，通过轻质滑轮（9）与砝码 m（14）相连。当 m 下落时，通过细线对刚体系施加外力矩。滑轮的支架可以借固定螺钉（10）升降，以保证当细线绕塔轮的不同半径时都可以保持与转动轴垂直。滑轮的固定架（11）上有一个指针（12），用来判断砝码的起始位置。实验开始时，取下塔轮，换上铅直准钉，通过调平螺钉（8）的调节可以使转轴竖直。当转轴调至竖直后，再换上塔轮，转动合适后，用螺丝固定。

【实验原理】

根据转动定律，当刚体绕固定轴转动时，有：

$$M = J\beta \tag{1}$$

因此，如果能测出系统的合外力矩 M 和转动时的角加速度 β，则转动惯量 J 可由上式求出。

系统的合外力矩 M 由两部分组成：细线施于塔轮上的力矩 $r \times T$（r 为塔轮半径，T 为细线的张力）和轴处的摩擦力矩 M_μ。如果选前者的方向为正，则 M 的大小为：

$$M = rT\sin\theta - M_\mu \tag{2}$$

式中 θ 为 T 与转轴之间的夹角。实验中保持细线的与转轴相垂直，则 $sin\theta = 1$，因此：

$$M = rT - M_\mu \tag{3}$$

若略去滑轮及细线的质量，并认为细线不可伸长，砝码 m 以加速度 a 下落，根据牛顿第二定律可得到：

$$T = m(g - a) \tag{4}$$

式中，g 为重力加速度，m 为砝码的质量。如果砝码在高度为 h 的地方从静止开始下落，到达地面所用的时间为 t，则有：

$$h = \frac{1}{2}at^2 \tag{5}$$

由于塔轮转动时边缘的切向加速度和砝码下落的加速度相等，即：

$$a = r\beta \tag{6}$$

根据（1）式，联合（3）～（6）式，可得刚体的合外力矩为：

$$M = rm(g - a) - M_\mu = J\beta = J\frac{2h}{rt^2} \tag{7}$$

实验中，若保持 $a \ll g$，则上式可简化为：

$$mgr = \frac{2hJ}{rt^2} + M_\mu \tag{8}$$

而利用（8）式，我们可得到如下三种测量转动惯量的方法。

方法一：实验中，保持 r、h 及 J 不变，我们可将（8）式改写为：

$$m = \frac{2hJ}{gr^2}\ \frac{1}{t^2} + \frac{M_\mu}{gr} = k_1\frac{1}{t^2} + c_1 \tag{9}$$

其中，$k_1 = \dfrac{2hJ}{gr^2}$，$c_1 = \dfrac{M_\mu}{gr}$。由式（9）可知，m 与 t^2 成反比。若设 M_μ 不随下落重物 m

的变化而变化，我们可以改变 m 而测出相应的下落时间 t，在坐标纸上作出 $m-\frac{1}{t^2}$图。如得到一直线，则由实验结果可证明转动定律成立，同时由斜率 k_1 可求得 J，由截距 c_1 可求得 M_μ。

方法二：实验中，保持 m、h 及 J 不变，我们可将（8）式改写为：

$$r=\frac{2hJ}{mg}\frac{1}{rt^2}+\frac{M_\mu}{mg}=k_2\frac{1}{rt^2}+c_2 \tag{10}$$

式中，$k_2=\frac{2hJ}{mg}$，$c_2=\frac{M_\mu}{mg}$。由式（10）可知，r 与 rt^2 成反比，我们可以改变 r 而测出相应的下落时间 t，在坐标纸上作出 $r-\frac{1}{rt^2}$图。如得到一直线，则由实验结果可证明转动定律成立，并由斜率 k_2 可求得 J，由截距 c_2 可求得 M_μ。

方法三：如果在转轴上添加适量润滑油，将轴的运转调整得较合适，可以认为 $M_\mu \ll mgr$，因而可将 M_μ 略去，则可将（8）式改写为：

$$m=\frac{2hJ}{gr^2}\frac{1}{t^2} \tag{11}$$

或：

$$r=\sqrt{\frac{2hJ}{mg}}\frac{1}{t} \tag{12}$$

仿照方法一和方法二也可测出刚体的转动惯量。

另外，实验中，若保持 r，h 及 m 不变，对称地改变重锤 m_0 的质心对转轴的距离 x，由刚体转动的平行轴定理可知，整个系统对转轴的转动惯量为：

$$J=J_0+J'_0+2m_0x^2 \tag{13}$$

式中，J_0 为杆和塔轮绕轴的转动惯量，J'_0 为两个圆柱体重锤对通过其质心且平行于塔轮转轴的转动惯量。将上式代入式（9）中可得到：

$$t^2=kx^2+c \tag{14}$$

其中，$k=\frac{4m_0h}{r(mgr-M_\mu)}$，$c=\frac{2h(J_0+J'_0)}{r(mgr-M_\mu)}$，在坐标纸上作 t^2-x^2 图即可证明平行轴定理。

【实验内容】

1. 验证刚体转动定律，测定刚体转动惯量

（1）取下塔轮换上铅直准钉，调节调平螺钉，使塔轮轴线取铅直位置（底座同时也达到水平），将塔轮重新装上，塔轮转轴固定合适，以转动时不晃动为限。

（2）将绕线均匀排列在塔轮的半径上，绕时不能重叠。调节滑轮支架的位置、高低，使塔轮绕线水平跨过滑轮。

（3）应用方法一或方法二测量刚体的转动惯量和摩擦力矩，验证转动定律。在实验中要注意砝码 m 每次都要从指针处下落，并且初速一定要是零，同时要保证 $g \gg a$。

2. 验证平行轴定理，观测转动惯量随质量分布的情况

取 $m=20g$，$r=2.5\text{cm}$ 不变，对称地改变重锤 m_0 的位置，测量对应的下落时间 t。

【数据处理】

根据式（9），我们固定 r（$r=2.0\text{cm}$），h，J，改变 m，测出相应的下落时间 t，可列出如下的数据（表 3－1）。

表 3－1　固定 r，h，J，改变 m 时，测得的下落时间 t

m（g）	5.00	10.00	15.00	20.00	25.00	30.00	35.00
t_1（s）							
t_2（s）							
t_3（s）							
$\bar{t}$（s）							
$(\bar{t})^{-2}$（s^{-2}）							

$r=2.0\text{cm}$，$h=$　　cm

根据式（14），我们固定 $m=20\text{g}$，$r=2.0\text{cm}$，改变重锤的位置，测量下落时间 t，可得表 3－2。

表 3－2　固定 r，h，m，改变 x 时，测得的下落时间 t

x	1，1′	2，2′	3，3′	4，4′	5，5′
t_1（s）					
t_2（s）					
t_3（s）					
$\bar{t}$（s）					
$(\bar{t})^2$（s^2）					

$r=2.0\text{cm}$，$m=20\text{g}$

表 3－1、表 3－2，t_i（$i=1$，2，3）为同一实验条件下的下落时间的三次测量值，$\bar{t}=(t_1+t_2+t_3)/3$。

我们以表格一为例，用两种方法来处理数据。

1. 作图法

我们在坐标纸上作 $m-t^{-2}$ 图（这里 t 指的是 $\bar{t}$），得一直线，则验证了刚体转动定律。从图线中求出直线的斜率 k_1 和截距 c_1，并根据测得的 h 值和式（9），有：

$$J=k_1gr^2/2h$$

$$M_{\mu}=c_1gr$$

2. 最小二乘法（线性回归法）

很多计算器可以在输入测得的数据后，直接选用线性回归法能得到直线方程 $y=a+bx$ 的截距 a 和斜率 b，并能得出相关系数 γ。当然也可通过手工计算。下面我们以手工计算来说明。本实验中砝码的质量 m 为自变量，可以认为是无误差的，因变量为时间 t。将式（9）改写为：

$$\frac{1}{t^2}=\frac{1}{k_1}m-\frac{c_1}{k_1}$$

令$\frac{1}{t^2}=y$，$m=x$，$-\frac{c_1}{k_1}=a$，$\frac{1}{k_1}=b$，上式写作 $y=a+bx$。按最小二乘法作直线拟合，求出 a、b：

$$b=\frac{S_{xy}}{S_{xx}}$$

$$a=\sum_i y_i/n-b\sum_i x_i/n$$

相关系数为

$$\gamma=\frac{S_{xy}}{S_{xx}\cdot S_{yy}}$$

式中

$$S_{xy}=\sum_i x_i y_i-\frac{1}{n}\sum_i x_i\sum_i y_i$$

$$S_{xx}=\sum_i x_i^2-\frac{1}{n}(\sum_i x_i)^2$$

$$S_{yy}=\sum_i y_i^2-\frac{1}{n}(\sum_i y_i)^2$$

要得到上述参数，我们可设计一个表格计算出 $\sum_i x_i$，$\sum_i y_i$，$\sum_i x_i y_i$，$\sum_i x_i^2$，$\sum_i y_i^2$。由 a 和 b 值即可求得 M_μ 和 J，并与作图法求得的值进行比较。

我们可以用同样的方法来处理表二中的数据，以验证平行轴定理。

【思考题】

1. 在转动惯量测量中，如果用钢制或胶木砝码来代替铝制砝码，其它条件不变，每一次测量的时间 t 会有什么变化？为什么？

2. 塔轮绕线重叠将产生什么影响？

3. 在验证转动定律和平行轴定理的实验数据处理中，为什么不作 $m-t^2$图和$t-x$ 图，而作 $m-t^{-2}$图和 t^2-x^2 图？

实　验　指　导

【预习要求】

1. 理解实验原理和目的。
2. 熟悉实验仪器的结构。
3. 写出重要的操作步骤。
4. 理解实验数据的处理方法。

【操作要点】

1. 实验过程中要保持转轴的竖直。

2. 塔轮必须调整至松紧合适。

3. 减少运动部件的摩擦。

【报告要求】

设计合适的数据记录表格，根据实验一的要求作图。

【实验后的思考】

本实验测量的偶然误差主要来源于时间的测量，通常约为百分之几的量级，如果 $M_\mu << mgr$，我们可根据 $J=\frac{mgr^2}{2h}t^2$ 来大致估计转动惯量测量的相对误差为

$$E_J=\sqrt{\left(\frac{\Delta m}{m}\right)^2+4\left(\frac{\Delta r}{r}\right)^2+\left(\frac{\Delta h}{h}\right)^2+4\left(\frac{\Delta t}{t}\right)^2}$$

式中 Δm 取天平仪器误差，Δr 取游标的最小读数，Δh 取米尺最小读数的一半，Δr 取秒表的最小读数，m 和 t 取每个测量值的平均。根据 $\Delta J=J\cdot E_J$，将测量结果写成标准形式：$(J\pm\Delta J)$（$g\cdot cm^2$）。

下面粗略分析一下系统误差对实验结果的影响。

（1）T 垂直于转轴的条件。如果其不能很好满足而有 5°（角度）的偏差，则 Tr 应为 $Tr\cos\alpha$，Tr 的相对误差为 $1-\cos\alpha\approx\frac{1}{2}\alpha^2\approx0.38\%$。本实验的基本误差为测量时间的误差，约为百分之几的量级，因此，0.38% 的误差就可以不考虑了。而 5°的偏差对一般人用目测就可以看出来，所以，在实验装置上没有作专门的考虑，调节时目测就可以了。

（2）$g>>a$ 的条件。先估计一下 a 的大小。本实验中，$h\approx90$cm，砝码 m 最大用 35g，对于砝码（铝制品）（m_0），t 最小约为 3.5s（这里指改变 m 测 t 的情况下），而 t 的最大值约为 20s。对于 $t=20$s，$a\approx0.45\text{cm/s}^2$，$a/g\approx4.5\times10^{-4}$；对于 $t=5$s，$a\approx7.2\text{cm/s}^2$，$a/g\approx7.3\times10^{-3}$；对于 $t=3.5$s，$a\approx15\text{cm/s}^2$，$a/g\approx1.5\times10^{-2}$。可见，$a/g$ 在某些情况下是要考虑的。

（3）滑轮质量 M_0 对结果的影响。考虑了 M_0 以后，影响了 T 的数值，即实际的拉力为 $T'=m[g-(1+M_0/2m)a]$，而不是 $T=m(g-a)$。仪器中 M_0 约为 15g，当 $m=35$g 时，$(1+M_0/2m)\approx1.2$，对于 $t=5$s，$(1+M_0/2m)a/g\approx1\%$；对于 $t=3.5$s，$(1+M_0/2m)a/g\approx2\%$。当 $m=5$g 时，$(1+M_0/2m)\approx2.5$，对于 $t=15$s，$(1+M_0/2m)a/g\approx0.2\%$。这可以看作是对上述讨论 a/g 影响的补充。

（4）h 不变的条件。由 $h=\frac{1}{2}at^2$，有：$\frac{dh}{h}=at\frac{dt}{t}$ 如果 $h=90$cm，$t=10$s，h 即使偏差 0.5cm，只相当于 t 的测量误差为小于 0.03s，因此，这个条件是很容易满足的。

（5）$M_\mu<<mgr$ 的条件。方法三中需要用该条件，而该条件可归结为 $M_\mu=m_\mu gr$ 中 m_μ/m 的大小。用顶尖轴承时，$m_\mu=1\sim3$g，如果仪器使用时间较久而没有定期重磨轴尖时，m_μ 甚至会大于 5g。因此，这时 m_μ 的影响是不可忽略的。所以使用方法三测量转动惯量，最好使用滚珠轴承，这时，$m_\mu\approx0.1\sim0.5$g。

3 The Determination of Moment of Inertia

Rigid – body rotation law is the law of motion of rigid body in external moment of force. The moment of inertia is the physical quantity which describes the physical characteristics of rigid body rotation. It's the measurement of the rotating inertia of rigid body; it is related to the shape of a rigid body, the mass of rigid – body, the mass distribution, as well as the location of the rotation axis. For example, dancers and figure skating athletes can change the moment of inertia of their own by changing their body location. Around a fixed axis of rotation, the moment of inertia of the rigid body with simple and regular shape can be calculated; but for complex or irregular ones, moment of inertia will be difficult to calculate. Usually we use the experimental approach. The determination of moment of inertia of specific objects, such as the flying shells, fly-wheel design, engine blade design, satellite design and other production work, has - great significance. There are many experimental approaches of rigid – body moment of inertia. This time, we choose rigid body rotation apparatus.

Objectives

1. Verifying of the law of rotation and theorem of parallel axes of rigid body.
2. Learn to handle data by graphic——making a curve straight.
3. Learn to handle data by least square method.

Equipments

the rigid body rotation apparatus, stop – watch and meter.

Rigid body rotation apparatus are shown in figure 1. The tower wheel (6) has wheels of different radius; two arms with scales are fixed symmetrically in the two sides of central cannula. There is a movable round weight m_0 (3) on every arm. All these three parts makes up of a rotation rigid body system. A line coiling around the tower wheel passes by the pulley (9) and links a weight m (14) in its end. When weight is falling, a moment of force is put on the rigid body system through the line. The pulley frame can be adjusted to a suitable position to make the line being parallel with the axes of tower wheel. There is indicating needle (12) to indicate the start position of falling weight. Before experiment, switch tower wheel with the vertical needle, then adjust the adjusting screw (8) to make the system axes vertically. When the tower wheel is put on in place, it should be able to rotate neatly.

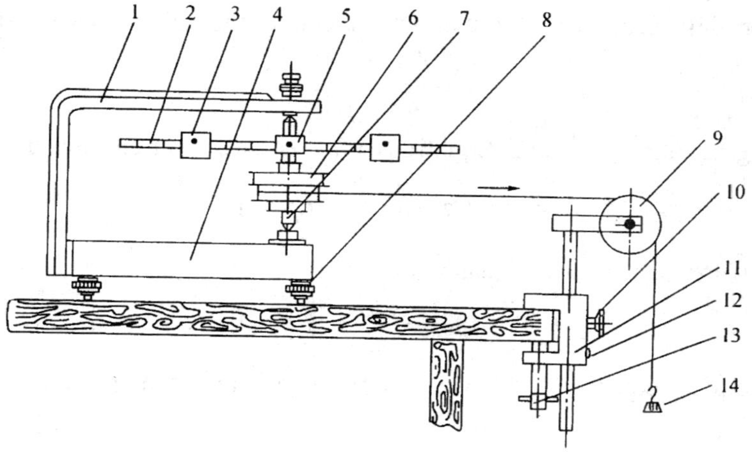

Figure 1 Rigid body rotation apparatus

1. Apparatus holder 2. Arm 3. Hammer 4. Basement
5. Central cannula 6. Tower wheel 7. Central axes
8. Adjusting screw 9. Pulley 10. Fixing screw 11. Stand
12. Indicating needle 13. Stand fixing screw 14. Weights

Principle

When a rigid body is rotating around fixed axes, the rotation law can be written as:

$$M = J\beta \tag{1}$$

If the torque M and angular acceleration β have been measured, the moment of inertia J can be determined.

The torque is made up of two parts: $r \times T$ (r is the radius of tower wheel, T is the tensile force), produced by the line; M_{μ}, friction torque in the axes. If we choose the former as the positive direction, the M will be:

$$M = rT\sin\theta - M_{\mu} \tag{2}$$

where θ is the angle between T and axes. Keeping the line perpendicular with the axes, and $\sin\theta = 1$, the equation (2) can be:

$$M = rT - M_{\mu} \tag{3}$$

We neglect the mass of pulley and line, and take for the line as non – elongate. When the weight m falling in accelerate a, according to Newton second law, we get:

$$T = m\ (g - a) \tag{4}$$

where g is gravity accelerate. Suppose the height of the weight falling down from rest state is h and takes time t to reach the ground, then

$$h = \frac{1}{2}at^2 \tag{5}$$

The tangential accelerate of tower wheel equals the falling accelerate of weight, that is:

$$a = r\beta \tag{6}$$

According equation (1) and equations (3) ~ (6), we can get the total torque

$$M = rm\ (g - a)\ - M_\mu = J\beta\frac{2hJ}{rt^2} \tag{7}$$

Keeping $a << g$, the above equation can be:

$$mgr = \frac{2hJ}{rt^2} + M_\mu \tag{8}$$

By the help of equation (8) we can measure the moment of inertia with three methods.

Method 1. If keep r, h, J unchanged, the (8) can be written as:

$$m = \frac{2hJ}{gr^2}\frac{1}{t^2} + \frac{M_\mu}{gr} = k_1\frac{1}{t^2} + c_1 \tag{9}$$

where $k_1 = \frac{2hJ}{gr^2}$, $c_1 = \frac{M_\mu}{gr}$. According to (9), m has a relation of inverse proportion with t^2, if keep M_μ unchanged, we can measure different falling times t of different weights m, and draw a graph of $m - 1/t^2$ on the plot paper. It's a verification of rotation law for getting a straight line. And also we can get J, M_μ from the slope k_1 and intercept c_1.

Method 2. Keep m, h, J unchanged; equation (8) can be written as:

$$r = \frac{2hJ}{mg}\frac{1}{rt^2} + \frac{M_\mu}{mg} = k_2\frac{1}{rt^2} + c_2 \tag{10}$$

where $k_2 = \frac{2hJ}{mg}$, $c_2 = \frac{M_\mu}{mg}$. From (10), r has a relation of inverse proportion with rt^2. We can measure the falling times t corresponding with different radius r, and draw the graph of $r - \frac{1}{rt^2}$. For the same reason it's a verification of rotation law for getting a straight line. And also we can get J, M_μ from the slope k_2 and intercept c_2.

Method 3. If we can keep M_μ unchanged and $M_\mu << mgr$, the M_μ can be neglected, equation (8) will be:

$$m = \frac{2hJ}{gr^2}\frac{1}{t^2} \tag{11}$$

or:

$$r = \sqrt{\frac{2hJ}{mg}}\frac{1}{t} \tag{12}$$

We can get the moment of inertia following the same steps of method 1 or method 2.

In addition, suppose r, h, m unchanged, the distance between the center of hammer m_0 and the system axes is x, the moment of inertia will be:

$$J = J_0 + J'_0 + 2m_0x^2 \tag{13}$$

where J_0 is the moment of inertia of arms and tower wheel, J'_0 is the moment of inertia for the axes passing through the hammer center and paralleling the system axes. Substitute J with the equation above in (9), we can get

$$t^2 = kx^2 + c \tag{14}$$

where $k = \dfrac{4m_0h}{r\,(mgr - M_\mu)}$, $c = \dfrac{2h\,(J_0 + J'_0)}{r\,(mgr - M_\mu)}$. Plot the graph of $t^2 - x^2$, and verify the law of parallel axes for a straight line.

Experiment

1. Verification the law of rotation and determination the moment of inertia

(1) Switch the tower wheel with the vertical needle and adjust the adjusting screw to make sure that the system axes are vertical. Then install the tower wheel, and the axes should not be too tight or too loose, rotation of the tower wheel should be neatly.

(2) Coil the line tidily on one of wheel with certain radius. The circles of line should not overlap each other. Adjust the frame of pulley to make the line passing through it horizontally.

(3) Determine the moment of inertia and friction torque by method 1 or method 2, and verify the rotation law of rigid body. Every time the weight should fall down from the indicating needle and start with zero speed. It should be certain that $g \gg a$.

2. Verification of the paralleling axes law and observing the changing of moment of inertia with different distribution of mass

Let $m = 20$g, $r = 2.5$cm, and keep them unchanged, change the places of the two hammers symmetrically, measure the falling time.

Data processing

According to (9), let $r = 2.0$cm, h and J fixed. Change m, measure the falling time t and we get the table 3-1.

Table 3 – 1 **r, h, J are fixed, the falling time t for different m**

m (g)	5.00	10.00	15.00	20.00	25.00	30.00	35.00
t_1 (s)							
t_2 (s)							
t_3 (s)							
$\bar{t}$ (s)							
$(\bar{t})^{-2}$ (s^{-2})							

$r = 2.0$cm, $h =$ cm

According to (14), fix $m = 20$g, $r = 2.0$cm, change the positions of hammers, then measure the falling time t, we get table 3 – 2.

Table 3 – 2 r, h, m are fixed, the falling time t for different hammer position

x	1, 1′	2, 2′	3, 3′	4, 4′	5, 5′
t_1 (s)					
t_2 (s)					
t_3 (s)					
$\bar{t}$ (s)					
$(\bar{t})^2$ (s^2)					

In the table 3 – 1 and table 3 – 2, t_i ($i = 1, 2, 3$) is the three measurement of falling time for the same condition and $\bar{t} = (t_1 + t_2 + t_3)/3$.

Now, we process the data from table 1 by two methods.

1. Drawing graph

Drawing the curve of $m \sim t^{-2}$ (t means $\bar{t}$), its a straight line. We can get the slope k_1, intercept c_1 of the line, and measure h. According to equation (9),

$$J = k_1 g r^2 / 2h$$

$$M_\mu = c_1 g r$$

2. Method of least – square (linear regression)

Many scientific calculators have the function of least – square. After inputting data, it can solve the a and b of equation $y = a + bx$ and the relative coefficient γ directly. It also can be processed manually as following. Let the mass of weight m be the independent variable, the falling time t be the dependent variable. The equation (9) can be written as:

$$\frac{1}{t^2} = \frac{1}{k_1} m - \frac{c_1}{k_1}$$

Let $\frac{1}{t^2} = y$, $m = x$, $-\frac{c_1}{k_1} = a$, and $\frac{1}{k_1} = b$, (15) will be in a form of $y = a + bx$.

The a and b can be solved by following steps of least - square method:

$$b = \frac{S_{xy}}{S_{xx}}$$

$$a = \sum_i y_i / n - b \sum_i x_i / n$$

And the relative coefficient will be

$$\gamma = \frac{S_{xy}}{S_{xx} \cdot S_{yy}}$$

Where

$$S_{xy} = \sum_i x_i y_i - \frac{1}{n} \sum_i x_i \sum_i y_i$$

$$S_{xx} = (\sum_i x_i^2) - \frac{1}{n} (\sum_i x_i)^2$$

$$S_{yy} = (\sum_i y_i^2) - \frac{1}{n} (\sum_i y_i)^2$$

A table including items such as $\sum_i x_i$, $\sum_i y_i$, $\sum_i x_i y_i$, $\sum_i x_i^2$, $\sum_i y_i^2$ will be helpful for calculating manually. Then we can get J, M_μ from a and b. The results getting from different methods should be compared with each other.

This procedure can also be applied to the data in table 3 - 2.

Problems

1. If we switch the aluminum weights with steel weights or bakelite weights, is there any change will happen to the measurement of falling time? Why?
2. Why should not let the coiling line overlapped?
3. In this experiment, why we do not draw the graph of $m - t^2$ or $t - x$, but $m - 1/t^2$ or $t^2 - x^2$?

Additional Instruction

Preparation requirement

1. Understand the principle and the purposes.
2. Be familiar with every part of the instrument.
3. Write down the important operation steps.
4. Understand the method of data procedure.

Operation key points

1. Keeping the system axes vertically.
2. The tower wheel should be able to rotate neatly.

3. Lessen friction of all moving parts.

Report requirement

Write down suitable data tables. See detail requirement of graph drawing in experiment 1.

Problems after experiment

The uncertainty of result mostly comes from the measurement of falling time. It will be about several percent. If $M_\mu << mgr$, we can estimate the relative uncertainty for the equation $J = \frac{mgr^2}{2h} t^2$. That is

$$E_J = \sqrt{\left(\frac{\Delta m}{m}\right)^2 + 4\left(\frac{\Delta r}{r}\right)^2 + \left(\frac{\Delta h}{h}\right)^2 + 4\left(\frac{\Delta t}{t}\right)^2}$$

According to $\Delta J = J \cdot E_J$, the final result should be in the standard form: $(J \pm \Delta J)$ $(\mathrm{g \cdot cm^2})$.

Here, we make some rough estimate for the system errors in this experiment.

(1) The tensile force T be perpendicular with the central axes. If there is a deviation of 5°, Tr will become $Tr\cos\alpha$, the relative error can be $1 - \cos\alpha \approx \alpha^2/2 \approx 0.38\%$, it's enough small comparing with the relative error of the measurement of time. And a deviation of 5° is enough big for bare eyes' judge.

(2) $g >> a$. In this experiment, $h \approx 90\mathrm{cm}$, t is in a range of 3.5 ~ 20s. For $t = 20\mathrm{s}$, $a \approx 0.45\mathrm{cm/s^2}$, $a/g \approx 0.045\%$; for $t = 3.5\mathrm{s}$, $a \approx 15\mathrm{cm/s^2}$, $a/g \approx 1.5\%$. Thus, the falling time should not be too short to bring in big error.

(3) Neglect the mass of pulley M_0. By considering the M_0, the tensile force will be in the form of $T' = m\,[g - (1 + M_0/2m)\,a]$, not $T = m\,(g - a)$. Actually the mass of pulley is about 15g. When $m = 35\mathrm{g}$, $(1 + M_0/2m)\,a \approx 1.2a$; $m = 5\mathrm{g}$, $(1 + M_0/2m)\,a \approx 2.5a$. Thus the effect of mass M_0 can be considered with the neglect of a wholly as $a << g$.

(4) Keep h as a constant. For $h = \frac{at^2}{2}$, we can get $\frac{\mathrm{d}h}{h} = at\frac{\mathrm{d}t}{t}$. If $h = 90\mathrm{cm}$, $t = 10\mathrm{s}$, and there is a deviation of 0.5cm for h, it is worth of an uncertainty of falling time of about 0.03s. So it is easy to keep h unchanged.

(5) $M_\mu << mgr$. We can write the M_μ as $M_\mu = m_\mu gr$, and it will become easy to compare the magnitude of m_μ/m. Usually for the tip shaft bearing $m_\mu = 1 \sim 3\mathrm{g}$, but it may be bigger for a rough bearing. It's better for a ball bearing, when $m_\mu = 0.1 \sim 0.5\mathrm{g}$.

实验四　液体黏滞系数的测定

液体的黏滞系数又称内摩擦系数或黏度，是描述液体内摩擦力性质的一个重要物理量，它表征液体反抗形变的能力，只有在液体内存在相对运动时才表现出来。在一定温度下，液体的黏度是一定的。测定液体药品的黏度可以检查其纯度，所以在药学上是鉴定液态药物的方法之一。黏度的测定有许多方法，如转筒法、落球法、黏度计法等。对于黏度较小的液体，如水、乙醇、四氯化碳等，常用奥斯瓦尔德黏度计（简称奥氏黏度计）测量其黏滞系数。

【目的要求】

1. 了解奥氏黏度计的结构原理。
2. 学会用奥氏黏度计测定液体黏度的方法。
3. 学会用韦氏天平测定液体的密度。

【仪器器材】

奥氏黏度计、韦氏天平、温度计、秒表、移液管、量筒、洗耳球、木架、蒸馏水及待测液体（乙醇）。

【实验原理】

1. 液体的黏度及奥氏黏度计

（1）液体的黏度　液体的黏度主要是由于分子力的作用而产生的，当液体在分层稳定流动时，不同液体层的运动速度是不同的，运动快的一层给运动慢的一层以拉力，而运动慢的一层又给运动快的一层以阻力，这一对作用力称为液体的内摩擦力，它是由流体分子之间相互作用而产生的。通过实验得知，液体流动时其内摩擦力的大小与两层液体的接触面积 S 成正比，与层间速度梯度 $\frac{dv}{dx}$（即沿垂直于速度方向上的单位长度的速度增量，z 方向平行于速度梯度的方向）成正比，如图 4-1 所示。

$$f = \eta \cdot S \cdot \frac{dv}{dx} \qquad (1)$$

式中的比例系数 η 就称为该液体的内摩擦系数（或黏滞系数、黏度）。液体不同，它的黏滞系数值也不同，η 是描述液体流动性质的一个重要物理量。它在数值上等于当速度梯度变化一个单位时，作用于液体层单位面积上的内摩擦力。在厘米-克-秒制中，η 的单位为 P（泊），以纪念法国科学家泊肃叶（Poiseuille），常用单位为 cP（厘泊）、mP（毫泊），在国际单位制中是 Pa·s（帕秒），它们的关系为：

$$1P = 10^2 cP = 10^3 mP = 10^{-1} Pa \cdot s$$

液体的黏滞系数 η，随着温度的改变会发生明显的变化，因此测出的 η 必须指明是在什么温度条件下测得的，当温度上升时，液体的黏滞系数要减小，而气体则正好相反，当温度上升时，气体的黏滞系数要增加，但液体的黏滞系数明显地比气体大。

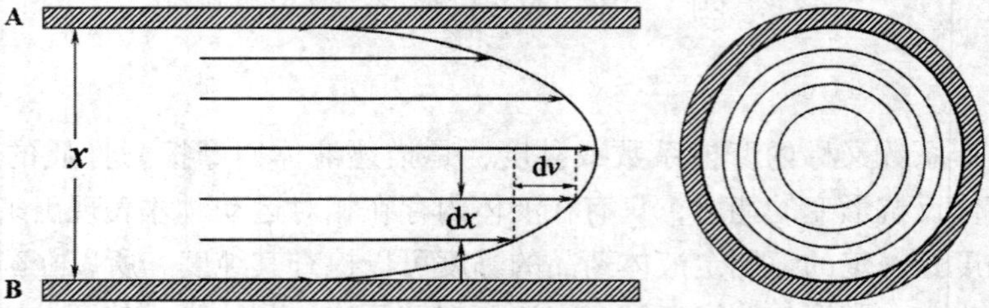

图 4－1　液体层流图

（2）泊肃叶定律　当液体在作层流的情况下，流过细长均匀的圆柱形毛细管时，流速相同的各层正好是和管壁同轴的若干圆柱面，如图 4－1 所示，假设维持液体在管子中的压强差是 $\Delta P = P_1 - P_2$，由式（1）可以推得 t 时间内流过液体的体积为

$$V = \frac{\pi R^4}{8\eta l} \cdot \Delta P \cdot t \tag{2}$$

此式即是泊肃叶定律，其中 R 为毛细管的内半径，l 为毛细管的长度，具体推导如下。

取一长为 l、内半径为 R 的粗细均匀的毛细管，水平放置，使液体在管中稳定流过，将管中的液体柱分为与管壁同轴的圆管状液层，参见图 4－1。各层受到来自内层与流动方向相同的内摩擦力 F 和来自外层与流动方向相反的内摩擦力 $-(F + \mathrm{d}F)$ 的作用。设半径为 r 的管状液体薄层的流速为 v，则由式（1）得

$$F = -\eta \cdot 2\pi r l \cdot \frac{\mathrm{d}v}{\mathrm{d}r}$$

负号表示 $\frac{\mathrm{d}v}{\mathrm{d}r} < 0$，而

$$\mathrm{d}F = \frac{\mathrm{d}F}{\mathrm{d}r} \cdot \mathrm{d}r = -2\pi l\eta \cdot \frac{\mathrm{d}}{\mathrm{d}r}\left(r \cdot \frac{\mathrm{d}v}{\mathrm{d}r}\right) \cdot \mathrm{d}r$$

设管两端的压强分别为 P_1 和 P_2，则此圆管状薄液层上作用的外力为 $2\pi r \cdot \mathrm{d}r \cdot (P_1 - P_2)$，由于液体作稳定流动，所以此外力必定和作用于此液层的内摩擦力 $-\mathrm{d}F$ 相互平衡，即：

$$2\pi r \cdot \mathrm{d}r \cdot (P_1 - P_2) + 2\pi l\eta \cdot \frac{\mathrm{d}}{\mathrm{d}r}\left(r \cdot \frac{\mathrm{d}v}{\mathrm{d}r}\right) \cdot \mathrm{d}r = 0$$

化简上式得：

$$\frac{\mathrm{d}v}{\mathrm{d}r} = -\frac{P_1 - P_2}{2l\eta} \cdot r$$

再将上式积分得

$$v = \frac{P_1 - P_2}{4\eta l} \cdot (R^2 - r^2)$$

设 t 时间内从管中流出的液体体积为 V，则

$$V=\int_0^R vt\cdot 2\pi r\mathrm{d}r=\frac{(P_1-P_2)\cdot\pi t}{2\eta l}\int_0^R(R^2-r^2)r\mathrm{d}r$$

此结果即为泊肃叶公式。

(3) 直接测量法　本实验采用的是自重型毛细管法测量，压强差为

$$\Delta P=P_1-P_2=\rho g\bar{h}$$

将其代入式 (2) 得

$$\eta=\frac{\pi R^4\rho g\bar{h}}{8Vl}\cdot t \tag{3}$$

如果 R、ρ、$\bar{h}$、t、V、l 各量均可测得，即可由上式计算出 η，这种测量方法称为直接测量法（或绝对测量法）。但由于高度差是一个变化的量，采用了取平均值（$\bar{h}$）来近似代替，而毛细管半径 R 又是以四次方形式计算，并且 R 和 V 又不容易精确测定，故其测量精度不高。

(4) 比较测量法　我们采用奥氏黏度计来测定液体的黏度，这是一种比较测量法。奥氏黏度计为 U 型管，参见图 4－2，它是由粗细玻璃管 A 和 E 组成，共有三个玻泡 B、C、D 和毛细管 no 组成。在 C 泡的上下两端的管壁上各有刻线 m 和 n，实验中要测定的就是 m 和 n 间的一定体积的液体流经毛细管所需的时间。玻泡 B 上端亦有刻线 p，以控制待测液体的体积。

如果同体积的两种液体（分别用下标 0 和 1 来标注）流经相同的毛细管所需的时间各为 t_0 和 t_1，由式 (2) 可得：

$$V_0=\frac{\pi R^4}{8\eta_0 l}\Delta P_0 t_0$$

$$V_1=\frac{\pi R^4}{8\eta_1 l}\Delta P_1 t_1$$

因为 $V_0=V_1$，所以

$$\frac{\eta_1}{\eta_0}=\frac{t_1}{t_0}\cdot\frac{\Delta P_1}{\Delta P_0}$$

由于液体在重力的作用下而流动，所以毛细管两端的压强差与流体的密度和黏度计两臂中的液面高度差成正比。在两次实验中，两臂中的液面高度差及其在流动时的变化相同，故：

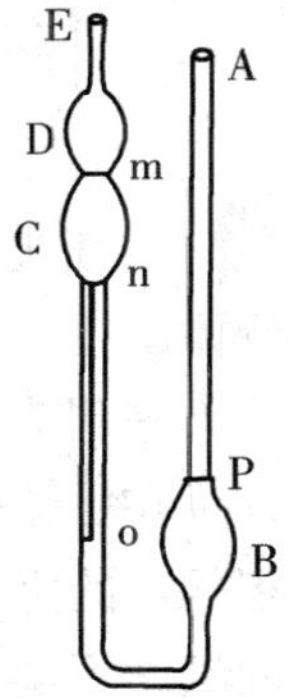

图 4－2　奥氏黏度计

$$\frac{\Delta P_1}{\Delta P_0}=\frac{\rho_1}{\rho_0}$$

ρ_1 和 ρ_0 分别为两种液体的密度，所以

$$\eta_1=\frac{\rho_1 t_1}{\rho_0 t_0}\eta_0 \tag{4}$$

一般以水作为标准液体，水的密度 ρ_0 及黏度 η_0 可以从表中查得；测出 t_0 和 t_1 后，再测出待测液体的密度 ρ_1，则可根据式 (4) 求得待测液体的黏度。

2. 用韦氏天平测定液体的密度

(1) 韦氏天平的结构　韦氏天平是根据阿基米德原理设计的用于测定液体密度的一

种特种天平，其装置如图4-3所示。LFC为一天平臂，由刀口F支于磨光的刀口架B上，可绕支点B向上下倾斜。臂的刻度分成10等分（包括C点）。臂的一端有重锤A及指针L，另一端有一挂钩，悬挂重锤E及玻泡P，在4℃时玻泡P的体积，恰等于同温度时5g纯水的体积。D是垂直支架，高度可用螺丝R调节。支柱底盘上附有螺丝S，用来调节天平水平。天平平衡时指针L应与指针K对齐。甲、乙、丙、丁四种砝码可以跨在天平臂凹槽中，甲砝码等于4℃时玻泡在纯水中减轻的重量。如果设甲砝码重量为1，则乙砝码为甲砝码的1/10，丙砝码为甲的1/100，丁砝码为甲的1/1000。所以四种砝码的真正重量各为5g、0.5g、0.05g、0.005g。。

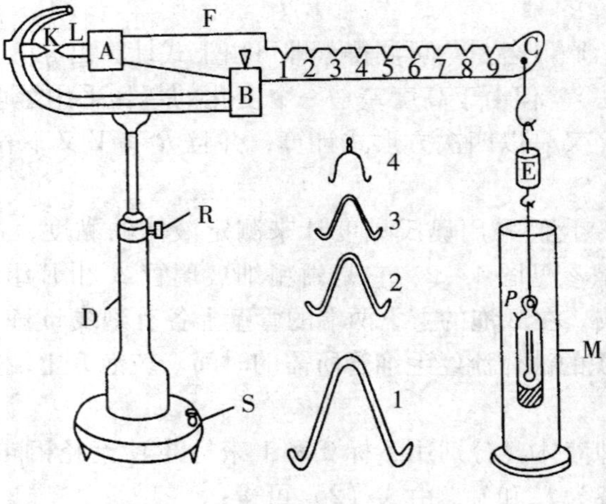

图4-3　韦氏天平的结构

（2）韦氏天平测定液体密度的原理　根据阿基米德定律，物体在液体中减轻的重量等于物体所排开液体的重量。如将重锤E及玻泡D悬挂于挂钩上（参见图4-3），在空气中将天平调节平衡。当玻泡D浸在水中后，因受浮力作用，天平平衡被破坏；要恢复平衡，可在称臂的挂钩上加重量W_0以补偿玻泡在水中所受到的浮力，则W_0即为与玻泡同体积的水的重量。若ρ_0为该温度下水的密度，则排开水的体积为$W_0/\rho_0 g$,，再将玻泡浸入同温度下的待测液体中，用重量W_1加在挂钩上与之平衡，则W_1即为与玻泡同体积的液体重量。如ρ_1是同温度下待测液体的密度，则排开液体的体积为$W_1/\rho_1 g$，因为同一温度下玻泡的体积是一定的，故

$$\frac{W_0}{\rho_0 g}=\frac{W_1}{\rho_1 g}$$

所以

$$\rho_1=\frac{W_1}{W_0}\rho_0 \qquad (5)$$

其中ρ_0可由表中查出，只要知道W_0及W_1的数值就可以求出待测液体在该温度下的密度。

W_0及W_1是用下法求出的。

当玻泡浸入水中后，用砝码跨在称臂的凹槽中使之达到平衡时，比如设甲砝码跨于9

处，乙砝码跨于9处，丙砝码跨于8处，丁砝码跨于5处。又设BC长为a，与玻泡同体积之水重为W_0，则根据力矩定律可得：

$$W_0 \times a = 5 \times \frac{9a}{10} + \frac{5}{10} \times \frac{9a}{10} + \frac{5}{100} \times \frac{8a}{10} + \frac{5}{1000} \times \frac{5a}{10}$$

即：

$$W_0 = 5 \times 0.9985\ (\mathrm{g})$$

这相当于在挂钩上加5×0.9985g的重量，其中“5”在式（5）中可消去，我们可把所跨的砝码读为0.9985。同法可得玻泡浸入待测液体中时W_1的读数。如果实验中不要求测出液体的密度，则可将（5）式代入（4）式，得：

$$\eta_1 = \frac{W_1 t_1}{W_0 t_0}\eta_0 \qquad (6)$$

【实验内容】

1. 测定乙醇的密度

（1）将韦氏天平如图4－3装置好，必须注意使螺丝S在称臂的平面内，以便调节。调节S使K、L尖端对齐，则天平达到平衡。

（2）在筒M内装入待测液体乙醇后，使玻泡刚好全部浸入待测液体中（注意悬丝尽量少浸入液体中）。加砝码于称臂上，使K、L重新对齐。记下砝码读数W_1。

（3）将筒M及玻泡P分别擦干，并用蒸馏水荡洗三次，再在筒内换上蒸馏水，将玻泡全部浸入水中，重新加砝码使天平平衡，记下砝码读数W_0。

（4）用温度计测出此时的水温。

2. 测定乙醇的黏度

（1）将黏度计铅直地固定于木架上。将待测液体用移液管从管口A注入B泡，参见图4－2，使液体满到刻度P。将洗耳球对准E管，小心将液体吸入C泡，使液面高于刻线m，体积约占D泡的1/4，放开洗耳球让液体稳定流下，用秒表记下液面依次通过m、n所需的时间，测量两次，取其平均值，作为t_1。

（2）用蒸馏水冲洗黏度计三次后，注入蒸馏水，重复（1）的操作，记下时间，测两次，取其平均值，作为t_0。

（3）查出该温度下水的黏度系数η_0，用式（6）算出待测液体的黏度系数η_1。

【注意事项】

1. 无论是用韦氏天平测密度或是用黏度计测黏度，都先测待测液体（乙醇），再测蒸馏水。

2. 注入M筒的待测液体和蒸馏水以及注入黏度计的乙醇和蒸馏水的体积应分别相等。

3. 韦氏天平系精密仪器，使用时要特别小心。尤应注意悬挂甲砝码时，勿使重锤E碰到筒内液体；悬挂玻泡时，谨防打碎，并勿使悬挂玻泡之悬线折断，使用砝码时必须用镊子，切勿使用手拿；谨防丁砝码遗失。

4. 黏度计一定要清洁，洗涤时及将液体从B泡吸入C泡时，要手执粗管A，以免损坏。

5. 韦氏天平使用后及时将玻泡 P 及筒 M 擦干、放好。黏度计使用后及时用自来水洗净、放好。

【思考题】

1. 为什么先测待测液体（乙醇），后测水？若反之，对实验有什么影响？
2. 用韦氏天平作实验时，为什么注入 M 筒的待测液体和蒸馏水的体积应相等？

【附表】

水在不同温度时的黏度

温度（℃）	黏度（$Pa \cdot s \times 10^{-3}$）	温度（℃）	黏度（$Pa \cdot s \times 10^{-3}$）	温度（℃）	黏度（$Pa \cdot s \times 10^{-3}$）
0	1.7921	14	1.1709	27	0.8545
1	1.7313	15	1.1404	28	0.8360
2	1.6728	16	1.1111	29	0.8180
3	1.6191	17	1.0828	30	0.8007
4	1.5674	18	1.0559	31	0.7840
5	1.5188	19	1.0299	32	0.7679
6	1.4728	20	1.0050	33	0.7523
7	1.4284	20.2	1.0000	34	0.7371
8	1.3840	21	0.9810	35	0.7225
9	1.3462	22	0.9579	36	0.7085
10	1.3077	23	0.9358	37	0.6947
11	1.2713	24	0.9142	38	0.6814
12	1.2363	25	0.8937	39	0.6685
13	1.2028	26	0.8737	40	0.6560

实 验 指 导

【预习要求】

1. 了解奥氏黏度计的测量原理；
2. 熟悉奥氏黏度计的使用方法；
3. 了解韦氏天平的测量原理；
4. 熟悉韦氏天平的结构及其使用方法。

【操作要点】

1. 测量时奥氏黏度计必须垂直放置；
2. 使用韦氏天平测量前，必须用螺丝 S 将天平调节水平，即指针 L 应与指针 K 对齐。

3. 无论是用韦氏天平测密度或是用黏度计测黏度，在换液体时要用蒸馏水荡洗干净。

【报告要求】

1. 测定乙醇的密度

水温（℃）	W_1	W_0

2. 测定乙醇的黏度

	1	2	平均值
乙醇			$t_1=$
水			$t_0=$

$$\eta_1=\frac{W_1t_1}{W_0t_0}\eta_0=$$

【实验后的思考】

实验中哪些因素影响乙醇黏度测定的准确性?

4 Determination of Viscosity

Viscosity is a scientific term describing the internal friction of a fluid. Fluids resist the relative motion of layers by different velocity among them. Viscosity is a measure of the resistance of a fluid which is being deformed by either shear stress or extensional stress. At the same temperature, different liquids have a different viscosity. Determination of viscosity is one way to exam liquid medicines. There are many ways to measure viscosity, including the method of falling balls, rotating cylinder, Ostwald viscometer and so on . We use the Ostwald viscometer in this experiment. It is suitable for the liquid with low viscosity, such as water, alcohol and carbon tetrachloride.

Objectives

1. Understand the measuring principle of Ostwald viscometer;
2. Learn to measure the liquid viscosity;
3. Learn to measure the liquid density by Westphal balance.

Equipments

Ostwald viscometer; Westphal balance; thermometer; stop – watch; pipette; rubber suction bulb; graduated cylinder; wood – frame; distilled water and alcohol.

Principle

1. Viscosity of liquid and Ostwald viscometer

(1) The viscosity of liquid

The viscosity of liquid depends on the magnitude of intermolecular forces. When two layers of fluids move past one another with a certain relative velocity, both layers experience a force which tends to oppose their relative motions. This force caused by intermolecular forces, is called internal friction. The magnitude of internal friction is depended on two factors, the area of the plane of contact between the layers and the velocity gradient normal to it. As shown in figure 4 – 1, there is a kind of liquid between the two boards A and B, now fix the board B and pull the board A with a constant force. The experiment has been demonstrated the internal friction f between two liquid layers is:

$$f=\eta \cdot s \cdot \frac{dv}{dx} \tag{1}$$

Where the coefficient η is the viscosity of the liquid, s is area of the liquid layer and $\frac{dv}{dx}$ is the velocity gradient of the liquid. The SI unit of viscosity is the pascal second (Pa · s). The most common unit of viscosity is the poise (P), which is given the name after the French physiologist Jean Louis Poiseuille. Ten poise equal one pascal second.

$$1P = 10^2 cP = 10^3 mP = 10^{-1} Pa \cdot s$$

In general, the viscosity of liquids decreases with increasing temperature. But viscosity of gases is opposite.

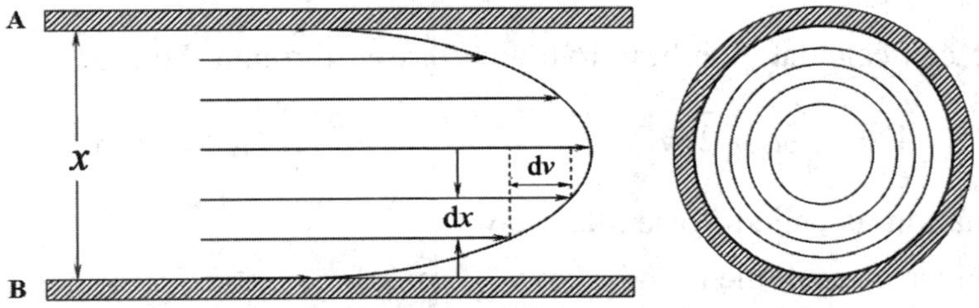

Figure4 – 1 Laminar flow

(2) Poiseuille's law

In a tube, the liquid in the center is moving fastest while the liquid touching the walls of the tube is stationary. To simplify the situation, let's assume that there are a bunch of circular layers of liquid, each having a velocity determined only by their radial distance from the center of the tube (in figure 4 – 1). When liquid is flowing through a small tube of radius R and length l under pressure difference of ΔP at two ends, we can calculate the liquid volume passed through the tube in time t:

$$V = \frac{\pi R^4}{8\eta l} \cdot \Delta P \cdot t \qquad (2)$$

In the above equation form of Poiseuille's Law, η is the viscosity of the flowing liquid, t is the time for which the liquid flows, V is the volume of liquid which flows in that time, R is the radius of the pipe, ΔP is the pressure difference between ends of the pipe, and l is the length of the pipe.

Derivation of Poiseuille's Law:

One layer of liquid has two frictions, F and $-(F + dF)$. $-(F + dF)$ is in the opposite the direction of movement of liquid and is negative. According equation (1):

$$F = -\eta \cdot 2\pi r l \cdot \frac{dv}{dr}$$

$$dF = \frac{dF}{dr} \cdot dr = -2\pi l \eta \cdot \frac{d}{dr}\left(r \cdot \frac{dv}{dr}\right) \cdot dr$$

Suppose there is no acceleration of liquid in the pipe. By Newton's first law, there is no net force.

$$2\pi r \cdot \mathrm{d}r \cdot (P_1 - P_2) + 2\pi l\eta \cdot \frac{\mathrm{d}}{\mathrm{d}r}\left(r \cdot \frac{\mathrm{d}v}{\mathrm{d}r}\right) \cdot \mathrm{d}r = 0$$

It is a differential equation:

$$\frac{\mathrm{d}v}{\mathrm{d}r} = -\frac{P_1 - P_2}{2l\eta} \cdot r$$

The velocity of liquid moving through the tube: $v = \frac{P_1 - P_2}{4\eta l} \cdot (R^2 - r^2)$

Then we can calculate the liquid volume passed through the tube in time interval t:

$$V = \int_0^R vt \cdot 2\pi r \mathrm{d}r = \frac{(P_1 - P_2) \cdot \pi t}{2\eta l}\int_0^R (R^2 - r^2) r \mathrm{d}r$$

This equation is called Poiseuille's law.

(3) Direct measurement

The pressure differential between ends of the pipe:

$$\Delta P = P_1 - P_2 = \rho g \bar{h}$$

We can get that:

$$\eta = \frac{\pi R^4 \rho g \bar{h}}{8Vl} \cdot t \qquad (3)$$

For the known quantities R, ρ, $\bar{h}$, t, V and l, we can calculate the viscosity of the flowing liquid η. But the relative accuracy of R and V are low, the error of η is large.

(4) Determine the viscosity by the Ostwald viscometer

Ostwald viscometer consists of a glass tube in the shape of U as shown in Figure 4-2. It has three bubbles B, C, D and a capillary no. Two marks (m and n) indicate a known volume. The time for the liquid in the glass bubble C between these marks passing through the capillary will be measured. If t_0 and t_1 are the time consuming by two kinds of liquid respectively, we can get that:

$$V_0 = \frac{\pi R^4}{8\eta_0 l}\Delta P_0 t_0$$

$$V_1 = \frac{\pi R^4}{8\eta_1 l}\Delta P_1 \cdot t_1$$

Because $V_0 = V_1$, we can get the following equation:

$$\frac{\eta_1}{\eta_0} = \frac{t_1}{t_0} \cdot \frac{\Delta P_1}{\Delta P_0}$$

Where ΔP_0 and ΔP_1 are the pressure difference of the two kinds of the liquid. And the pressure difference is proportional to the liquid density ρ under the condition of same height, that is:

$$\frac{\Delta P_1}{\Delta P_0}=\frac{\rho_1}{\rho_0}$$

Thus, we get:

$$\eta_1=\frac{\rho_1 t_1}{\rho_0 t_0}\eta_0 \tag{4}$$

Usually, water is taken for standard liquid for its density ρ_0 and viscosity ρ_0 are known, and we can determine the viscosity of a certain liquid by the t_0, t_1 and ρ_1.

2. Determine the liquid density by Westphal balance

(1) The structure of Westphal balance

Westphal balance is based on the Archimedes' Principle and used to measure the specific density of liquid as shown in figure 4 – 3. LFC is the arm of balance, with nine equally spaced notches on it. There are a glass plummet P and a hammer E on the end of the arm. The plummet has a built – in thermometer and has a known volume (equal to 5 g of water at 4°C). D is the frame of balance adjustable in height. The screw S on the foot is used to adjust the balance. With no weight on the arm of the balance, the two points K and L must be aligned. There are four different weights 1#, 2#, 3# and 4#, loaded onto the notches on arm. Their values correspond to 5g, 0.5g, 0.05g and 0.005g.

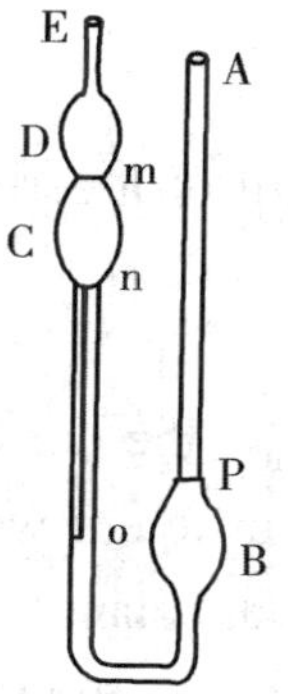

Figure 4 – 2 Ostwald viscometer

(2) The principle of Westphal balance

First, balance the plummet in air (Figure4 – 3) by turning the screws on the foot until the pointer L heading on the point K on the frame. Then immerse the plummet completely in water, and the system becomes imbalanced. Placing four different weights on the notches, the system is rebalanced. By Archimedes' Principle, the weight of the water displaced by the plummet is equal to the mass loss of the plummet when it is submerged. Thus the weight W_0 hung at the end of arm, is equal to the weight of water displaced by the plummet. At last, the plummet is placed in an unknown liquid. Adjust the weights so that the system restores balance. W_1 is the weight of this liquid displaced by the plummet. In the same temperature, water and unknown liquid have equivalent volume (equal to the volume of plummet). It can be written as:

$$\frac{W_0}{\rho_0 g}=\frac{W_1}{\rho_1 g}$$

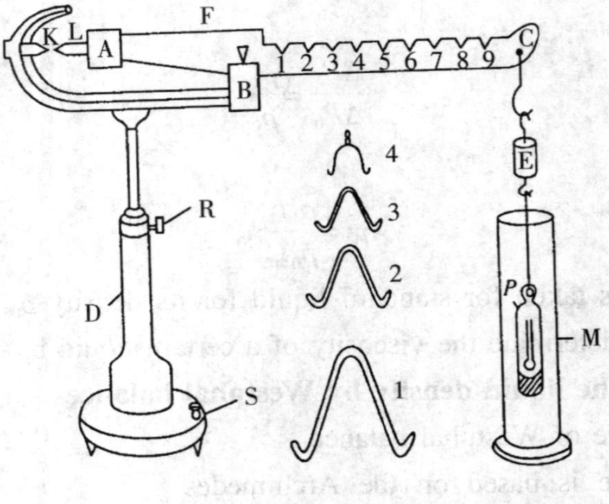

Figure4 −3 The structure of Westphal balance

The density of a certain liquid can be measured by the following equation:

$$\rho_1 = \frac{W_1}{W_0}\rho_0 \tag{5}$$

Where, ρ_0 is the density of water. W_0 and W_1 can be measured by this following method. Four different weights have the relative values of 1.0, 0.1, 0.01 and 0.001. Suppose that a balance is obtained when the glass plummet is immersed in water with weights of 1.0 and 0.1 at notch 9, the 0.01 weight at notch 8 and the 0.001 weight at notch 5. Thus, we get:

$$W_0 \times a = 5 \times \frac{9a}{10} + \frac{5}{10} \times \frac{9a}{10} + \frac{5}{100} \times \frac{8a}{10} + \frac{5}{1000} \times \frac{5a}{10}$$

$$W_0 = 5 \times 0.9985 \text{ (g)}$$

Now, we can get the equation for determining the liquid viscosity in this experiment:

$$\eta_1 = \frac{W_1 t_1}{W_0 t_0}\eta_0 \tag{6}$$

Experiment

1. Determine the density of alcohol

(1) Place the Westphal balance as the figure 4 −3.

(2) Handle the Westphal balance according to its instruction book. First measure the density of alcohol by reading the weight W_1, then the reading W_0 of water.

(3) Get the temperature of water by the thermometer.

2. Determine the viscosity of alcohol

(1) Place the Ostwald viscometer on the wood – frame vertically. Inject alcohol into the port A by the pipette until the liquid level reaching the scale line P, as shown in

Figure 4 −2. Suck the liquid by the rubber suction bulb at the end of E until the liquid level higher then the line m, then release the liquid and record the interval t_1 of the liquid level passing through the lines m and n. Do this process for two times.

(2) Wash the viscometer by distilled water for three times, and then record the interval t_0 for water to times.

(3) Get the viscosity of water η_0 under the lab temperature according to the table in appendix, and then calculate the viscosity of alcohol by the equation (6).

Notice

1. Determine the density of alcohol by Westphal balance first. Later measure the water. Measure the viscosity in the same order.

2. Make sure the alcohol introduced into the viscometer or balance has the same volume as water.

3. Be careful with Westphal balance. It is precision instrument.

4. Make sure that the viscometer is cleaned after each run. Don't break the glass tube when wash it.

5. Clean and dry the glass plummet of Westphal balance after use. And wash the Ostwald viscometer in time.

Problems

1. Why we must measure the alcohol first, then the water? Can we measure the water first?

2. Why should we keep the volume of alcohol and water same for pouring into the Westphal balance?

Appendix

Viscosity of water under different temperature

Temperature (℃)	Viscosity (Pa · s×10^{-3})	Temperature (℃)	Viscosity (Pa · s×10^{-3})	Temperature (℃)	Viscosity (Pa · s×10^{-3})
0	1.7921	14	1.1709	27	0.8545
1	1.7313	15	1.1404	28	0.8360
2	1.6728	16	1.1111	29	0.8180
3	1.6191	17	1.0828	30	0.8007
4	1.5674	18	1.0559	31	0.7840
5	1.5188	19	1.0299	32	0.7679
6	1.4728	20	1.0050	33	0.7523
7	1.4284	20.2	1.0000	34	0.7371

Temperature (℃)	Viscosity (Pa · s×10^{-3})	Temperature (℃)	Viscosity (Pa · s×10^{-3})	Temperature (℃)	Viscosity (Pa · s×10^{-3})
8	1.3840	21	0.9810	35	0.7225
9	1.3462	22	0.9579	36	0.7085
10	1.3077	23	0.9358	37	0.6947
11	1.2713	24	0.9142	38	0.6814
12	1.2363	25	0.8937	39	0.6685
13	1.2028	26	0.8737	40	0.6560

Additional Instruction

Preparation Requirements:

1. Understand the principle of Ostwald viscometer;
2. Be familiar with the use of Ostwald viscometer;
3. Understand the principle of Westphal balance;
4. Be familiar with the structure of Westphal balance and the use of Westphal balance.

Operation Key Points

1. Make sure that the Ostwald viscometer is vertical.
2. Adjust the arm of the balance by the screw S, so that the balance is horizontal. When the system is balance, the two pointers K and L are in line.
3. Be careful to see that the glass plummet and the viscometer are carefully cleaned each time before a new liquid is used.

Report Requirement:

1. Determine the density of alcohol

temperature of water (℃)	W_1	W_0

2. Determine the viscosity of alcohol

	1	2	the average time
alcohol			t_1 =
Water			t_0 =

$$\eta_1 = \frac{W_1 t_1}{W_0 t_0}\eta_0 =$$

Problem after experiment

What are the factors affecting the accuracy of viscosity measurement?

实验五　电学基本实验

电学实验常用的基本仪器有：电源、电表、电阻箱、变阻器、万用表、开关等。本实验通过介绍常用电学仪器的结构、原理、特征及使用时的注意事项，学会基本电学线路的连接和数据处理。

【目的要求】

1. 了解电学实验的要求、操作规程和安全知识。
2. 了解几种电学基本仪器的规格、用途和使用方法。
3. 学会简单的电路连接，掌握用滑线变阻器作为分压器和制流器的方法。
4. 练习使用伏特计、毫安计和多用电表。

【基本仪器】

1. 电源

电源有直流和交流两种。

（1）直流电源　用符号“DC”或“－”表示，常用的直流电源有蓄电池、直流稳压电源和直流稳流电源。蓄电池需要经常充电，维护比较麻烦，直流稳压电源的电压稳定、内阻小，使用方便，其输出电压连续可调，直流稳流电源在一定的负载阻值范围内，能给出一定的恒定电流，电流大小连续可调。

选用电源时除了应注意其输出电压是否符合需要外，还须注意取用的电流或负载电阻是否在电源的额定值之内。如果电流过载，电源将急剧发热而损坏，例如直流稳压电源标明 15V，1A，则我们使用时，电流不能超过 1A。

此外使用电源时特别要防止其短路（即外电路电阻为 0），因此连接电路时必须最后接电源。

直流稳压电源的输出端都有正负两接线柱，接电表时应加以注意。

（2）交流电源　用符号“AC”或“～”表示，常用的交流电有 220 伏和 380 伏两种，频率为 50Hz，根据需要还可使用变压器升高或降低电压。使用高压电，要防止触电，注意用电安全。

2. 电表

实验室通常用磁电式电表。永久磁铁的两个极上连着带圆筒空腔的极掌，极掌之间装有圆柱形软铁芯，其作用是使极掌和铁芯间的空隙中磁场很强，并且磁力线是以圆柱的轴为中心呈均匀辐射状。在圆柱形铁芯和极掌间空隙处放有长弓形线圈，线圈上固定一根指针。当有电流流通时，线圈就受电磁力矩而偏转，直到与游丝的扭力矩平衡。线圈偏角的大小与所通入的电流成正比，电流方向不同，偏转方向也不同。

（1）检流计　其特征是指针零点在刻度中央，指针可以向左或向右偏转，所以接线时可以不考虑接线柱的正负。但检流计只能通过微小的电流，使用时必须串联一只大电阻加以保护。

（2）安培计和伏特计　安培计（符号Ⓐ）和伏特计（符号Ⓥ）都是由电流计改装而成，在表头线圈上串联一个高电阻就成了伏特计，在表头线圈上并联一个低阻值电阻就成了安培计。安培计（包括毫安计、微安计）用来测量电路中电流的大小，串联在被测电路中，伏特计用来测量电路中某两点间的电势差，应与被测电路并联。

使用电表应注意的事项：

①要在仪器规定的条件下使用；

②仪表正、负极不能接反　直流电表的接线柱上标有“+”、“-”号，电流从“+”接线柱流入电流，从“-”接线柱流出电表，故“+”接线柱接电路中电势较高的一端，“-”接电势较低的一端；

③选择适当量程：量程（测量范围）指电表可测量出的最大电流值或电压值，最好选择使仪表指针在满刻度的2/3左右位置；

④读数时视线应垂直于指针处表面，如有镜面，最好用单目观察，应使指针与其在镜面中的象重合。

3. 多用电表

多用电表是常用的电工测量仪表，其主要部分是一个电流计，还装配着一套大小不同的电阻和整流器等。利用一个多挡选择开关的控制，可以测量交直流电压、电流和电阻，还可以对半导体器件进行简单的测试。

多用电表的使用方法如下：

（1）测量直流电流和电压　与使用安培计和伏特计一样，多用电表有两个表笔：红色表笔从“+”接出来，黑色表笔从“-”接出来，注意红表笔接电路中电势高的一端，黑表笔接电势低的一端。

（2）测量交流电压：由于多用电表内部有整流器（通常是一个二极管），整流器将正弦交流电变成脉动直流电，从而多用电表可用来测量交流电。用两表笔（不考虑红黑）与被测电路并联，即可测出交流电压的有效值。

（3）测量电阻——欧姆计　将选择开关旋在测量电阻处，此时启用多用电表内接电源（原理详见实验十），黑表笔接电源正极，红表笔接负极，以保证流过待测电阻的电流仍从红表笔流入多用电表。

每次测量电阻前及每次欧姆挡换挡后，都要进行电阻调零：将两表笔短路，调节调零旋钮，使指针满格对准0Ω，测量时将两表笔置于待测电阻两端，表针偏转，表面电阻刻度尺上指针所指数值乘以该挡倍率，即得待测电阻阻值。读数时应注意电阻刻度尺上刻度是自右向左逐渐增大的，而且刻度不均匀。

使用多用电表应注意：

（1）执表棒时，手不能接触表棒上任何金属部分。

（2）测试时，电表先选用大量程，观察电表指针偏转，再选用适当量程测量，并采用跃接法，即在用表棒接触测量点的同时，注意电表指针偏转情况，并随时准备在出现不正常现象时，使表棒离开测量点。

（3）用作欧姆计时，不得测带电的电阻和额定电流极小的电阻（如灵敏电流计的内阻）。

（4）用作欧姆计时，不得双手同时接触两表笔的金属尖端，测高阻时尤须注意。

（5）使用完毕，务必将万用电表选择开关拨离欧姆挡，应拨到空挡或交流最大电压挡，数字表要选择“关”，以保安全。

电表的主要规格除量程外，还有准确度等级和内阻。电表内阻一般在仪表说明书上已给出，或由实验室测出。电表的标称误差是其主要技术特性，将电表用一个标准电表进行校准，而得到电表各个刻度的绝对误差，取其中最大的绝对误差除以量程，即为该电表的标称误差。我国国标（GB）规定，按标称误差的大小，电表可分为七级（0.1，0.2，0.5，1.0，1.5，2.5，5.0），电表出厂时一般已将级别标在表盘上。

例如伏特计量程为10V，等级为1.5级，则绝对误差最大为：

$$\Delta U=\frac{1.5}{100}\times 10=0.15\approx 0.2\text{V}\quad \text{（注意：误差取一位有效数字）}$$

图5-1为伏特表的面板。

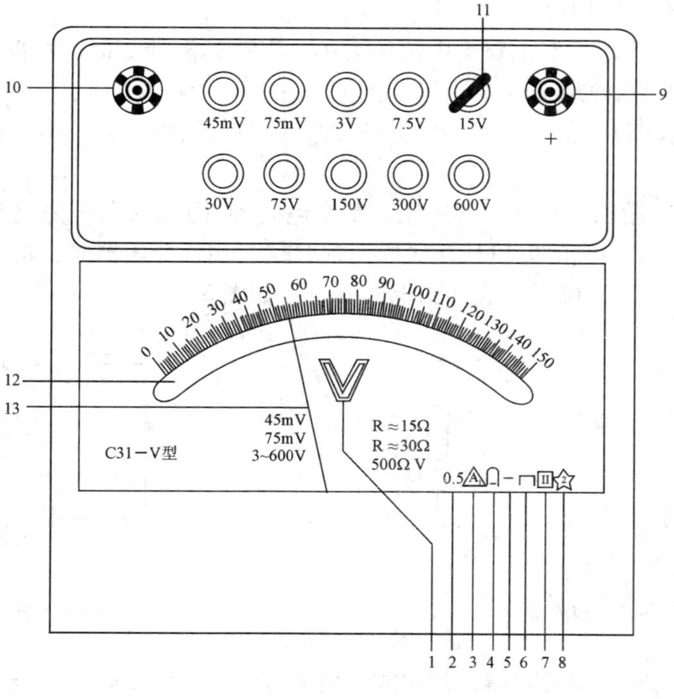

图5-1　伏特计面板

1. 伏特计；2. 电表（等级为0.5级）；3. 电表使用条件属 A_1 组（在周围气温为0～+40℃及相对湿度不大于85%的条件下工作）；4. 磁电式仪表；5. 直流电表（～为交流单相）；6. 标度尺位置为水平放置（T或↑为竖直放置）；7. Ⅱ级防外磁场及电场；8. 电表的绝缘强度试验电压为2kV；9. 正负接线柱；11. 插塞（用于仪表的量程转换）；12. 镜面（读数用。读数时使指针和镜中的像重合，以避免视差）；13. 指针

4. 滑线变阻器

如图 5－2 所示，把均匀的电阻丝绕在瓷筒上，绕线的两端分别与接线柱 a、b 连接，因此 a、b 间的电阻值即为全部电阻线的电阻。绕线上方有条形金属杆，装有滑键 d 并与电阻丝有良好接触，杆的一端装有接线柱 c，因此电阻线通过 d 与接线柱 c 连通，移动滑键 d 时，其与电阻丝的接触点随之改变，即可改变 a、c 或 b、c 之间的电阻值。滑线变阻器的规格在出厂铭牌上标出：

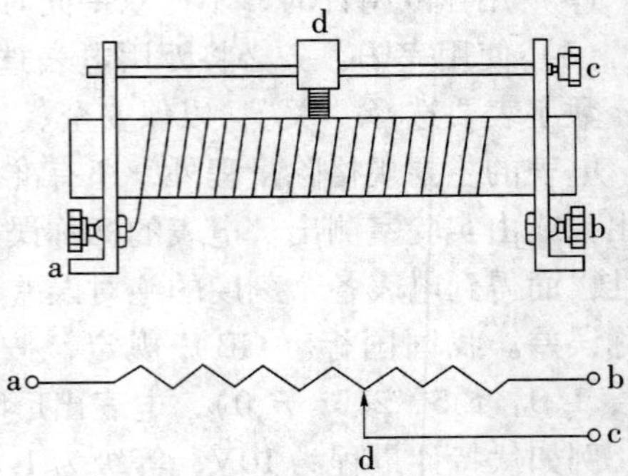

图 5－2　滑线变阻器

（1）阻值　指 ab 间电阻值。

（2）额定工作电流　指长时间连续工作时允许通过的最大电流值。使用时注意选择合适规格的变阻器，以免过载。

滑线变阻器有两种用途。

（1）限流器　用以控制电路中电流的大小，当移动滑片 d 时，电路中的阻值就改变，从而电流也发生改变，如图 5－3。

接通电路前及断开电路前，电阻调至最大，以使电路中电流最小，确保电路中电表和用电器的安全。

（2）分压器　用以获得变化的电压，整个电阻 ab 两端的电压为电源电压 E。当滑键 d 由 b 滑向 a 时，bc 两端的电压从 0 伏增大到 E 伏。用作分压器时，接通电路及断开电路前，应使分压 bc 为零，确保电表和用电器安全，如图 5－4。

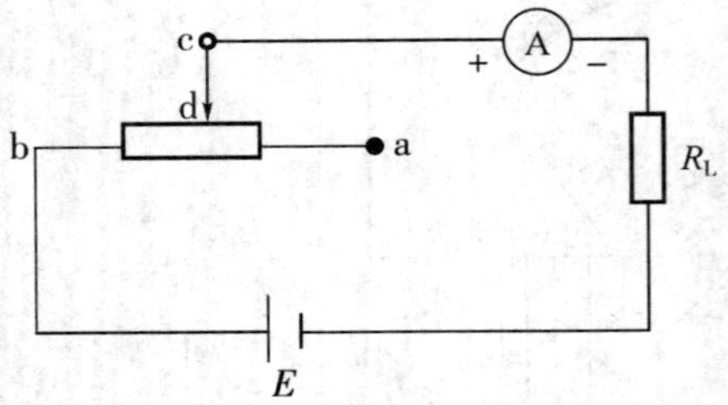

图 5－3　制流器

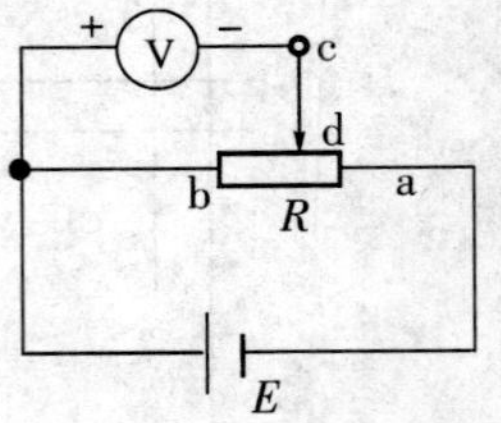

图 5－4　分压器

电子仪器中用的小型变阻器，其功能和使用方法与滑线变阻器完全一样，唯体积与通过的电流较小。

5. 旋钮电阻箱

旋钮电阻箱外形如图 5－5。

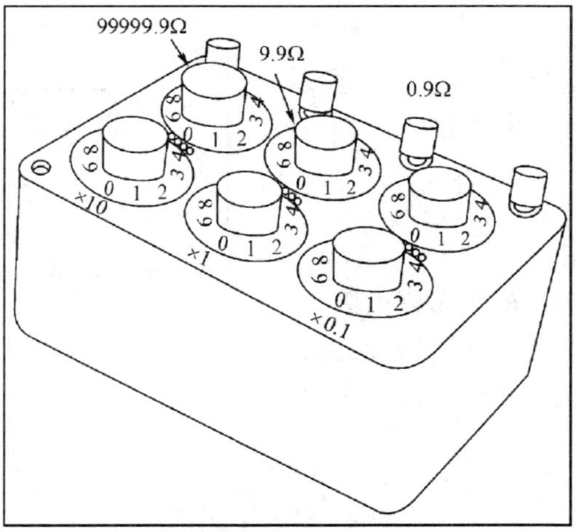

图 5－5　电阻箱外形

箱面上有六个旋钮，其上刻有 0，1，2……9 等数字，每旋钮旁箱面上分别刻有“×10000”，“×1000”……“×0.1”等标记，分别表示该旋钮对着标记的字数的数位。各旋钮相互串联，其内部构造如图 5－6，所以总电阻值为各旋钮读数之和。当用99999.9 Ω和 0 这两个接线柱时，意味着将六个旋钮的电阻全部串联接入了电路，当需用电阻阻值小于 10　Ω 时可用接线柱 9.9　Ω 和 0，此时只将“×1”及“×0.1”这两个旋钮的电阻串联接入了电路，避免了高数位旋钮的接线电阻对低电阻阻值的影响。

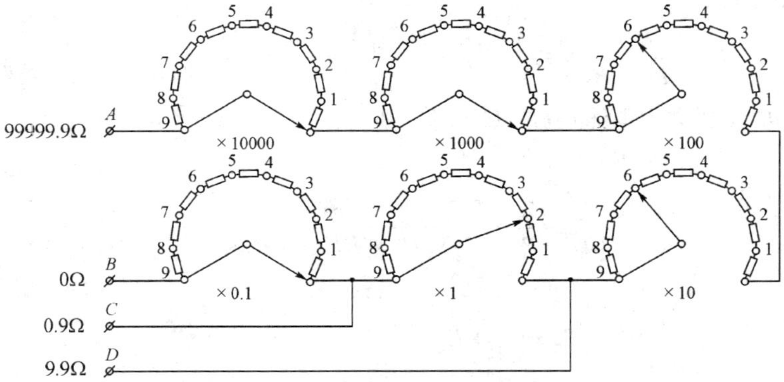

图 5－6　电阻箱各旋钮内部构造

在使用电阻箱前，应将每个旋钮转动几次，以免内部接触不良。要防止流过电阻箱的电流超过该挡的额定电流。电阻箱每挡电阻允许通过的电流不同，由额定功率可计算出各挡的额定电流，如实验室常用的 ZX21 型电阻箱的额定功率为 0.25W，如用 ×0.1 挡的电阻时，允许通过的电流为：

$$I=\sqrt{\frac{P}{R}}=\sqrt{\frac{0.25}{0.1}}=1.6\ \text{(A)}$$

【实验原理】

滑线变阻器作为制流器，如图 5－7 所示。

当 c 移至 a 端，电路中电阻最小，相应电流最大，即

$$I_{max}=\frac{\varepsilon}{R_L}$$

式中 ε 为电源电动势，R_L为负载电阻。

当 c 移至 b 端，变阻器全部串入回路，回路电阻最大，电流最小。令 R_0 表示变阻器的总电阻，

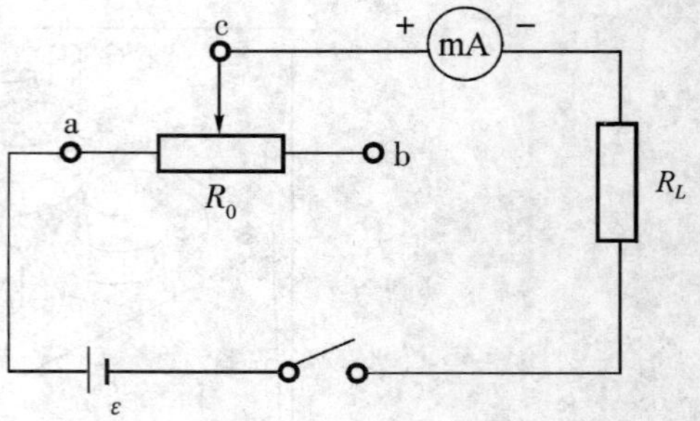

图 5－7　制流电路

$$I_{min}=\frac{\varepsilon}{R_L+R_0}$$

所以该制流电路电流调节范围是$\frac{\varepsilon}{R_L+R_0}\sim\frac{\varepsilon}{R_L}$，其范围与变阻器阻值 R_0 有关，R_0 越大，则调节范围越大。

滑线变阻器作为分压器，如图 5－8 所示。

滑键将滑线电阻 R_0 分为 R_1 和 R_2 两部分，负载电阻 R_L 与 R_1 并联，电路的总电阻为：

$$R=R_2+\frac{R_LR_1}{R_L+R_1}$$

c、a 两点间电阻为：

$$R_{ca}=\frac{R_LR_1}{R_L+R_1}$$

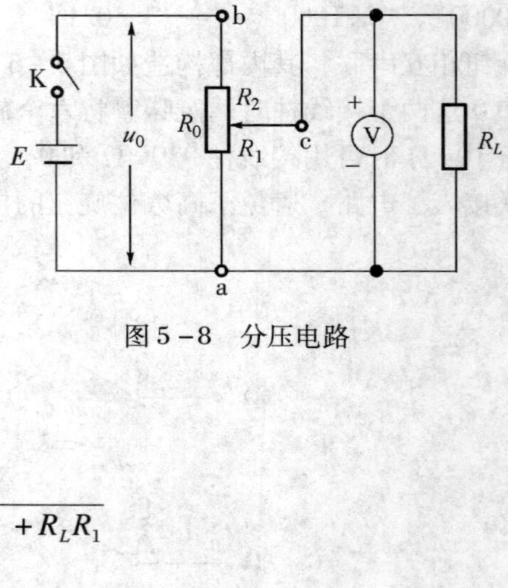

图 5－8　分压电路

负载电阻 R_L 上的电压降即分压 U 与总的电压 U_0 之比为：

$$\frac{U}{U_0}=\frac{R_{ca}}{R}=\frac{R_LR_1}{R_2(R_L+R_1)+R_LR_1}$$

$$\because R_0=R_1+R_2$$

$$\therefore \frac{U}{U_0}=\frac{R_LR_1}{R_0R_L+R_0R_1-R_1^2}=\frac{\frac{R_L}{R_0}\cdot\frac{R_1}{R_0}}{\frac{R_L}{R_0}+\frac{R_1}{R_0}-\frac{R_1{}^2}{R_0{}^2}}$$

令 $k=\frac{R_L}{R_0}$是负载电阻相对于滑线电阻器阻值大小的参数。

令 $x=\frac{R_1}{R_0}$是滑线电阻器滑动头所在位置的参数。

$$\frac{U}{U_0}=\frac{kx}{k+x-x^2}$$

对于不同参数 k 及 x 和分压比$\frac{U}{U_0}$的关系如图 5－9 所示。

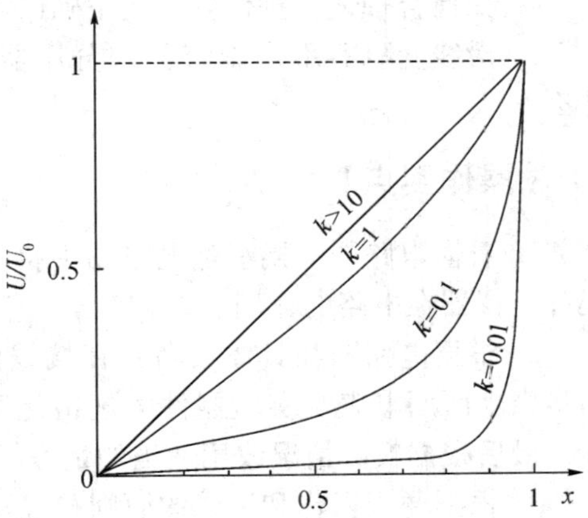

图 5－9　分压比随参数的关系曲线

【实验内容】

1．制流电路的连接

用 100Ω（或 90Ω）的滑线变阻器作制流器，负载电阻用 50Ω，按图 5－7 接线，接通稳压电源 E（取 3V），将 c 分别移动到 10，20，30……100 分度处，记录相应电流值。

2．分压电路特性的研究

按图 5－8 接线，令 k 分别等于 0.1，10（注：R_L分别调到相应的值，$R_L = kR_0$），改变滑线变阻器滑动头的位置参数 x（如 x＝0.1，0.2……1），测出相应的分压 U，并作出 U/U_0-x关系曲线。

【思考题】

1. 如图 5－8 的分压电路，取滑动端 c 和固定端 b 作为分压输出端接至负载，哪端电位高？分压输出为零时 c 端应在什么位置？

2. 如果分压电路误接成图 5－10 所示，将会发生什么问题？

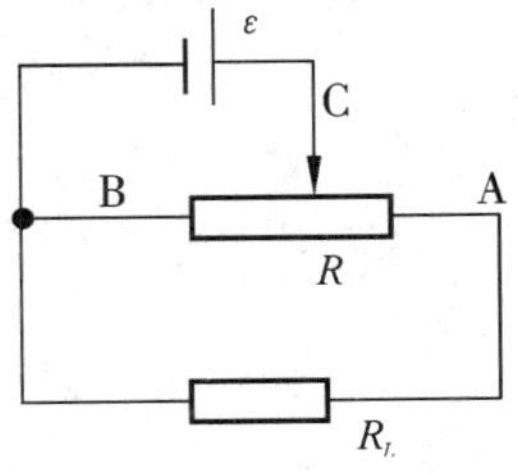

图 5－10　分压电路的错误接法之一

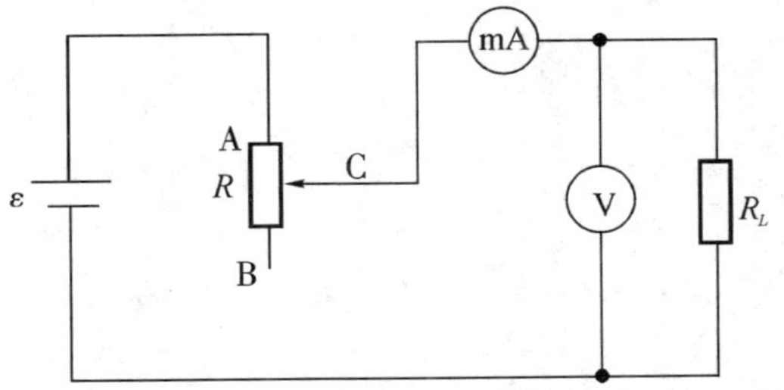

图 5－11　分压电路的错误接法之二

3. 根据图 5－11 所示线路（R 作为一个可调限流电阻）测量伏安特性曲线，较图 5－8所示的用分压器接法的线路有什么区别？各有什么特点？

实　验　指　导

【预习要求】

1. 使用直流电流计、直流电压计要注意什么？
2. 如何正确地使用多用电表，用多电表测电阻时如何读数？
3. 已知电表的等级和量程，怎样求绝对误差？

4. 如何合理地安排仪器以及正确连接电路？要注意哪些问题？

5. 滑线变阻器用作制流器和分压器时分别有几个接线柱接入电路？使用时要注意什么？

【操作要点】

1. 看懂电路图：搞清楚电路图中符号所代表的仪器、元件，会分析电路，弄清各个元件、仪器在电路的作用。

2. 根据电路图和具体设备：在接线前首先估计电路中可能出现的电流和电压大小，初步判断所用电表及实验器件的规格是否适用。在把握不大的情况下，尽可能先用大量程，最后根据实际情况改用适当的量程。

3. 根据便于操作和读数的原则布置好仪器。

4. 连接电路：连接电路一定要切断电源，先接包括电源在内的主要电路，接线顺序从电源正极开始按回路法接线；接线时要充分利用电路中的等位点，避免在一个接线柱上集中过多的接线片，并注意电线要接牢、防止断路。

5. 接好线后，对照电路图检查、核对，确定是否正确，然后检查所用仪器（如电源、电表等）的正负极、量程是否正确，电阻箱、滑线变阻器所放阻值是否正确。

6. 一切无误后，用跃接法接通电源，一切正常后方可闭合开关进行实验。

7. 实验过程中需要更换仪器或电路，或改变电表量程时，都必须切断电源，不能带电作业。要养成不用手和身体接触电路中的导体部分的好习惯，以免发生危险。

8. 实验结束后，先切断电源，然后拆线，整理好仪器。

5 Elementary Experiment of Electricity

There are a lot of basic equipments for electrical experiment, such as power source, electric meters, resistance, multimeter, switch, and so on. Here is a simple introduction for some of them. We can learn to join the simple circuit and to deal the data.

Objectives

Knowing the operating rules and safety knowledge of electrical experiment.

To learn how to use some elementary electricity apparatus such as voltmeter, ampere meter and multimeter.

To learn how to join the simple circuit, master the method of slide rheostat as potential divider and current restrictor.

Equipments

1. Power source

There are two kinds of power sources: direct current source and alternating current source.

(1) Direct Current, symbol "DC" or " – ". Accumulator is an ordinary kind of DC power, but it needs recharging regularly and is not very easy for maintenance. The DC regulated power supply is a common device in labs. Usually it can provide stabilized electric voltage and has a very small internal resistance; also its output can be adjusted continuously and conveniently. The direct stabilized current supply can supply steady current, and current can be changed continuously.

It is very important for using the power source device that not let the output current beyond the rating value, or the source device may be damaged for over – heating. For example: for a direct regulated power supply marking with "15V, 1A", its output current can't exceed 1A.

There are two output connectors: positive and negative in the DC power source, not let them be short circuit (the external resistance equals zero). And it is suggested that link the power source at last when joining an electrical circuit.

(2) Alternating current source, symbol "AC" or " ~ ". The common AC source is of 220V or 380V with a frequency of 50Hz. Potential transformers are often used for

increasing or reduction voltages if necessary. Always keeps safeguard in mind when handling with high voltage power.

2. Electric meters

The common electric meters in physics lab are magnetoelectric meters. There is a cylinder space between the two magnetic poles of permanent magnet, the column soft iron in the space can make the magnetic field strong enough and be equally radiative. A long arc coil is placed on the soft iron and a needle is fixed on the coil. The coil will deflect when electric current flows in it. The deflection angle is proportional to the current and for different current direction the deflection is also different.

(1) Electric current meter (galvanometer). The zero scale is in the middle of the scale window, therefore, usually, it is not necessary to consider the positive or negative connector. Because galvanometer is only permitted a small amount of electric current, it is often joined in series with a large resistance for protection.

(2) Ampere meter (Ⓐ) and voltmeter (Ⓥ). Both are refitted from galvanometer. If a large resistance is arranged in series with the meter coil, then, a voltmeter is formed. And an ampere meter is made up with arranging a small resistance in parallel with the meter coil. Ampere meter is for measuring electric current, it should be in series in the circuit. Voltmeter is for measuring potential, it should be in parallel in the circuit.

Notes about DC electric meters:

(1) Be sure to use them under right conditions.

(2) The positive and negative connectors should be linked right. The two connectors are marked by " + " and " – ". It is defined that the electric current flows into " + " and flows out " – ", so " + " connector should be linked with the higher potential point and the " – " connector lower potential.

(3) Select proper range. A certain measurement range means the largest current or potential the electric meter can indicate. It's better to select a measurement range with its reading near the 2/3 of the scale.

(4) Make sure that the line of sight be vertical to the scale surface or the pointer. If there is a mirror under the pointer, it's better to read with single eye for making sure the pointer coincide with its own image in the mirror.

3. Multimeter

Multimeter is a kind of commonly used electric meter. Its main unit is a galvanometer, servicing by a set of several resistors and rectifier. The multimeter can measure voltage, current, and resistance by turning a switch, also it can measure semiconductor transistor simply.

(1) Measuring DC current or voltage. Like amperemeter and voltmeter, the multimeter has two connectors or pens: the red pen connects with " + ", and the black con-

nects with " – ", where the red one links higher potential point.

(2) Measuring alternating voltage. There is a rectifier in multimeter (usually a diode), which can invert alternating current into pulsating direct current. So, when the multimeter is in parallel with circuit (here it needn't consider the red or black connector), it can measure effective alternating current voltage.

(3) Measuring resistance——ohmmeter. Turning the switch to "Resistance" step, the internal electric battery is started, for the black and red connector links with the positive and negative ends of the battery respectively. And now the electric current flowing through resistance still flows into the red connector.

Before measuring resistance, the ohmmeter has to be zero adjusted. Make the black and red connectors short circuit, turn the zero – adjusting knob to make pointer indicating at the "0Ω" position. When measuring, the number that the needle pointer showing should be multiplied by a relevant ratio for getting a right measurement of the resistance. Pay attention that the number of the ohmmeter scale is not uniform and increasing from right to left side.

Here are some points need pay attention。

(1) Don't touch any metal part of measure pens during measuring.

(2) It's better to select a large range at the beginning of measurement, then, decide a proper measuring range according to the first deflection. The jumper connection is a suitable method for using a multimeter: it's easy to make the measuring pens left the measuring point at anytime when abnormal conditions occur.

(3) As an ohmmeter, multimeter can't be used to measure the live resistance or the resistance of small rated current, such as the internal resistance of galvanometer.

(4) After measuring, the switch must be deviated from the ohmmeter steps to the largest range of "alternate voltage" (signal multimeter must be turned off).

Except measuring range, the main specification of an electric meter includes degree of accuracy and internal resistance. The internal resistance is usually indicated in the instruction book or can be measured by lab. Nominal error of electric meter is a chief technical characteristic. Calibrating electric meter with standard meter can get absolute error of each scale, and the nominal error of each range is the quotient of max absolute error divided by measuring range. In China, the National Standard (GB) defines that electric meters are classified by seven grades (0.1, 0.2, 0.5, 1.0, 1.5, 2.5, and 5.0) according to the nominal error, and usually the grade is marked on the dial by manufacturer.

For example, a meter's measuring range is 10V, its grade is 1.5, for any measured value of this range, and the maximum absolute error is:

$$\Delta U = \frac{1.5}{100} \times 10 = 0.15 \approx 0.2\text{V}$$

Figure 5 – 1 is a picture of the panel of voltmeter.

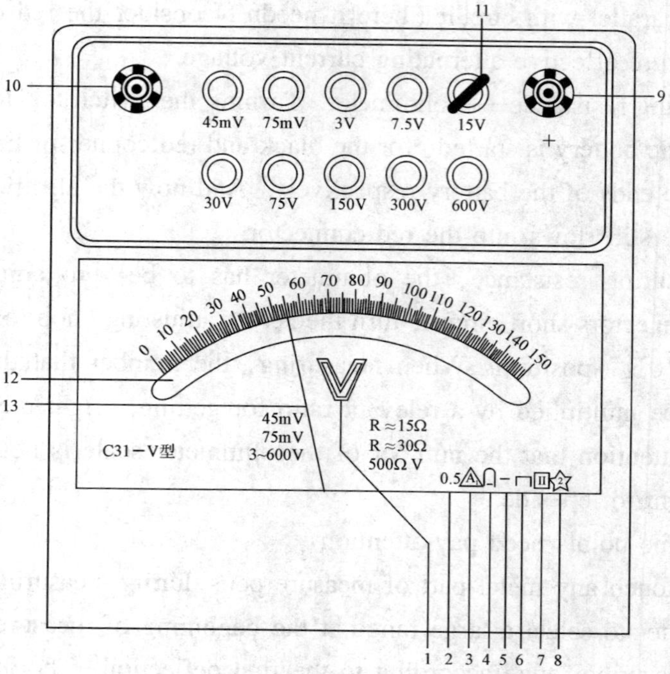

Figure 5 – 1　Panel of voltmeter

1. voltmeter; 2. electric meter (which level is 0.5); 3. using condition is belong to group A1 (work with temperature about 0℃ ~ +40℃ and relative humidity not more than 85%); 4. moving – coil instrument; 5. DC meter (~ AC single phase); 6. marking scale to be hold horizontal (T or ↑ to be hold upright); 7. to defend outer magnetic field and electric field with Ⅱ level; 8. test voltage of insulating strength is 2 kV; 9. positive post head; 10. negative post head; 11. jack plug (to choose the range); 12. mirror (make the clock hand have its mirror symmetry to avoid parallax while reading); 13. Clock hand.

4. Slide rheostat

The main part of a slide rheostat is made up by coiling uniform resistance wire onto the surface of porcelain pipe, (see figure 5 – 2), the two ends of the resistance wire connecting with two connectors, a or b separately. The resistance between the two points is the whole resistance. There is a metal rod equipping with a slide key "d" over the resistance wire, and "d" connects with resistance wire well. There is a connector "c" at one end of the metal rod, so either part of the resistance wire separating by "d" can communicate with "c" through "d". By sliding "d", the resistance between "a" and "c" or "b" and "c" will have being changed accordingly. Usually, the characteristic, such as max resistance and nominal current of the slide rheostat, is marked on the rating plate with manufacturer. So please pay attention to select proper rheostat avoid overload.

Usually, the two purposes of slide rheostat are:

(1) As a current - limiting resistor. See figure 5 - 3, by sliding "d", the resistance of circuit is changed; therefore, the current of circuit will be changed.

Remember to adjust the rheostat to the max resistance position before switch - on or switch - off the circuit, so that the current is the min for protecting electric meter and other electrical appliance.

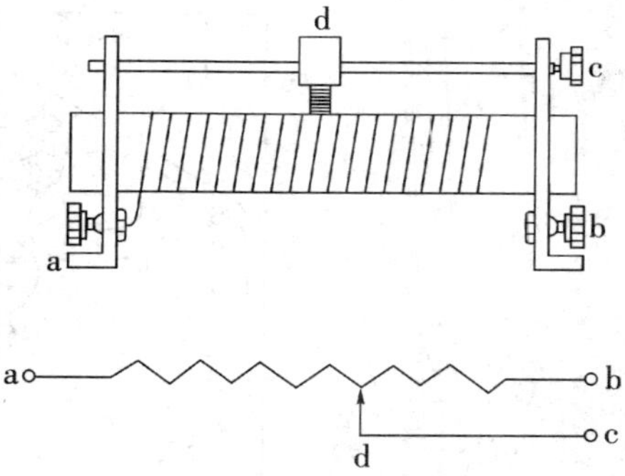

Figure 5 - 2　Slide rheostat

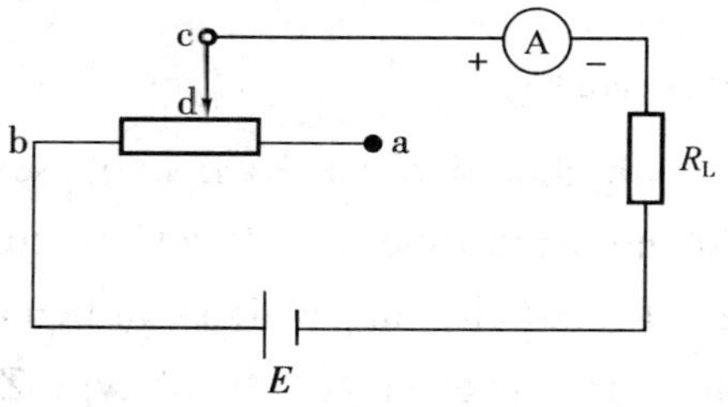

Figure 5 - 3　Current - limiting resistor

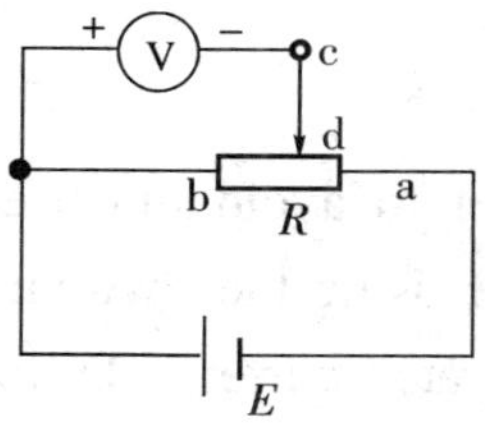

Figure 5 - 4　Voltage - divider

(2) As a voltage - divider. The rheostat is joined into circuit as in Figure5 - 4. The voltage between "a" and "b" equals to source voltage E, and the voltage between "b" and "c" increases from 0 to E when slide key "d" slides from "b" to "a", so, it is got a variable voltage. For protection of electric meter and appliance, we should make sure the voltage between "b" and "c" is 0 before turning on or off the circuit. The miniature rheostat with smaller size in the electronic appliance has same function as slide rheostat, except its nominal current.

5. Revolving resistance box

The picture of revolving resistance box is shown in figure 5 - 5, there are six knobs on the surface, and each knob marked with "×10000", "×1000", ···, "×0.1" respectively. Figure5 - 6 shows the diagram of internal structure of the box, so the total resistance equals to the sum of each knob's reading. For example, when the two connectors 0Ω and 99999.9Ω are joined in circuit, the six knobs are all in series in circuit.

When the resistance needed is less than 10Ω, we can joined the two connectors, 0Ω and 9.9Ω, now only the two knobs " ×1" and " ×0.1" are in series in circuit. Here, we avoid the influence of connecting – resistance of high – digits knobs for the small resistance value.

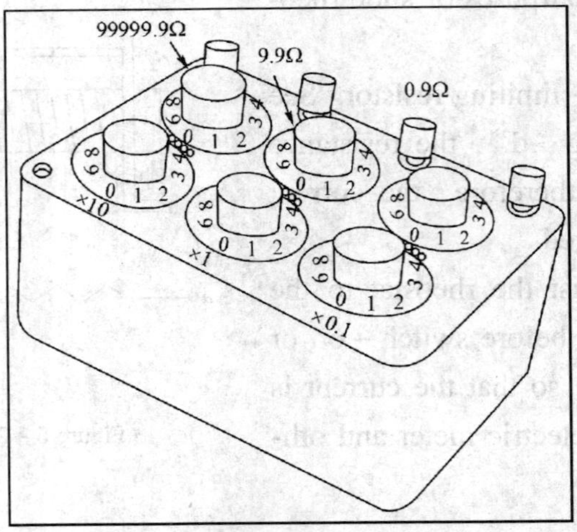

Figure 5 – 5 Revolving resistance box

In order to avoid imperfect internal contact, we should rotate each knob several times before using the revolving rheostat box. At the same time, don't let the current exceed the rated current. Different step has different rated current, which can be calculated according to power capability. For example: the power capability of type ZX21 revolving rheostat box is 0.25W, when using " ×0.1" step, the max current is

$$I = \sqrt{\frac{P}{R}} = \sqrt{\frac{0.25}{0.1}} = 1.6\ (A)$$

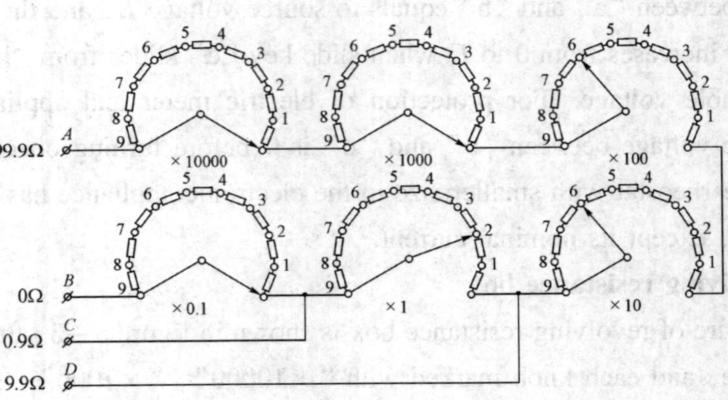

Figure 5 – 6 Internal structure of revolving resistance box

Principle

In figure5 – 7, the slide rheostat is taken for current – limiting resistor: when "c" is moved "a" position, the resistance of circuit is min, and the current is max:

$$I_{max} = \frac{\varepsilon}{R_L}$$

Where, "ε" is the source voltage and R_L is the load resistance.

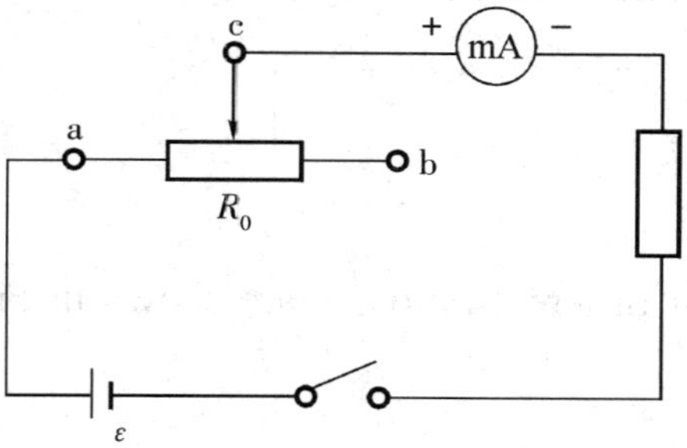

Figure 5 – 7 Current – limiting circuit

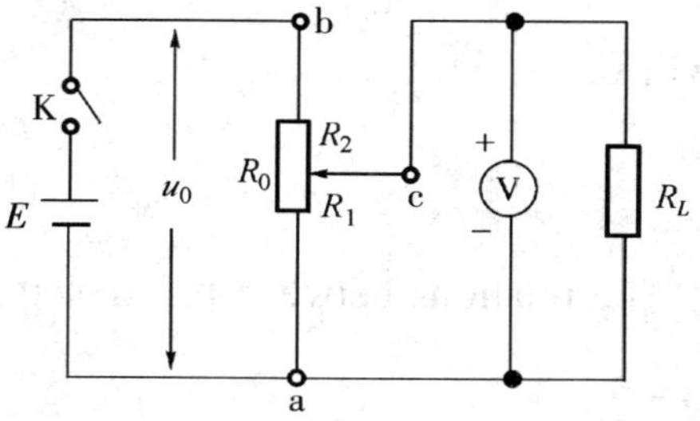

figure 5 – 8 Voltage – divider circuit

When "c" is moved "b" position, the total rheostat is in series, the resistance is max, and the current is min:

$$I_{min} = \frac{\varepsilon}{R_L + R_0}$$

Therefore, the current of the circuit is controlled in the range: $\frac{\varepsilon}{R_L + R_0} < \frac{\varepsilon}{R_L}$, the range is related to the resistance of rheostat R_0, the bigger R_0 is, the bigger of the range.

In figure 5 – 8, rheostat is regarded as a voltage – divider, slide key "c" divides the rheostat R_0 into two parts: R_1 and R_2, the load resistance R_L parallels with R_1, the circuit's total resistance is $R = R_2 + \frac{R_L R_1}{R_L + R_1}$. The resistance between "c" and "a" is:

$$R_{ca} = \frac{R_L R_1}{R_L + R_1}$$

The ratio of the potential U on R_L and the total voltage U_0 is:

$$\frac{U}{U_0} = \frac{R_{ca}}{R} = \frac{R_L R_1}{R_2(R_L + R_1) + R_L R_1}$$

$$\because R_0 = R_1 + R_2$$

$$\therefore \frac{U}{U_0} = \frac{\frac{R_L}{R_0} \cdot \frac{R_1}{R_0}}{\frac{R_L}{R_0} + \frac{R_1}{R_0} - \frac{R_1^2}{R_0^2}}$$

Let $k = \frac{R_L}{R_0}$, as the parameter of the ratio of load resistance and slide rheostat resistance. And let $x = \frac{R_1}{R_0}$, as the parameter of the position of the slide key. The prior formula will be:

$$\frac{U}{U_0} = \frac{kx}{K + x - x^2}$$

The relations between parameters k and x for different ratio $\frac{U}{U_0}$ are shown in Figure5 -9.

Experiment

1. Current – limiting circuit

Arrange a 100Ω (or 90Ω) rheostat as a current – limiting resistor, a 50Ω resistor as a load, join a circuit according to figure 5 – 7. Take the electric source 3V, move the slide key "c" to 10, 20, 30, …, 100 marks respectively, and record the corresponding current values.

2. Voltage – divider circuit

Arrange the rheostat as voltage – divider and join the circuit according to figure 5 -8. Let K be 0. 1 and 10 respectively, and adjust R_L according to the formula . Now, slide key "c" to different position x, such as $x = 0.1$, 0. 2, …, 1, record the potential U correspondingly, and draw the curve $\frac{U}{U_0} - x$.

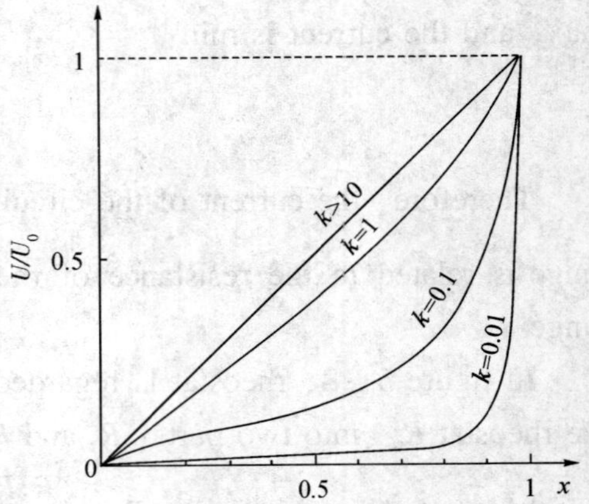

Figure 5 -9 The relation between $U/U_0 - x$

Problems

1. For the circuit in figure 5 -8, let the potential between "c" and "b" as output voltage of the load resistance, now, which terminal, "c" or "b" has higher potential? Where the position of "c" will be for output voltage is 0?

2. What will happen if the voltage – divider circuit is connected as figure 5 – 10?

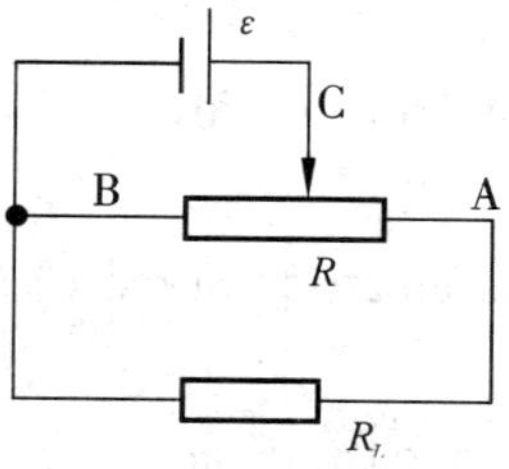

Figure 5 – 10 One of the wrong nnections for voltage – divider circuit

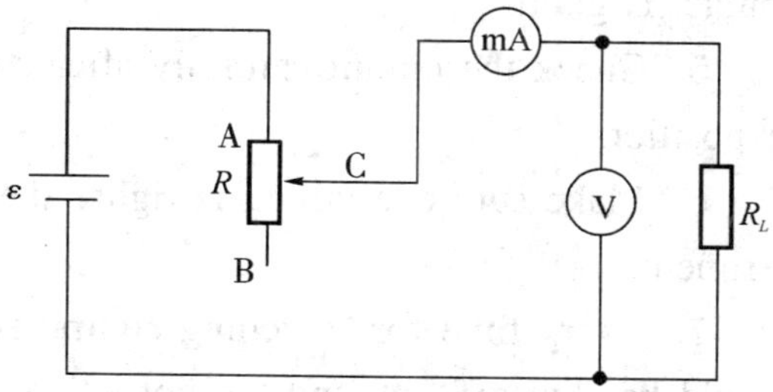

Figure 5 – 11 A wrong connection for voltage – divider circuit

3. What will be the difference of the $U \sim V$ curve between the circuit in figure 5 – 11 and the circuit in figure 5 – 8? And what's characteristic each circuit?

4. Why a multimeter cannot be used for measuring an electric meter's internal resistance or battery's internal resistance?

Additional instruction

Preparation Requirements

1. What kind of things should pay more attention to for a DC amperemeter or a voltmeter?

2. How to use the multimeter correctly? And how to read resistance when it is used as an ohmmeter?

3. How to calculate the absolute error of a reading of an electrical meter?

4. How to arrange the device properly and to connect circuit correctly? What kind of things should be paid more attention to?

5. How many connecters are connected in circuit of current – limiting and voltage – divider, respectively?

Operation key points

1. Read the circuit diagram carefully, understand meanings of every symbol.

2. Estimate voltages and currents according to equipment requirements. For uncertain or changeable value, please select a bigger measure range as possible firstly, and change to the proper measure range according to practical situation.

3. Arrange the device in a way that it is convenient for operation and reading.

4. When connecting circuit, keep power source off and connect the circuit in loop – way. Usually, link the main loop first and start from positive terminal of source.

We can make use of the concept of equipotent point to arrange wires. Be sure every contact is good.

5. Check the circuit carefully after connecting. Confirm the devices in right states or position.

6. Make sure everything is right, then, turn on the electric source to start the experiment.

7. Every time for switching equipment or changing measure range, electric source should be shut off first and the hot – line work is not suggested. And it's a good habit of safety that not touching any conductor part of a circuit.

8. When experiment is over, shut off the source first, then, disconnect the wires and rearrange the devices.

实验六　伏安法测电阻

电阻的测量方法有很多种，常见的有伏安法，补偿法，电桥法等。用伏特计、安培计根据欧姆定律测量电阻的方法称为“伏安法”。虽然伏安法测电阻不如补偿法和电桥法精确，但其测量方式便捷，测量范围宽，是研究各种元件和材料导电特性的基本方法，广泛应用于工业生产和实验室科研工作。

【目的要求】

1. 掌握用伏安法测电阻时电流表内接、外接的条件。
2. 掌握电表量程的选择对测量结果的影响。

【仪器器材】

伏特计、安培计、稳压电源、滑线变阻器、电阻箱、单刀双掷开关、单刀单掷开关。

【实验原理】

根据欧姆定律 $R_x = \frac{U}{I}$ 测出通过未知电阻 R_x 的电流 I 及 R_x 两端的电压 U，便可以计算出 R_x。伏安法测电阻的缺点是测量的准确度不高，一方面是由于电表有一定内阻，在测量电流或电压时会产生系统误差（也叫电表的接入误差）；另一方面是由于电表的准确度等级使电流和电压的测量有误差。

1. 系统误差的产生与修正

图 6－1 所示的是用伏安法测电阻的两种电路连接方法，将 K 倒向 1 时，安培计为内接，简称内接法；将 K 倒向 2 时，安培计为外接，简称外接法。不管哪种接法，由于电表有一定的内阻，对测量结果都会产生系统误差。

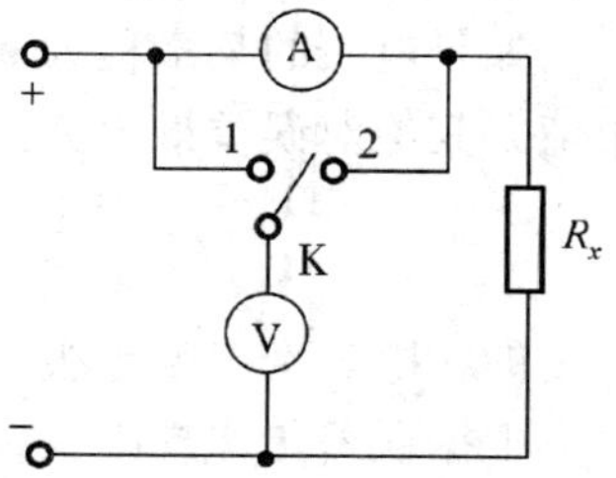

图 6－1　伏安法测电阻

（1）内接法　将 K 倒向 1，安培计的读数 I 是通过待测电阻 R_x 的电流 I_x，即 $I = I_x$，伏特计的读数 U 不是 R_x 两端的电压 U_x，而是 U_x 与电流表两端的电压 U_A 之和，即 $U = U_x + U_A$。

设电流表内阻为 R_A，待测电阻的测量值为 R，则

$$R = \frac{U}{I} = \frac{U_x + U_A}{I_x} = R_x + R_A \qquad (1)$$

$$R_x = R - R_A \qquad (2)$$

即测量值比真实值偏大，由此带来的相对误差 E 为：

$$E=\frac{\Delta R_x}{R_x}=\frac{R-R_x}{R_x}=\frac{R_x+R_A-R_x}{R_x}=\frac{R_A}{R_x}\times 100\% \tag{3}$$

可见只有当 $R_x >> R_A$ 时，才能保证测量有足够的准确度。

（2）外接法　将 K 倒向 2，这时伏特计的读数 U 就是 R_x 两端的电压 U_x，即 $U=U_x$，但安培计的读数 I 不是 I_x，而是 I_x 与通过伏特计的电流 I_V 之和，即 $I = I_x + I_V$。

设伏特计内阻为 R_V，则

$$R=\frac{U}{I}=\frac{U_x}{I_x+I_V}=\frac{1}{\frac{I_x}{U_x}+\frac{I_V}{U_x}}=\frac{1}{\frac{1}{R_x}}+\frac{1}{\frac{1}{R_V}}=\frac{R_xR_V}{R_x+R_V} \tag{4}$$

$$R_x=\frac{R}{1-\frac{R}{R_V}} \tag{5}$$

即测量值比真实值偏小，由此带来的相对误差 E 表示为：

$$E=\frac{\Delta R_x}{R_x}=\frac{R-R_x}{R_x}=\frac{\frac{R_xR_V}{R_x+R_V}-R_x}{R_x}=-\frac{R_x}{R_x+R_V}\times 100\% \tag{6}$$

可见当待测电阻较小（$R_V >> R_x$）时，宜采用外接法，式中负号表示当电流表外接时电阻测量结果偏小。

综上讨论，伏安法测电阻，测得的电阻值总是偏大或偏小，具体采用哪种接法，主要根据 R_x、R_A、R_V 三者的相对大小确定。

当 $R_x >> R_A$，采用内接法；

当 $R_x << R_V$，采用外接法；

当 $R_x >> R_A$ 且 $R_x << R_V$，两种接法均可。

若要得到待测电阻的准确值，则必须分别用式（2）、式（5）加以修正。

2. 电表量程的选择对测量结果的影响

如果电表的某量程为 A，准确度等级为 p 级，则对于用该量程进行测量所得的任意测量值，其绝对误差最大为：

$$\Delta A=A\times\frac{p}{100} \tag{7}$$

即在电表准确度等级一定的条件下，选择的量程越小，测量的准确度越高。

根据误差传递规则，在用伏安法测电阻时，由于电表本身准确度造成的相对误差可以表示为：

$$E=\frac{\Delta U}{U}+\frac{\Delta I}{I} \tag{8}$$

其中，U、I 分别表示电压和电流的测量值，ΔU、ΔI 分别表示某一量程下的最大绝对误差。因此，要使测量的结果更加准确，就需要选择合适的量程，使得电表在读数时指针在 2/3 满量程以上。

【实验内容】

按图 6－2 连接线路。

（1）当待测电阻 $R_x >> R_A$（如 R_x 取 $2k\Omega$），观察开关 K_2 在 e 处和在 f 处时伏特计和安培计分别有何变化。选择合适的量程，记下开关在 e 处时的伏特计和安培计的数值以及安培计的内阻 R_A。改变量程再测一次，比较两次测量的误差。

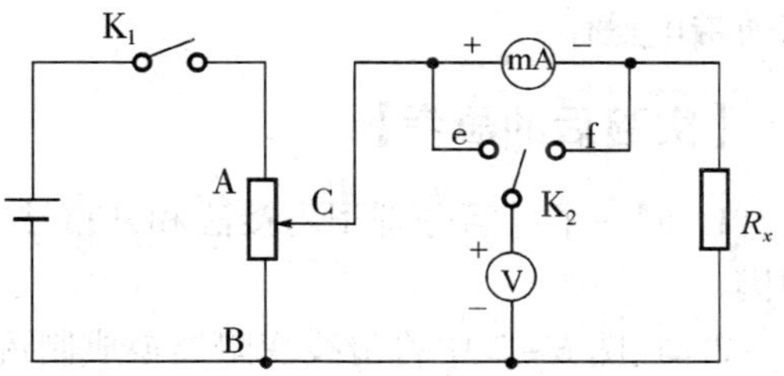

图 6－2　伏安法测电阻线路图

（2）当待测电阻 $R_x << R_V$，R_x 又不过分大于 R_A（如 R_x 取 20Ω），观察开关 K_2 在 e 处和在 f 处时伏特计和安培计分别有何变化。选择合适的量程，记下开关在 f 处时的伏特计和安培计的数值以及伏特计的内阻 R_V。改变量程再测一次，比较两次测量的误差。

【数据处理】

将待测电阻 R_x 用式（2）、式（5）修正，相对误差要用误差传递公式计算，再计算平均绝对误差 ΔR_x，测量结果采用 $\overline{R_x} \pm \Delta R_x$ 表示。如表 6－1 所示。

表 6－1　内接法数据处理

U_0 (V)	I_0 (mA)	U (V)	I (mA)	$R=\frac{U}{I}$ (Ω)	$R_x = R - R_A$ (Ω)	$E_R = \frac{\Delta U}{U} + \frac{\Delta I}{I}$	$\Delta R = R_x E_R$ (Ω)	$R = R_x \pm \Delta R_x$ (Ω)

注：伏特计最大绝对误差 ΔU = 伏特计量程 U_0 × 伏特计标称误差（%）；
安培计最大绝对误差 ΔI = 安培计量程 I_0 × 安培计标称误差（%）。

【思考题】

1. 内接法、外接法适用的条件分别是什么？如何进行系统误差的修正？
2. 为什么选择不同的量程会影响测量结果的准确性？如何选择合适的量程？

实　验　指　导

【预习要求】

1. 掌握伏安法测电阻安培表内接或外接的条件。
2. 掌握电表量程的选择对测量误差的影响。

【操作要点】

1. 待测电阻用变阻箱表示，按实验要求设定阻值，并根据电表内接外接的条件选择适当的线路连接方法。

2. 无论是外接法还是内接法，首先用合适的电表量程进行测量，即尽量使电表的指针在满刻度的 2/3 以上，并正确读数。改换量程后，注意读数的有效数字的改变。

3. 实验过程中，改换电表量程、改变待测电阻阻值、内外接法转换时均要切断电源，切勿带电操作。

【实验后的思考】

1. 同一个电阻分别用内接法和外接法测量，伏特计和安培计分别有何变化，解释其原因。

2. 在图 6－2 中将滑线变阻器换成制流器接法是否可以？为什么？

6 Measuring Resistance by Volt – ampere Method

There are many ways to measure resistance such as volt – ampere method, compensation method and electrical bridge method and so on. Volt – ampere method is defined as measuring resistance by voltmeter and ampere meter according to Ohm's law. Volt – ampere method is not as precious as compensation method and electrical bridge method, but it is convenient and has wide measuring range. Therefore volt – ampere method is the basic method to examine electrical components and materials, and is widely applied in industrial and lab practice.

Objectives

1. Understand the conditions for amperemeter inner – link or outer – link while measuring resistance.

2. Analyze the influence of the accuracy of electric meters on the results of measurements.

Equipments

Voltmeter, amperemeter, regulated power supply, slide rheostat, revolving resistance box, single – pole double – throw (SPDT) switch, single – pole single – throw (SPST) switch.

Principle

According to Ohm's law $R_x = \frac{U}{I}$, we can get the value of unknown resistance R_x by measuring the potential U and current I of resistance R_x. Measuring resistance with the volt – ampere method have the shortcoming of low accuracy. One reason is the system error caused by internal resistances of electric meters. The other is the instrument error owing to the limitation of the meter's accuracy.

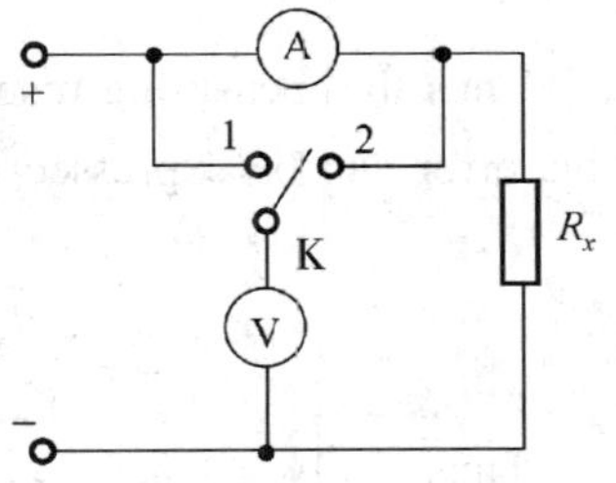

Figure 6 – 1 Measuring resistance with volt – ampere method

1. System error and its correcting

Figure 6 – 1 shows the two ways of joining the amperemeter into the measuring resistance circuit by volt – ampere method. Inner – link: by turning switch K to "1" position, the amperemeter is connected internally. Outer – link: by turning switch K to "2" position, the amperemeter is connected externally.

For each way, the measurement has system errors, which are discussed as following:

(1) Inner – link

The reading of amperemeter I shows the current I_x passing through the resistance R_x, that is $I = I_x$. The reading of voltmeter U does not shows the potential difference U_x on the two terminals of R_x, but it is the sum of U_x and the potential difference U_A on the terminals of amperemeter, i. e., $U = U_x + U_A$. Suppose the internal resistance of amperemeter is R_A and the measuring result of the unknown resistance is R, then:

$$R = \frac{U}{I} = \frac{U_x + U_A}{I_x} = R_x + R_A \tag{1}$$

That is:

$$R_x = R - R_A \tag{2}$$

Thus the measuring result R will be bigger than the actual value R_x. The relative error will be:

$$E = \frac{\Delta R_x}{R_x} = \frac{R - R_x}{R_x} = \frac{R_x + R_A - R_x}{R_x} = \frac{R_A}{R_x} \times 100\% \tag{3}$$

Only when $R_x >> R_A$, the result of inner – link will be accurate enough.

(2) Outer – link

The reading of voltmeter U shows the potential difference U_x of R_x, that is $U = U_x$. The reading of amperemeter I is not the current I_x flowing through R_x, but the sum of I_x and the current I_V flowing through the voltmeter, i. e., $I = I_x + I_V$. Suppose the internal resistance of voltmeter is R_V, then:

$$R = \frac{U}{I} = \frac{U_x}{I_x + I_V} = \frac{1}{\frac{I_x}{U_x} + \frac{I_V}{U_V}} = \frac{1}{\frac{1}{R_x} + \frac{1}{R_V}} = \frac{R_x R_V}{R_x + R_V} \tag{4}$$

That is:

$$R_x = \frac{R}{1 - \frac{R}{R_V}} \tag{5}$$

Thus the measuring result R will be smaller than the actual value R_x. The systematic relative error can be expressed as:

$$E = \frac{\Delta R_x}{R_x} = \frac{R - R_x}{R_x} = \frac{\frac{R_x R_V}{R_x + R_V} - R_x}{R_x} = -\frac{R_x}{R_x + R_V} \times 100\% \tag{6}$$

Thus, only when $R_V >> R_x$, the result of outer – link will be accurate enough, where the negative sign that measurement result is smaller than actual value with amperemeter inner – link.

Now, we can conclude that all of the measured values by volt – ampere method are bigger or smaller than actual values. The way of amperemeter linking in the circuit should be selected depending on the relative magnitude of R_x, R_A, R_V:

If $R_x >> R_A$, the amperemeter can be inner – linked;

If $R_x << R_V$, the amperemeter can be outer – linked;

If $R_x >> R_A$ 且 $R_x << R_V$, both work.

Formula (2) or (5) can be adopted for getting a more accurate measured value of resistance.

2. The influence of electric meter's measuring range on measuring result

If a measuring range of a electric meter is A, the electric meter's grade is p, the maximum value of absolute error is:

$$\Delta A = A \times \frac{p}{100} \tag{7}$$

So the measuring range of electric meter is smaller, the measuring result is more accurate.

According to the formula of propagation of error, the relative error caused by the accuracy of electric meter itself can be written as:

$$E = \frac{\Delta U}{U} + \frac{\Delta I}{I} \tag{8}$$

Where, U and I are the readings of voltmeter and amperemeter respectively, ΔU and ΔI are the maximum absolute errors corresponding to the measuring ranges of voltmeter and amperemeter. As shown above, the suitable measuring range of electric meter must be chosen to get more accurate measuring result. It's better to read electric meter when the value is above 2/3 measuring range.

Experiment

Join the circuit as shown in figure 6 – 2.

(1) When $R_x >> R_A$ (for example R_x = 2kΩ), compare the changes happening to amperemeter and voltmeter by turning K_2 switch from "e" position to "f" position. Then turn K_2 switch to "e" position, chose the suitable measurement range of voltmeter and amperemeter and record the readings of them correspondingly. The internal resistance of amperemeter R_A should be recorded also. Change the measurement range of voltmeter and amperemeter to measure again, analyze how the measurement range affects the accuracy of final measurement result.

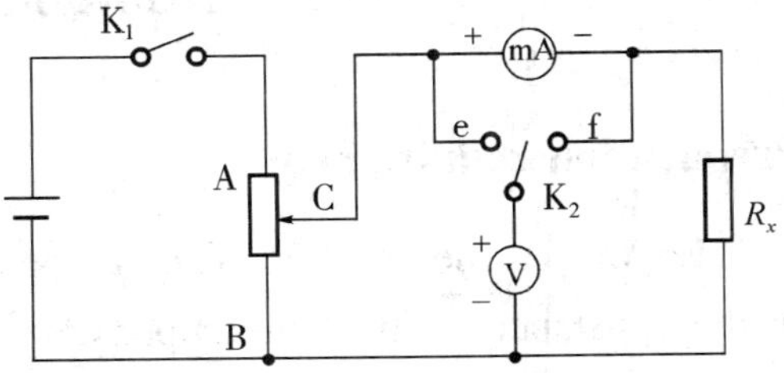

Figure 6 – 2 The circuit of the experiment

(2) When $R_x << R_V$, and R_x is not bigger than R_A very much (for example $R_x = 20\Omega$), compare the changes happening to amperemeter and voltmeter by turning the K_2 switch from "f" position to "e" position. Then turn K_2 switch to "f" position, chose the suitable measurement range of voltmeter and ampermeter and record the readings of them correspondingly. The internal resistance of voltmeter R_V should be recorded also. Change the measurement range of voltmeter and ampermeter to measure again, analyze how the measurement range affects the accuracy of final measurement result.

Data processing

The actual result of resistance R_x can be calculated by formula (2) and (5). The relative error E_R and absolute error $\Delta R_x = R_x E_R$, should be calculated with propagation formula. Table 6 – 1 is an example for data handling of inner – link.

Table 6 – 1 Table for data handling of inner-link

U_0 (V)	I_0 (mA)	U (V)	I (mA)	$R=\frac{U}{I}$ (Ω)	$R_x = R - R_A$ (Ω)	$E_R = \frac{\Delta U}{U} + \frac{\Delta I}{I}$	$\Delta R_x = R_x E_R$ (Ω)	$R = R_x \pm \Delta R_x$ (Ω)

Where, ΔU = voltmeter' s measurement range U_0 × voltmeter' s nominal error (%);

ΔI = amperemeter' s measurement range I_0 × amperemeter' s nominal error (%).

Problems

1. What is the conditions for ampermeter inner – link or outer – link? How to correct the system error?

2. Why the measurement range affects the accuracy of final measurement result? How to select the proper measurement range?

Additional Instruction

Preparation requirement

1. Master the conditions for amperemeter inner – link or outer – link while measuring resistance with volt – ampere method.

2. Master the influence of electric meter's measurement range on measuring result.

Operation key points

1. The revolving resistance box is used as the resistance R_x, which should be set as

the experiment requirement. Choose the appropriate method to join the circuit in accordance with the conditions for amperemeter inner – link or outer – link.

2. Whether inner – link or outer – link, first of all, measure the resistance with the appropriate measurement range, that is, try to make the pointer of meter above 2/3 of full scale, and read the meter rightly. After change the measurement range, pay attention to the change of the effective number of readings.

3. Every time for changing measurement range, changing revolving resistance box, or changing ampermeter link method, electric source should be shut off first and the hot – line work is not suggested.

Problems after experiment

1. Measure a resistor with inner – link and outer – link method respectively, if the voltmeter and ammeter change? Explain the cause.

2. For the circuit in figure 6 – 2, join the slide rheostat for a current – limiting resistor, not a volt – divider, does it also work? Why?

实验七　用惠斯通电桥测定电阻

电桥电路是电学中一种基本的电路连接方式，具有非常广泛的应用。电桥按其电源可分为直流电桥和交流电桥两大类。惠斯通电桥是典型的直流电桥，其四个桥臂都是纯电阻元件，探测器一般用灵敏检流计，可以精确测量电阻。交流电桥用的桥臂可以是电容、电感或电阻，以及它们的组合。由于影响交流电桥平衡的因素很多，因而使用范围更为广泛：可以用来测量电容、电感、互感、磁性材料的磁导率、电容器的介质损耗等。

【目的要求】

1. 掌握惠斯通电桥测量电阻的原理和方法。
2. 了解电桥的灵敏度与元件参量之间的关系。

【仪器器材】

检流计、电阻箱、待测电阻二个、开关、电源、滑线变阻器。

【实验原理】

1. 电桥工作原理

惠斯通电桥原理电路图如图 7－1 所示。R_1、R_2、R_s均是可调电阻，R_x是待测电阻，BD 间接检流计。适当调节 R_1，R_2、R_s的阻值，可使检流计指针指零（即 $I_g=0$），则 BD 两点的电势相等，此时电桥平衡。则有：

$$I_1R_1=I_xR_x$$

$$I_2R_2=I_sR_s$$

因 $I_g=0$，故 $I_1=I_2$，$I_s=I_x$，得：

$$\frac{R_1}{R_2}=\frac{R_x}{R_s}$$

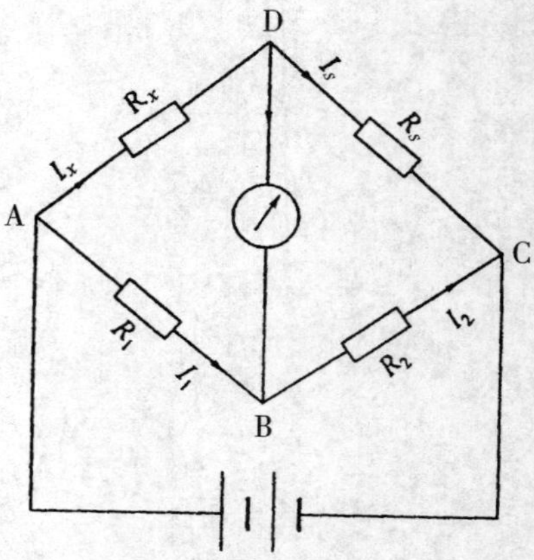

图 7－1　惠斯通电桥原理电路图

或：

$$R_x=\frac{R_1}{R_2}R_s=K_rR_s$$

式中，$K_r=\dfrac{R_1}{R_2}$，叫做比率系数或倍率。

2. 电桥的灵敏度

当电桥有微小不平衡时，我们希望知道检流计如何偏转的。通常情况下人眼所能判断的最小偏转为 0.1mm，若偏转小于 0.1mm，则认为电桥是平衡的。这必然会带来测量上的误差。这种误差与哪些因素有关呢？为此引入电桥灵敏度 S：

$$S=\frac{\Delta n}{\dfrac{\Delta R_s}{R_s}} \tag{1}$$

上式中 ΔR_s是电桥平衡后 R_s的微小变化量，Δn 是因电桥不平衡，引起检流计的偏转格数。S 越大灵敏度越高，所带来的误差就越小。

可以证明电桥的灵敏度为：

$$S=S_iU\frac{R_1/R_2}{R_g^*\left(1+\dfrac{R_1}{R_2}\right)^2} \tag{2}$$

其中 S_i为检流计的灵敏度，U 为电源电压，$R_g^*=R_g+(R_1//R_2)+(R_X//R_s)$，$R_g$为检流计的内阻。

由上式可知：

（1）增加检流计的灵敏度可以提高电桥的灵敏度。

（2）增加电源电压 U（注意不能超过电阻的额定功率）可提高电桥的灵敏度。

（3）电桥的灵敏度还与倍率或桥臂比率有关，当$\dfrac{R_1}{R_2}=1$ 时，S 为最大。

【实验内容】

1. 用自组电桥测电阻

按图 7－2 连接线路。R_1、R_2、R_s均为电阻箱，R 和 R_H为滑线变阻器，K_1和 K_2为单刀单掷开关。

选择适当倍率 $K_r=R_1/R_2$（10 进倍率如 0.1、1、10、100 等），测量未知电阻 R_x 的阻值。

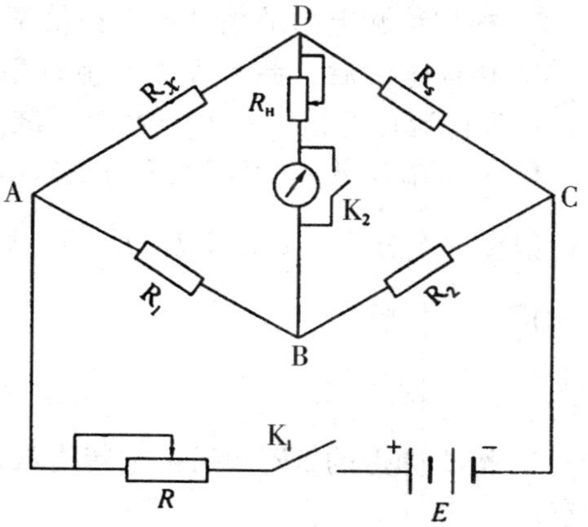

图 7－2　线路图

2. 电桥灵敏度的研究

（1）选择不同的倍率 K_r值，进行测量。

（2）改变电源电压，进行测量。

（3）选择灵敏度不同的检流计，进行测量。

（4）测量 R_s的调节误差 δ_{Rs}。

【注意事项】

1. R_H为保护电阻，测量前应将 R_H值调到最大。

2. 为了保护检流计，应注意开关次序，接通时先合 K_1，后合 K_2；断开时先断 K_2后断 K_1。

3. 开关应短时间接通，即采用跃接法。避免通电时间过长，使额定功率较小的电阻发热，引起阻值的变化。

【思考题】

1. 电桥由哪些部分组成？电桥平衡条件是什么？

2. 用电桥法测电阻比用欧姆计测电阻，准确的原因是什么？

3. 在电桥能测量的范围内，为了提高测量的准确度，桥臂比率应如何选择？

4. 在图 7－2 中，当电桥平衡时，交换电源和检流计的位置，电桥是否仍保持平衡状态？试证明之。

实 验 指 导

【预习要求】

1. 熟悉惠斯通电桥的结构及各部分的作用，了解电桥的使用方法。

2. 简要写出实验操作步骤。

【操作要点】

1. 测量前，先调准检流计的机械零点。

2. 根据待测电阻的估计值选择倍率 K_r。若 R_x未知，则应选倍率为 1。

3. 接通电源后，调节 R_s使检流计指针指零，逐渐减小 R_H的数值，并反复调节 R_s的数值，直到检流计的指针在 $R_H=0$ 时，准确地停止在零点为止。

4. R_s调节误差 δ_{Rs}可用如下方法得到。当电桥平衡后，使 R_s值改变（R_s，并使检流计指针偏转一个分度（1 分度为 1mm），则相应$\frac{1}{10}$分度的 R_s值的改变量即为 R_s的调节误差 δ_{Rs}，即：

$$\delta_{Rs}=\frac{1}{10}\Delta R_s$$

5. 按 4 测得的 ΔR_s和指针偏转格数 Δn（为 1 分度），由定义式（1）可计算电桥的灵敏度。

【报告要求】

本实验中误差主要是仪器误差，可分为两部分：

（1）电桥灵敏度引入的误差：

由公式$\frac{R_1}{R_2}=\frac{R_x}{R_s}$可得：

$$\frac{\delta R_x}{R_x}=\frac{\delta R_s}{R_s}$$

（2）桥臂元件的级别引入的误差，其最大可能误差为：

$$\frac{\Delta R_x}{R_x}=\frac{\Delta R_1}{R_1}+\frac{\Delta R_2}{R_2}+\frac{\Delta R_s}{R_s}$$
$$\geqslant 0.3\%$$

总的仪器误差应为这两部分之和。

【实验后的思考】

1. 若用检流计 G 取代电阻 R_x，BD 间只连接一个开关。能否测得检流计的内阻。如能测，如何判断电桥是否平衡？

2. 互易桥臂对测量结果有何影响？并说明互易桥臂的作用。

7 Determination of Resistance by Wheatstone Bridge

As one of basic wiring method, electric bridge circuit has a widely application in the electric measurement. There are two kinds of electric bridges, direct current bridge and alternative current bridge, which are classified by its source. Wheatstone bridge is a typical direct current bridge, which has four bridge arms of resistance components and a detector of galvanometer. It can measure resistance with high precision. Meanwhile, an alternative current bridge has arms of capacitors, inductors, resistances, or their combination. Because there are a lot of factors working on the balance of the bridge, an alternative bridge can measure capacitance, inductance, mutual inductance, permittivity of magnetic material, dielectric loss, and so on.

Objectives

1. Learn the measuring principle and method of Wheatstone bridge;
2. Understand the relationship between bridge sensitivity and its elements' parameters.

Equipments

Galvanometer, resistance boxes, resistor to be measured, switches, electric source, slide rheostat.

Principle

1. Principle of Wheatstone bridge

The electric circuit of Wheatstone bridge is shown in Figure 7-1. R_1, R_2 and R_s are all changeable resistances; R_x is the resistance to be measured. Join a galvanometer between point B and D. Adjust R_1, R_2 and R_s until the Galvanometer indicating zero ($I_g = 0$), so the point B and D are equipotential, and we call the bridge is balanced.

We can get:

$$I_1 R_1 = I_x R_x,$$

$$I_2 R_2 = I_s R_s.$$

Because $I_g = 0$, it means $I_1 = I_2$, $I_s = I_x$, and we get:

$$\frac{R_1}{R_2} = \frac{R_x}{R_s},$$

That is: $R_x = \frac{R_1}{R_2} R_s = K_r R_s$. The $K_r = \frac{R_1}{R_2}$, is called ratio parameter or ratio.

2. Sensitivity of the bridge

What will be the deflection of the pointer of the galvanometer for a slight unbalance of the bridge? Usually we can see a deflection of not less than 0.1mm, so that if the deflection is smaller than 0.1mm, we consider the bridge is in equilibrium. Now, we introduce the sensitivity of the bridge:

$$S = \frac{\Delta n}{\frac{\Delta R_s}{R_s}}, \qquad (1)$$

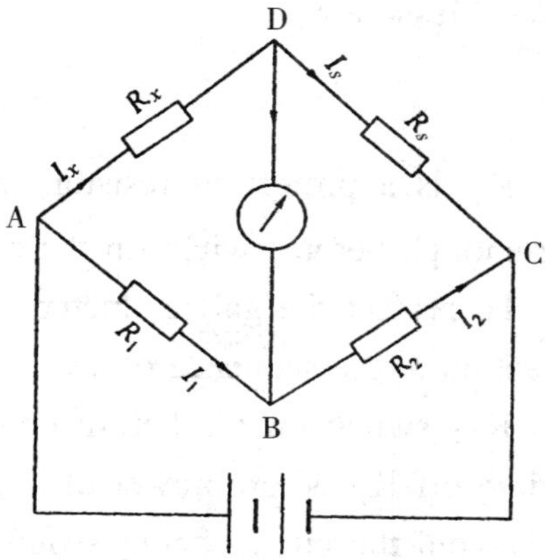

Figure 7 – 1 The electric circuit of Wheatstone bridge

The R_s is slight changed amount of R_s under the condition of balance, Δn is the deflection of the hand causing by the ΔR_s. The bigger of the S means the bridge is more sensitive and smaller error.

It has been demonstrated that the S can be written as:

$$S = S_i U \frac{R_1/R_2}{R_g^* \left(1 + \frac{R_1}{R_2}\right)^2}, \qquad (2)$$

S_i is the sensitivity of the galvanometer, U is the voltage of source, $R_g^* = R_g + (R_1//R_2) + (R_X//R_s)$, and R_g is the internal resistance of the galvanometer.

According to the formulas above, we can conclude that:

(1) The bridge sensitivity can be improved for a more sensitive galvanometer;

(2) The bridge sensitivity can be improved by increasing the voltage of source U;

(3) The bridge sensitivity has a relation with the ratio, when $\frac{R_1}{R_2} = 1$, the S will get its maximum value.

Experiment

1. Joint a bridge to measure resistance

Joint the circuit as shown in figure 7 – 2. R_1, R_2 and R_s are resistance boxes, R and R_H are slide rheostats, K_1 and K_2 are switches (SPST).

For different ratios of R_1/R_2, such as 1, 0.1, 10, 100, measure the R_x.

2. Study the sensitivity of the bridge

(1) Measure R_x under different ratios.

(2) Measure R_x under different source voltages.

(3) Measure R_x with different galvanometers of different sensitivities.

(4) Measure δ_{Rs}.

Notice

1. R_H is a protecting resistor. Adjust R_H big enough before switch on the circuit.

2. To protect the galvanometer, please pay attention to the sequence of switching of K_1 and K_2: switch on K_2 before power on and switch off K_1 before power off.

3. Avoid the circuit being switched on for such a long time that the resistance being over heated.

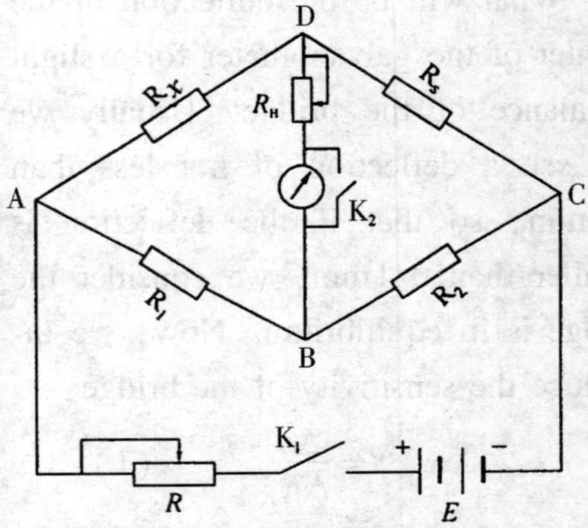

Figure 7 − 2 The circuit diagram

Problems

1. What are the main parts of the bridge? What is the judgement for a balanced bridge?

2. Why it is a more accurate measurement of Wheatstone bridge than that of an ohmer?

3. How to select a correct ratio?

4. For a balanced bridge, if exchange the positions of the power source and galvanometer, the bridge will keep in balance, prove it.

Additional Instruction

Preparation Requirements

1. Be familiar with the structure of the Wheatstone bridge.

2. Write down the experiment steps briefly.

Operation key points

1. Adjust the mechanical zero of the galvanometer before measuring.

2. Select a suitable value for the ratio K_r according to an estimated value of R_x. For an unknown R_x, K_r should be 1.

3. After connect to the electric source, adjust R_s to make the galvanometer pointing zero. And then reduce the value of R_H, adjust R_s again to make the galvanometer

pointing zero until $R_H = 0$.

4. The adjustment error δ_{Rs} of R_s can be gotten according to the following process. For the balanced bridge, change R_s by a small value of ΔR_s to make the pointer deflect just one scale (1mm), thus the $\frac{1}{10}$ of the ΔR_s will be δ_{Rs}:

$$\delta_{Rs} = \frac{1}{10}\Delta R_s$$

5. According to formula (1), the bridge sensitivity can be calculated from ΔR_s above step and the deflection of pointer Δn (1 scale) .

Report Requirements

The errors in this experiment can be divided into two parts:

(1) The sensitivity of the bridge:

Because $\frac{R_1}{R_2} = \frac{R_x}{R_s}$, we can get:

$$\frac{\delta R_x}{R_x} = \frac{\delta R_s}{R_s}$$

The δR is the adjusting error of resistance of R.

(2) The errors caused by the resistors:

$$\frac{\Delta R_x}{R_x} = \frac{\Delta R_1}{R_1} + \frac{\Delta R_2}{R_2} + \frac{\Delta R_s}{R_s}.$$
$$\geqslant 0.3\%$$

The total error should be the sum of the two parts.

Problems after experiment

1. Can we determine the internal resistance of a galvanometer by the bridge?
2. What is the result if we exchange the places of R_x and R_s?

实验八　用电势差计测电池的电动势

电势差计是利用补偿原理测量电动势的一种精密仪器。它用途广泛，不但可以直接测量电源电动势或电压，还可与标准电阻配合间接测量电流、电阻和功率等，不仅适用于直流电路，也适用于交流电路。因此电势差计在工业测量、自动控制等系统中得到普遍的应用。

【实验目的】

1. 掌握用补偿法测量电动势的原理。
2. 掌握 UJ－25 型直流电势差计的结构、原理和使用方法。
3. 用补偿法测量电源电动势。

【实验原理】

电势差和电动势是电学实验中经常碰到的物理量，一般情况下可用伏特计进行测量。但是，用伏特计测量电势差时，由于测量支路的分流作用（图 8－1），所测得的电势差必小于未测量时的真实值。为了准确测量电势差，必须使分流到测量支路上的电流等于零。直流电势差计就是为了满足这个要求而设计的。

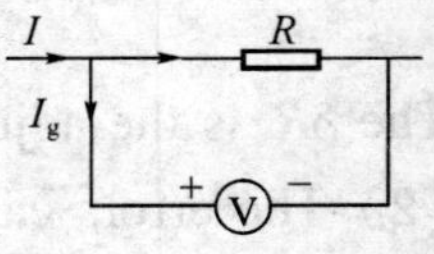

图 8－1　用伏特计测量电势差

电势差计以补偿原理为设计基础，我们用图 8－2 简单说明补偿原理：把已知的可调电动势 ε_0 和未知电动势 ε_x 同极相联，并和检流计Ⓖ串联成图 8－2 所示的闭合电路。$\varepsilon_0 \neq \varepsilon_x$时，回路中必然有电流，检流计偏转。调节 ε_0，使检流计Ⓖ指零，此时回路中 ε_0和 ε_x两电动势必然大小相等，方向相反，即 $\varepsilon_0 = \varepsilon_x$，我们称电路得到了补偿。又 ε_0是可知的，这样 ε_x即被求出。这种测量方法称为补偿法。从上述可知，补偿原理测电动势需要一个大小可调，电压稳定并能准确读数的电动势 ε_0和一个用来判断补偿与否的灵敏检流计Ⓖ。

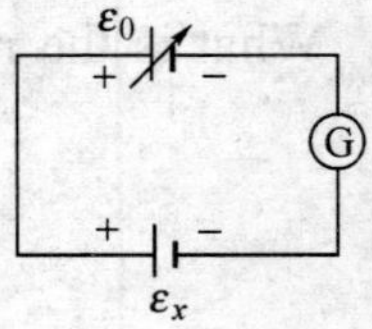

图 8－2　补偿原理测电源电动势

电势差计的原理线路如图 8－3 所示。这个电路图可分成三部分：$\varepsilon R_a R_N R\varepsilon$ 称为辅助回路；ε_NⒼ$R_N \varepsilon_N$称为校准回路；$\varepsilon_x R_k$Ⓖε_x称为测量回路。

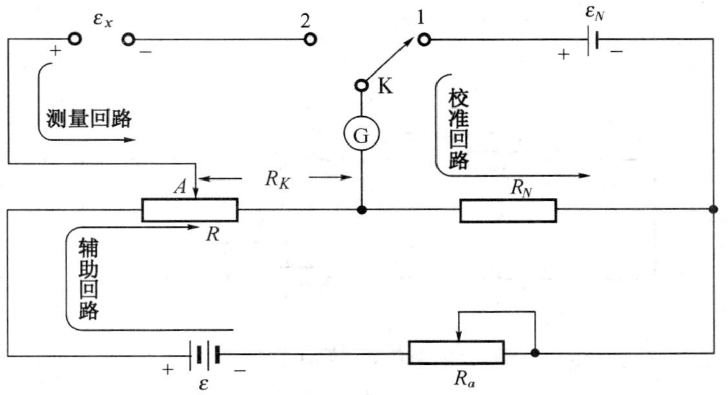

图 8－3　UJ－25 型直流电势差计原理图

电势差计根据补偿原理测量未知电动势 ε_x 的方法如下：

如图 8－4 所示，在辅助回路中，电源 ε（工作电池组）提供工作电流 I，改变可变电阻 R_a 的阻值大小，可调节 I 的大小。

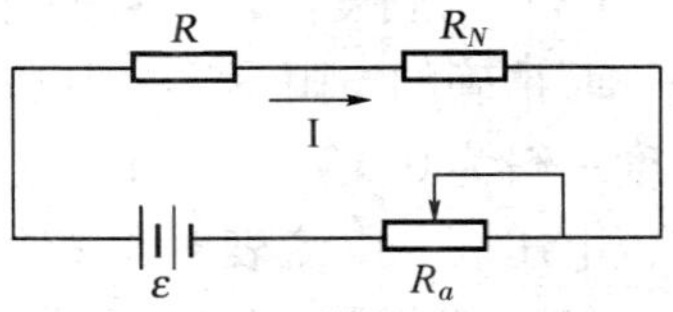

图 8－4　辅助回路工作原理

校准回路是利用补偿原理从而在辅助回路中获得标准工作电流 I_0。

我们将校准回路接通之前，先要做一个确定 R_N（标准电池 ε_N 的补偿电阻）阻值大小的工作。假设我们希望辅助回路中获得的标准电流为 I_0，我们先得知标准电动势 ε_N 的大小（可根据环境温度得知），则 $R_N=\dfrac{\varepsilon_N}{I_0}$，将电阻 R_N 调节成此数值。再将开关 K 合向“1”位置，此时校准回路接通（图 8－5）。调节 R_a，当 R_N 两端的电压降 IR_N 与 ε_N 完全补偿，Ⓖ指零，此时由 $IR_N=\varepsilon_N$ 可知 $I=I_0$。即辅助回路中的电流 I 已等于我们想要的标准电流 I_0。

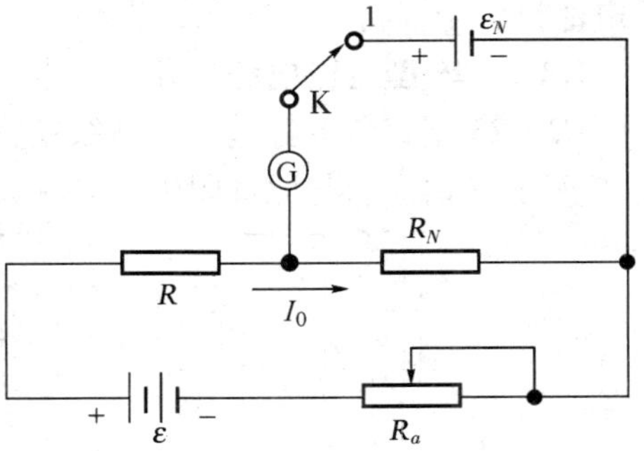

图 8－5　校准回路工作原理

将开关 K 合向“2”位置，连通测量回路 ε_xⒼ$R_K\varepsilon_x$（图 8－6）。此时电阻 R_K（R 的一部分）中已有标准电流 I_0 流过，即 R_K 两端有一电压降 I_0R_K。此电压与 ε_x 的极性相反。适当调节 R_K 值，即滑动头 A 的位置，使测量电路完全补偿，检流计Ⓖ再次指零。此时有 $I_0R_K=\varepsilon_x$，又因为 $I_0R_N=\varepsilon_N$，可得 $\varepsilon_x=\dfrac{R_K}{R_N}\varepsilon_N$。

若要测任一电路两点间的电势差，只需将待测的两点接入测量回路代替 ε_x 即可。

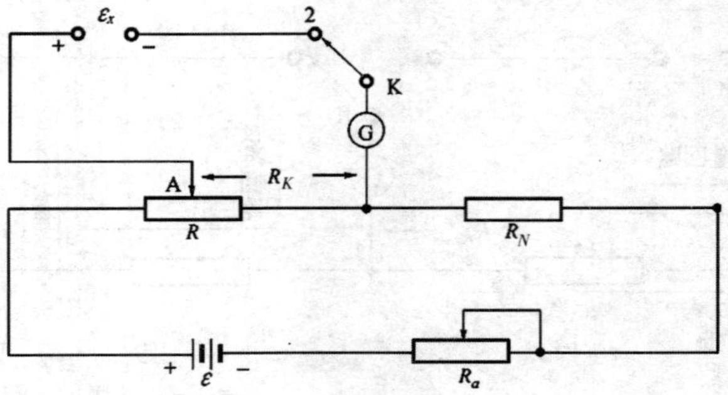

图 8－6　测量回路工作原理

补偿法测量电源电动势或电压的优点是：

（1）准确度高。当电路完全补偿时，Ⓖ指零，无电流流过测量电路，测量电路不消耗待测电路的能量，故可准确测出被测电动势。

（2）精确度高。电势差计是一种电阻仪器，它的精确度主要取决于自身所使用的电阻，而精确的电阻是容易制造的，故电势差计也就成了精确测量中常用的电工仪器，其精确度一般为 0.1%，高的可达 0.005% 乃至 10^{-7}。因此这种高精度的电势差计被广泛用于校正电压表、电流表、欧姆表及功率表。

（3）能测量多种物理量。电势差计除了可以用于测量电势差、电动势以外，还可以用于间接测量电流、电阻、电功率等，不仅用于直流电路，也可用于交流电路，加上各种换能器，还可以进行非电学量的测量，因此使用范围十分广泛。

【仪器器材】

UJ－25 型直流电势差计、AC15 型检流计、多用电表、标准电池、工作电池（1.5V 干电池）二节、待测干电池二节。

1. UJ－25 型直流电势差计

UJ－25 型直流电势差计是高级精密仪器，使用时必须正确连接线路，它的面板如图 8－7 所示（请注意与原理图中的对应关系）。

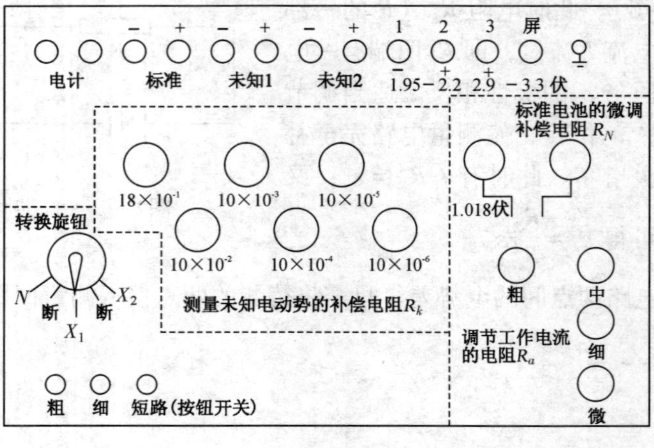

图 8－7　UJ－25 型直流电势差计面板

面板中：

“电计”——接检流计Ⓖ。

“标准”——接标准电池 ε_N。

“未知 1”、“未知 2”——接待测电源 ε_{x_1}、ε_{x_2}。

“1”、“2”、“3”——接工作电池 ε_0，“2”、“3”接线柱下注有所接工作电池电动势的范围。若用 3V 电源，应接在“1”和“3”接线柱上。

“转换旋钮”——“N”为标准挡，用于工作电流的校准；“断”为检流计与线路断开；“X_1”、“X_2”为待测电源挡，分别用于测量两个独立的待测电源。

“粗”、“细”、“短路”按钮——用来将检流计接通或将检流计本身短路。按下“粗”时，在检流计回路中接入 500kΩ 的保护电阻，以限制流过检流计的电流。按下“细”时，检流计直接接入回路中。这两个按钮按下后，稍加转动即可锁住，再转动即可弹起。按下“短路”时，检流计本身被短路，通过检流计的电流快速消失，但此钮只能按下一会儿便应放开，切不可锁住，否则无法实验。

“粗”、“中”、“细”、“微”旋钮——调节工作电流的电阻 R_a。电势差计的工作电流为 1×10^{-4}A。

“R_N”旋钮——标准电池的微调补偿电阻。标准电池电动势的尾数部分可由两个“R_N”旋钮分别调节。其中左边旋钮指 1 时代表万分之一伏特；右边旋钮指 1 时代表十万分之一伏特。与 1.01800V 相应的补偿电阻值已在仪器内部串联好了。

“R_K”旋钮——待测电源的补偿电阻。待测电动势的数值，可以从六个十进位的转盘电阻 R_K 旋钮旁窗孔中所示数字的总和直接读出。

2. AC15 型检流计

AC15 型检流计是高灵敏度的指零仪器，由光标的偏转来显示电流的数值，每分度为 3×10^{-10}A，其使用方法及注意事项如下。

（1）电源开关置于 220V 一侧，即接通电源。

（2）调零点。在检流计未接入回路前应先进行零点调节，即旋转“零点调节”旋钮，使光标落在标尺的零刻度线上。为方便调节，除使用“零点调节”旋钮外，若指针与零刻线相差在 ±5 个小格的范围内，可以利用“标尺活动调零器”（刻度标尺盘上的一个金属小柱）移动刻度盘的办法来调零。

（3）检流计的“分流器”共有五挡，如图 8-8所示。测量使用时，“分流器”应从最低灵敏度挡“0.01”开始，调到光标偏转不大时，则可逐步提高灵敏度至“0.1”、“1”和“直接”挡。灵敏度越高，检流计所连接的保护电阻就越大。测量完毕，分流器必须旋到“短路”挡。

（4）检流计上有两个接线柱用来连接测量电路，如按标明的“+”和“-”连接时，电流从“+”流入，则检流计的光标向右偏转，反之左偏。

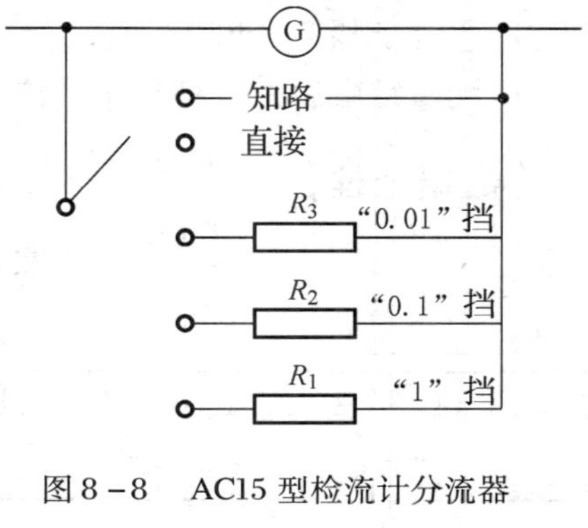

图 8-8　AC15 型检流计分流器

3. 标准电池

电势差计需要标准电池配合使用。标准电池的电动势受温度影响，在某一温度下标准电池的电动势可按下式计算：

$$\varepsilon_t = \varepsilon_{20} - [\alpha (t-20) + \beta (t-20)^2 - \gamma (t-20)^3] \times 10^{-6}\text{V}$$

式中 ε_t为 t℃时的标准电动势，ε_{20}为 20℃时的标准电动势，大小为 1.01860V，t 为测量时的环境温度，α、β、γ 的数值分别为 40.6、0.95、0.01。

由于标准电池内部为液体，因此使用时不能倾斜。另外，标准电池只能作为电动势的标准，不能用作电源，也不能用伏特计或多用电表去测量其电动势，更不得短路。

【实验内容】

1. 熟悉 UJ－25 型直流电势差计面板上各旋钮的作用。然后用多用电表粗测待测干电池的电动势。

2. 按图 8－7 所示位置把工作电池 ε、标准电池 ε_N、两只待测干电池 ε_{x_1}、ε_{x_2}及 AC15 型检流计与 UJ－25 型直流电势差计连接起来，检流计背面接地接线柱应和电势差计的接地接线柱相连接。注意，各电池的正负极切勿接错!

3. 将检流计电源开关置于“220V”、分流器置于“0.01”挡，接上电源，调节“零点调节”旋钮，使光标指零。

4. 校准电流 I_0。将“转换旋钮”拨至“N”；“R_N”的两个旋钮分别调到按式（1）算出的标准电动势的最后两位数（也可以查表获得）。用 R_a旋钮按粗细顺序来调节工作电流的大小，直至检流计分流器置于“直接”挡时，按下电势差计“细”按钮时，检流计光标仍在零点处且无偏转为止。此时，$I = I_0$已调至 1×10^{-4}A。

5. 测量待测电动势 ε_{x_1}。将“转换旋钮”拨至“X_1”；六个大旋钮（其总阻值为 R）的前三个拨至“未知 1”电动势的粗测值（即用多用电表测出的的数值）。按粗细顺序仔细调节六个大旋钮，直至检流计分流器置于“直接”挡，按下电势差计“细”按钮时，检流计光标停在零点处且无偏转。此时六个大旋钮组成的阻值即为补偿电阻 R_K，它们旁边的窗孔中示数的总和即为待测电动势 ε_{x_1}的值。

6. 重复步骤 4、5，再测一遍 ε_{x_1}。

7. 重复步骤 4，然后将“转换旋钮”拨至“X_2”处，重复步骤 5，测出 ε_{x_2}。

8. 重复步骤 7，再测一遍 ε_{x_2}。

9. 测量结束后，电势差计“转换旋钮”拨至“断”，检流计分流器拨至“短路”。

【数据处理】

表 8－1　测量待测电动势

T = ________℃　　　　ε_T = ________V

待测电池	第一次测量	第二次测量
ε_{x_1} (V)	ε_{x11} =	ε_{x12} =
ε_{x_2} (V)	ε_{x21} =	ε_{x22} =

电势差计的仪器误差（U 为量程）：$\Delta U_x = Ua\% =$

测量误差：$\Delta\varepsilon_{x_1} = \max\left(\Delta U_x, \dfrac{|\varepsilon_{x_{11}} - \varepsilon_{x_{12}}|}{2}\right) =$

$\Delta\varepsilon_{x_2} = \max\left(\Delta U_x, \dfrac{|\varepsilon_{x_{21}} - \varepsilon_{x_{22}}|}{2}\right) =$

测量结果：$\varepsilon_{x_1} = \overline{\varepsilon_{x_1}} \pm \Delta\varepsilon_{x_1} =$

$\varepsilon_{x_2} = \overline{\varepsilon_{x_2}} \pm \Delta\varepsilon_{x_2} =$

【思考题】

1. 用电势差计测量电池的电动势有何优点？
2. 为什么工作电池组 ε 的电动势必须大于标准电池 ε_N和待测电池 ε_x？
3. 使用 UJ－25 型电势差计和 AC15 型检流计必须注意哪些事项？

实　验　指　导

【预习要求】

1. 掌握补偿法测电动势或电压的原理。
2. 掌握 UJ－25 型电势差计的电路结构原理及各回路的作用。
3. 熟悉 UJ－25 型电势差计和 AC15 型检流计面板上各旋钮的作用。
4. 了解标准电池的使用方法、温度补偿公式及使用注意事项。
5. 熟练掌握 UJ－25 型电势差计和 AC15 型检流计的操作要求和调节顺序。

【操作要点】

1. 连接线路时要注意，各电池正负极与电势差计上相应位置的接线柱相连时，正正相连，负负相连，不可接反，否则无法进行实验。

2. 所有电源都不得短路，因此连线时须十分注意连线顺序，先由电势差计一端引出导线，再把导线的另一端逐一接到指定的电池上。拆线时反之。

3. AC15 型检流计在使用前应先调零点。检流计每提升一个灵敏度挡都需要进行零点调节，但在校准和测量过程中，不得再随意旋转“零点调节”旋钮或移动刻度盘。

4. 为获得标准工作电流 I_0需要调 R_a。R_a有四个旋钮“粗”、“中”、“细”、“微”，是按电阻值的改变从大到小排列的，使用时应按从“粗”经“中”、“细”到“微”的顺序来用，若前一旋钮旋转一格已可使Ⓖ中的光标从零线的某一侧偏转到另一侧，则说明此旋钮过粗，应改用下一个旋钮。但在检流计已经达到较高灵敏度挡时，一般只调节“细”、“微”旋钮。同理，对测 ε_x时所需调节的 R_K，也有精度高低之分，用法类似。

5. 整个实验的过程就是使检流计Ⓖ光标指零的过程，而且是在调节精度不断提高的条件下，使Ⓖ光标指零。在 UJ－25 型直流电势差计上有两个按钮式开关“粗”和“细”，表示粗调和细调。注意检流计光标远离平衡点时这两个按钮应跃按，只有接近平衡时才可按下锁住，仔细调节。在实验中，这两个开关要先后与分流器的各挡配合使用，才能使检流计在精度逐渐提高的过程中指零。配合使用顺序如表 8－2 所示。

表 8-2　电势差计及检流计的调节步骤

<table>
<tr><th>调节精度</th><th>Ⓖ分流器</th><th>电势差计开关</th><th>为使 Ⓖ指零而需调节的旋钮</th></tr>
<tr><td rowspan="8">低
↓
高</td><td rowspan="2">0. 01</td><td>粗</td><td rowspan="8">1. 若校准电流 I_0，则调 R_a，调节顺序由“粗”、“中”、“细”至“微”

2. 若测 ε_x，则调 R_K，其使用顺序由 10×10^{-1}V，逐渐至 10×10^{-6}V。在此过程中 R_a 不得再变，以保证工作电流不变</td></tr>
<tr><td>细</td></tr>
<tr><td rowspan="2">0.1</td><td>粗</td></tr>
<tr><td>细</td></tr>
<tr><td rowspan="2">1</td><td>粗</td></tr>
<tr><td>细</td></tr>
<tr><td rowspan="2">直接</td><td>粗</td></tr>
<tr><td>细</td></tr>
</table>

6. 在上表所示的各调节步骤中，为加快实验速度，可在第一步Ⓖ的“0. 01”挡与电势差计的“粗”、“细”开关的配合使用时，仔细调节 R_a（或 R_K），但可不必使Ⓖ的光标十分准确地指零刻线，因为这只是粗调；在“0. 1”挡及以后的各步调节中，可进行适当的省略，条件是：只要分流器旋钮及电势差计的按钮开关处于该步的位置时，检流计Ⓖ的光标偏转在刻度盘内，且距零点较小，则此步调节可省略，直接晋升至下一步调节。注意：最后一步，即分流器的“直接”挡与电势差计的“细”按钮配合时，一定要仔细调节 R_a（或 R_K），直至使Ⓖ准确指零，电路才完全补偿。

7. 电势差计上的开关系统由两部分构成：转换旋钮（分五挡：“N”、“断”、“X_1”、“断”、“X_2”）及按钮式开关（分“粗”、“细”、“短路”三种）。使用时，除“转换旋钮”应放在相应的指定位置上外，“粗”或“细”两钮必须按下其中一个，否则电势差计与检流计Ⓖ的电路就未连通。“短路”按钮是为减少光标回平衡位置所用时间、加快实验速度而设置的；当检流计光标来回大幅度偏转而无法在短时间内停止下来时，可按下“短路”按钮，使光标立即停止摆动。但此钮只能按下一会儿便应放开，切不可锁住，否则无法实验。

8. 电势差计测电动势或电压的仪器误差可写成 $\Delta U_x = Ua\%$，式中 a 为电势差计的准确度级别，U 为测量盘最大值，单位为伏特（V）。

【实验后的思考】

1. 实验中如果发现检流计光标总往一边偏，无法调到零点，试分析可能有哪些原因？
2. 为什么电阻 R_N 的数值要随温度的变化而作适当的调节？
3. 如何利用电势差计和一个标准电阻来测量一个未知电阻 R_x？

8 Measuring Electromotive Force with Compensation Method

Potentiometer is a precise instrument for measuring electromotive force with compensation method. It is widely applied not only in measuring electromotive force or voltage directly but also in measuring current, resistance and power with the help of standard resistance. It can be applied to DC circuit as well as AC circuit. So it has extensive use in industrial instrument and automatic control system.

Objectives

1. Understand the principle of compensation method.
2. Master the structure, the principle and the using method of potentiometer (type UJ – 25).
3. Measure the electromotive force with compensation method.

Principle

In electrical experiments we often need to measure potential difference and electromotive force, which are usually measured with voltmeter. But the internal resistance of any electromotive force is finite. As shown in figure 8 – 1, the reading of the voltmeter, therefore, must be less than the electromotive force. So it is necessary to make the current flowing through the voltmeter zero. An useful and accurate device called potentiometer is designed for measuring potential difference and electromotive force.

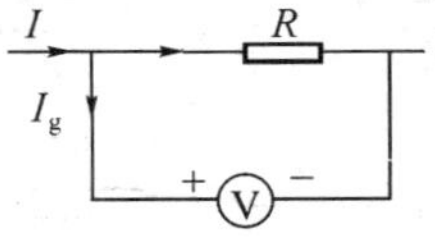

Figure 8 – 1 Measure potential difference with voltmeter

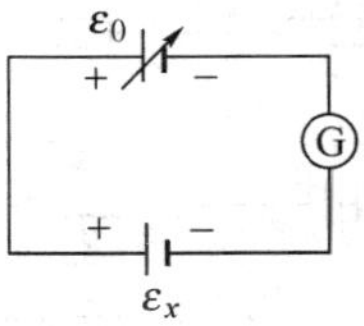

Figure 8 – 2 Measure electromotive force with compensation method

The design principle of potentiometer bases on compensation theory. See figure 8 – 2, ε_x is an unknown electromotive force, and ε_0 is an adjustable electromotive force whose value can be read out. If the two electromotive forces does not equal, there must be current in the circuit, and the galvanometer will not indicate zero. Adjust the ε_0 till the galvanometer indicate zero. Now, ε_x equals ε_0 and they have inverse polarities. It means that the unknown electromotive force is compensated, and we get the value of ε_x, for it is the reading of ε_x. We call this method compensation method. If the electromotive force ε_0 is adjustable, steady and accurate, it can be used to measure an unknown electromotive force with the compensation method. Of course a galvanometer is necessary to judge the circuit is compensated or not.

Figure 8 – 3 shows a schematic drawing of a potentiometer. It divides into three parts: where, the loop of $\varepsilon R_a R_N R \varepsilon$ is accessorial loop; the loop of ε_N Ⓖ $R_N \varepsilon_N$ is calibration loop; the loop of $\varepsilon_x R_K$ Ⓖ ε_x is measuring loop.

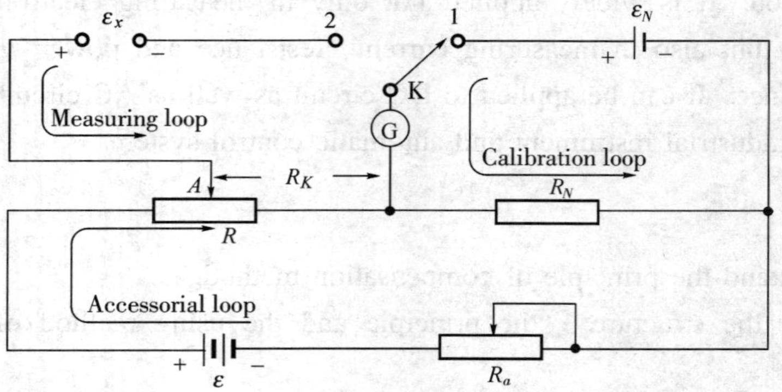

Figure 8 – 3 The working principle of potentiometer

The compensation method is illustrated as following:

Let's see the accessorial loop. As shown in figure 8 – 4, the power ε (batteries) in accessorial loop provides electric current I for working, adjust the value of the resistance R_a, the current can be modified.

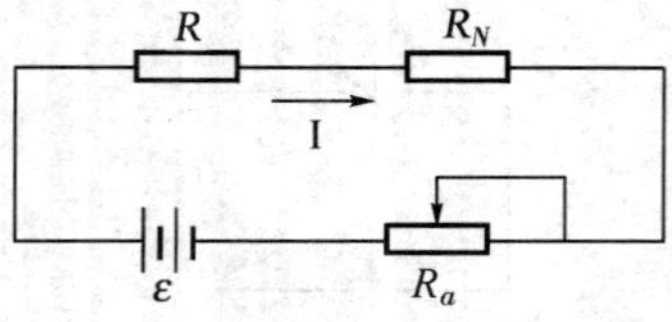

Figure 8 – 4 Accessorial loop

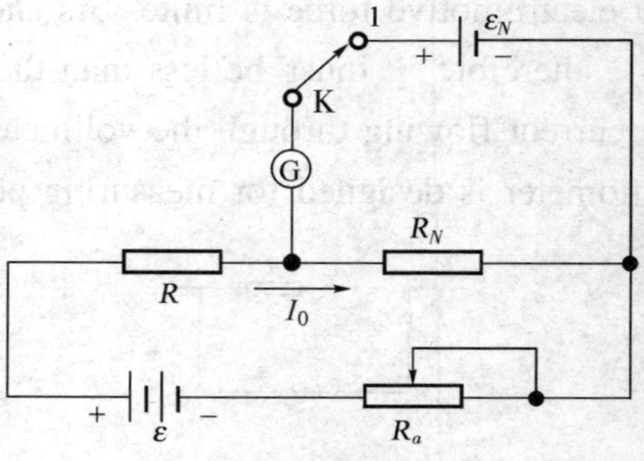

Figure 8 – 5 Calibration loop

The calibration loop is to get standard current I_0 in accessorial loop with compensation method.

The value of resistance R_N (the compensation resistance of the standard battery ε_N) must be calculated before we close the calibrating loop. Suppose the standard current we designed is I_0, we can calculate R_N for $R_N = \frac{\varepsilon_N}{I_0}$, and standard battery ε_N can be gotten according to environment temperature. Adjust the standard resistance R_N to this value. Now, throw the switch K to "1" to close the calibrating loop (Figure 8 – 5).

Modify R_a till the galvanometer Ⓖ indicates zero, then the potential difference between the terminals of R_N equals ε_N, that is $IR_N = \varepsilon_N$, we get $I = I_0$. In this way, the current in accessorial loop comes to the standard current I_0.

Throw the switch K to "2" to close the measuring loop (As shown in figure 8 – 6). There is the standard current I_0 flowing through resistance R_K (a part of R), also there is the potential drop I_0R_K across R_K, whose polarity is inverse to ε_x. Move the sliding contact A to adjust R_K until the galvanometer Ⓖ shows no deflection again. At this moment, $I_0R_K = \varepsilon_x$. Since $I_0R_N = \varepsilon_N$, therefore $\varepsilon_x = \dfrac{R_K}{R_N}\varepsilon_N$.

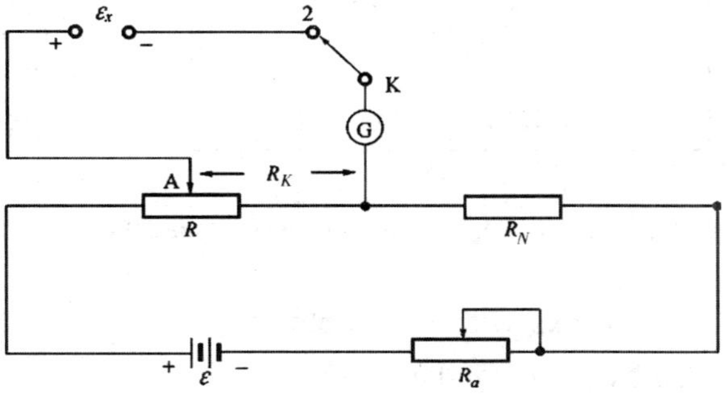

Figure 8 – 6 Measuring loop

By replacing ε_x with potential difference between two points of a certain circuit we can also measure the potential difference by the compensation method.

The compensation method has the advantages as follows:

(1) Great accuracy. When the circuit is compensated, the galvanometer Ⓖ points to zero. The measuring loop hasn't any current, therefore no energy of the circuit is consumed and the electromotive force can be determined with great accuracy.

(2) High precision. The potentiometer is a resistant instrument, its precision depends on the resistors it using. As we know, precise resistor can be made easily, so the potentiometer is commonly used in electrical measurements with high precision. Its precision is high from 0.1% to 10^{-7}, and it is widely used in calibrating voltmeter, current meter, ohmer meter and power meter.

(3) It is also quite versatile. It may be used for measuring voltage, current, resistance, power, etc., not only in DC circuit but also in AC circuit. In addition, it can measure non – electrical value with the help of different transducer.

Equipments

D. C. Potentiometer (type UJ – 25), galvanometer (type AC15), multi – meter, standard battery, working batteries (1.5V ×2), batteries to be determined

1. UJ – 25 Potentiometer

UJ – 25 potentiometer is a kind of fine instrument, and it should be handled carefully. Its panel is shown in figure 8 – 7.

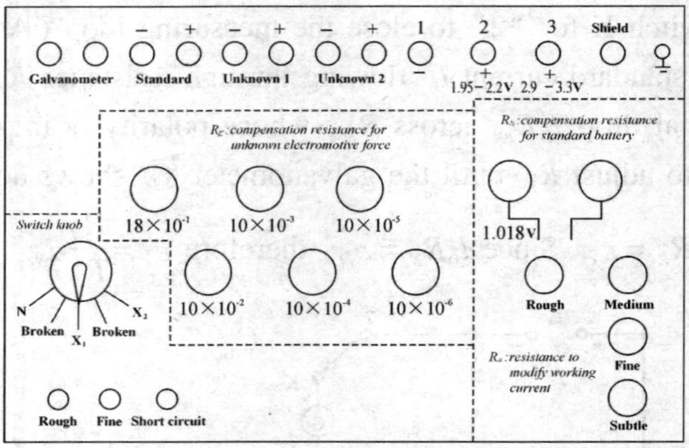

Figure 8 – 7 D. C. Potentiometer (type UJ – 25)

"Galvanometer" ——Links with galvanometer Ⓖ.

"Standard" ——Links with standard battery ε_N.

"Unknown 1" and "Unknown 2" ——Link with batteries to be determined, ε_{x_1} and ε_{x_2} respectively.

"1", "2", "3" ——Connectors for working batteries ε_0. The voltage range is indicated below the connectors. A battery with 3 voltage electromotive force should be connected to "1" and "3"

"Switch knob" —— "N" is for "Standard", used when calibrating; "Off" is pointed to make the loop break; "X_1", "X_2" is for "Unknown", used when measuring two independent electromotive forces.

"Rough", "Fine" and "Short circuit" buttons——If pressed down, they can make the galvanometer link to the relevant loop or short – cut. The difference between the "Rough" and "Fine" is that there is a 500kΩ resistance linked with the galvanometer when "Rough" button is pressed. If the buttons are pressed down and turned for a small angle, they will be locked. Samely, they will be unlocked by turning for a small angel again. "Short circuit" can be pressed when you want the current flowing through the galvanometer disappear quickly but it must be unlocked, otherwise it is impossible to do the experiment.

"Rough", "Medium", "Fine" and "Subtle" knobs——Adjust rheostat R_a for modifying working current. The working current for UJ – 25 potentiometer is 1×10^{-4} A.

"R_N" knobs——Compensation resistance for standard battery. Adjust the last two digits of standard resistance R_N. Resistance of compensation for standard battery with electromotive force of 1.01800V has installed in series beforehand.

"R_K" knobs——Compensation resistance for unknown battery. The result of measurement is the sum of the readings showing in the six windows besides the six knobs of R_K. The six figures are arranged in decimalism.

2. AC15 Galvanometer

AC15 galvanometer is a kind of high sensitive zero – indicating instrument. Its deflection of light cursor with 1 scale indicates an electric current of 3×10^{-10}A. The following is its operation points.

(1) Switch on the power to 220V electric source firstly.

(2) Adjust zero point. Before the galvanometer is linked to the circuit, the light cursor must be adjusted to the zero point preciously. The knob "Zero – point adjusting" can adjust zero – point finely. The "Movable scale" can be adjusted in the range from left five scales to right five scales for adjusting zero point conveniently.

(3) The diverter of galvanometer must be used from the step of lowest sensitivity "0.01". When the deflection of light cursor is not big around zero – point, the higher sensitivity step of diverter can be taken in. The order of using the diverter must be "0.01", "0.1", "1" and "Direct". The inner structure of galvanometer is shown as figure 8 – 8. Different steps are linked with different protective resistances, the sensitivity is higher, the protective resistance is larger. At the end of operation, the diverter should be turned to the step of "Short – circuit" for keeping it from electric shock.

(4) There are two connectors, "+" and "–", on the back of AC15 galvanometer. When measuring, if the electric current flows into the "+", the light cursor will deflect to right side, or it will deflect to left.

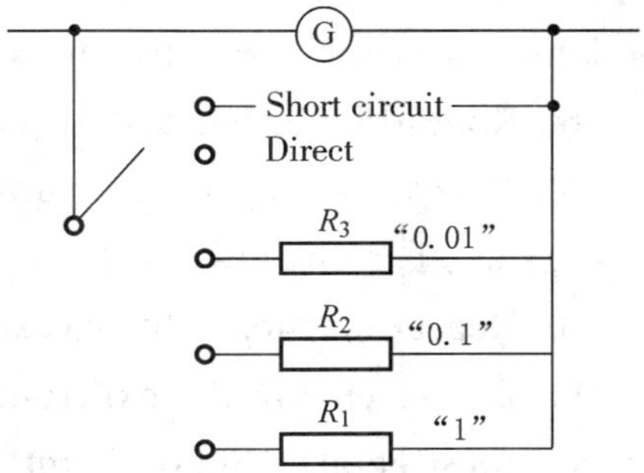

Figure 8 – 8 the structure of diverter

3. Standard battery

Potentiometer must be operated under the help of standard battery. The value of standard battery can be changed under different temperature. The formula for the electromotive force of standard battery is:

$$\varepsilon_t=\varepsilon_{20}-(\alpha(t-20)+\beta(t-20)^2-\gamma(t-20)^3)\times10^{-6}\text{V}$$

Where ε_t is the value of t℃ and $\varepsilon_{20}=1.01860$V is the value of 20℃. The values of α, β, γ are 40.6, 0.95 and 0.01 respectively.

Standard battery can not be inclined for the electrolyte in it. In addition, the standard battery can provide a standard of electromotive force only, which cannot be used as an electric power source and cannot be measured with voltmeter or multi – meter.

Experiment

1. Be familiar with the functions of knobs on the panel of UJ – 25 potentiometer. Then, get the electromotive force of the batteries to be determined by multi – meter for roughly measurement.

2. Connect the working batteries, standard battery, batteries for determining and AC15 galvanometer with the potentiometer. The earth terminal on the back of galvanometer also should be connected with the corresponding terminal on the potentiometer. Be sure the poles of batteries are connected correctly.

3. Power on the galvanometer and set the diverter to the step of "0. 01", then turn the "Zero – point adjusting" knob to adjust the light cursor to zero point.

4. Calibrate current I_0. Turn the "Switch knob" to "N"; set the two knobs of "R_N" to the figure of the last two digits of the value of standard battery. Modify the working current by adjusting the R_a in the order from "Rough" to "Subtle", until the light cursor stays at zero point while the diverter is on the step of "Direct" and the "Fine" button is pressed. Now, the working current is $I = I_0 = 1 \times 10^{-4}$ A.

5. Measure the value of the battery ε_{x_1} to be determined. Turn the "Switch knob" to "X_1", set the first three knobs of R_k to the value of the numbers of roughly measurement. Adjust the six knobs in the order from "$\times 10^{-1}$" to "$\times 10^{-6}$" until the light cursor can stay at zero point while the diverter at step of "Direct" and the "Fine" button is pressed. Now the value made up by the numbers showing on the six windows beside the knobs is the electromotive force ε_{x_1} of the battery for measuring.

6. Repeat the step 4 and step 5 above for measuring ε_{x_1} another time.

7. Repeat the step 4, then turn the "Switch knob" to "X_2" and repeat the step 5 to measure another unknown battery ε_{x_2}.

8. Repeat the step 7 for measuring ε_{x_2} another time.

9. Before ending the experiment, "Switch knob" must be turned to "Off" and diverter must be at "Short circuit".

Data Processing

Table 8 – 1 Measuring unknown electromotive force

$T =$ ________ ℃ $\varepsilon_T =$ ________ V

Readings for unknown batteries	First measuring	Second measuring
ε_{x1} (V)	$\varepsilon_{x11} =$	$\varepsilon_{x12} =$
ε_{x2} (V)	$\varepsilon_{x21} =$	$\varepsilon_{x22} =$

Instrument error of potentiometer (U is the measuring range): $\Delta U_x = Ua\% =$

Measuring error: $\Delta\varepsilon_{x1} = \max\left(\Delta U_x, \frac{|\varepsilon_{x11} - \varepsilon_{x12}|}{2}\right) =$

$\Delta\varepsilon_{x2} = \max\left(\Delta U_x, \frac{|\varepsilon_{x21} - \varepsilon_{x22}|}{2}\right) =$

Measuring result: $\varepsilon_{x1} = \overline{\varepsilon_{x1}} \pm \Delta\varepsilon_{x1} =$

$\varepsilon_{x2} = \overline{\varepsilon_{x2}} \pm \Delta\varepsilon_{x2} =$

Problems

1. What are the advantages of measuring electromotive force by a potentiometer?
2. Why the value of working batteries has to be greater than ε_N or ε_x?
3. What are the special attentions for operation when using UJ – 25 potentiometer and AC15 galvanometer?

Additional instruction

Preparation requirements

1. Understand the principle of compensation method.
2. Master the structure and the principle of UJ – 25 potentiometer and understand the functions of different loops.
3. Be familiar with the panel of UJ – 25 potentiometer and AC15 galvanometer.
4. Know about the standard battery and the formula for electromotive force of it.
5. Keep in mind the operation rules and adjustment order of UJ – 25 potentiometer and AC15 galvanometer.

Operation key points

1. Link the wires of same sign from the potentiometer to the batteries.
2. All the batteries can not be short circuited. Link the wires from potentiometer to the batteries when connecting the circuit and disconnect the circuit in reverse order.
3. Zero – point must be adjusted before using galvanometer. Zero – point adjusting must be done again when upgrading the sensitivity of the galvanometer, but "Zero – point adjusting" knob and "Movable scale" can not be used while calibrating and measuring at the same sensitivity step.
4. R_a is adjusted for the standard working current I_0. The four knobs of "Rough", "Medium", "Fine", "Subtle" are arranged from lower sensitivity to higher sensitivity orderly and they must be used in this order. Generally, when the diverter of galvanometer is at the higher sensitivity step of "1" and "Direct", "Rough" and "Medium" knob

can not be adjusted . Similarly, R_K knobs must be adjusted in the same way.

5. The central work of the experiment is to make the galvanometer stay at the zero point. There are two buttons “Rough” and “Fine” on the potentiometer denoting rough adjustment or fine adjustment. Press the button “Rough” or “Fine” on the potentiometer and release at once for checking the balance of circuit. When the light cursor can stay near the zero point, the button can be locked for further adjustment. The two buttons must co – work with the diverter during the experiment. The order of the steps is shown in table 8 – 2.

Table 8 – 2 Adjustment steps of the Potentiometer and Galvanometer

<table>
<tr><th>Sensivity</th><th>Galvanometer diverter</th><th>Potentiometer buttons</th><th>Knobs need to be adjusted</th></tr>
<tr><td rowspan="8">Low
↓
High</td><td rowspan="2">0. 01</td><td>Rough</td><td rowspan="8">1. Modify working currentI_0, by adjusting R_a, in the order from “Rough” to “Subtle”
2. Measureε_x, by adjusting R_K, in the order from 10×10^{-1} V, to 10×10^{-6} V。 R_a must keep unchanged during the process of adjusting R_K</td></tr>
<tr><td>Fine</td></tr>
<tr><td rowspan="2">0. 1</td><td>Rough</td></tr>
<tr><td>Fine</td></tr>
<tr><td rowspan="2">1</td><td>Rough</td></tr>
<tr><td>Fine</td></tr>
<tr><td rowspan="2">Direct</td><td>Rough</td></tr>
<tr><td>Fine</td></tr>
</table>

6. As shown in the table above, in order to speed up the experimental speed, when the divider of potentiometer is “0. 01”, regulate R_a (or R_K) carefully but do not have to make the light cursor of galvanometer refer to zero – point very accurately, because it is only a rough tune; when the diverter is “0. 1” or above, some steps can be omitted on the condition that: as long as the light cursor of galvanometer deflect near the zero point, then this step can be omitted and promoted to the next step directly. Note: At the final step, namely the diverter is “Direct” and the potentiometer is “Fine”, we must carefully regulate R_a (or R_K) until the pointer of galvanometer is at zero point accurately, the circuit is fully compensated.

7. The switching system of potentiometer consists of two parts: “Switch knob” (five sub – files: “N”, “Broken”, “X_1”, “Broken”, “X_2”) and press – button (including “Rough”, “Fine””, “Short circuit”) . When used, the “Switch knob” should be placed on the corresponding location, the “Rough” or “Fine” must be pressed, or the potentiometer and the galvanometer are not connected. “Short circuit” button must be unlocked, otherwise it is impossible to do the experiment.

8. The instrumental error of potentiometer can be written as $\Delta U_x = Ua\%$, a is the grade of potentiometer, U is the measuring range and its unit is volt.

Problems after experiment

1. If the light cursor always deflects to one side of the galvanometer, what is the reason?

2. Why the R_N should have to be modified with different temperature?

3. How to measure an unknown resistance R_x by using a potentiometer and a standard resistance?

实验九　直流电流表和电压表的扩程

常用的直流电流表和电压表一般由表头、调零电阻、测量电路和测量转换开关组成，常用来测量直流电流和电压，属于常见的电工仪表。表头通常是一只具有磁电式测量机构的微安表或毫安表，它所允许通过的电流很小（微安级或毫安级），同时它的满偏电压也很小（一般只有零点几伏），因此只能用它来测量很小的电流或电压。如果要用它来测量较大的电流或电压，就必须对其进行改装，以扩大其量程。若在表中配以整流电路将交流变为直流，则扩程改装后的表头还可以用来测量交流电的有关参量。日常接触到的各种电表几乎都是经过扩程改装的，因此学习电表的扩程改装在电学实验部分是非常重要的。本实验主要是将表头改装扩程以满足实际需要并对扩程表进行校验。

【实验目的】

1. 了解直流电流表和电压表的构造与原理。
2. 掌握直流电流挡和电压挡的扩程设计方法。
3. 掌握测量电压降的概念。
4. 掌握直流电流挡和电压挡的装配及检验，并学会正确使用电表。

【实验仪器】

表头（直流微安计或毫安计）、直流电流表、直流电压表、双路稳压电源、电阻箱（3 只）、滑线变阻器、单刀双掷开关、单刀单掷开关、导线若干。

【实验原理】

表头有两个重要参数：表头内阻 R_g 和满偏电流 I_g（表头指针满刻度偏转时所通过的电流强度值，也称量程），这两个参数一般均为已知量（测量方法参见实验指导部分）。表头符号如图 9－1 所示。

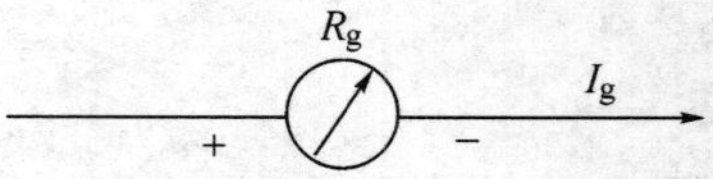

图 9－1　表头图示

1. 直流电流挡的扩程

（1）单量程直流电流表的扩程设计　一般来说，表头量程 I_g 较小。怎样使量程为 I_g 的表头能测量大于 I_g 的电流，即如何扩大表头的量程，这就是直流电流挡设计所要解决的问题。下面对此问题进行分析：

由电路知识可知，并联电路具有分流作用，根据这个作用，我们在表头两端并联一个分流电阻 R_s，则通过干路中的电流将增大，只要分流电阻 R_s 的大小选择合适，就可将表头设计成一个新量程为 I 的电流计（此时表头用作指示干路中的电流，其本身的量程仍保持不变），如图 9－2 所示。

由电路知识可知：

$$I=I_g+I_S$$

I_s表示表头满偏时流过分流电阻R_s的电流强度。由于并联电路中各支路两端的电压相等，即满足：

$$I_gR_g=I_SR_S$$

则有：

$$I_S=\frac{I_gR_g}{R_S}$$

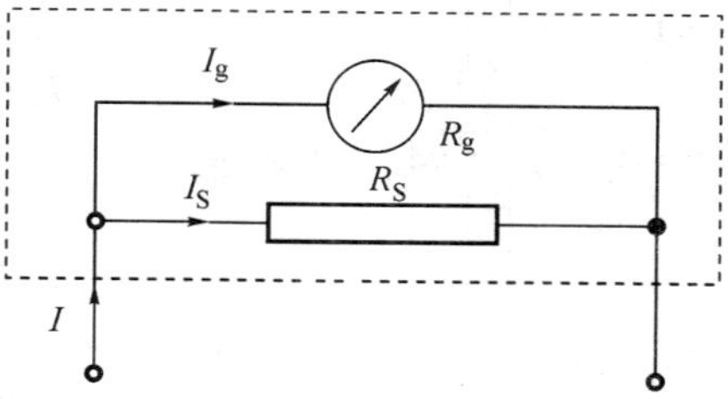

图9-2　单量程直流电流表的扩程设计

因此：

$$I=I_g+I_S=I_g+\frac{I_g\,R_g}{R_S}=I_g\left(\frac{R_S+R_g}{R_S}\right) \qquad (1)$$

由式（1）可知，并联分流电阻R_s后，电流计的量程扩大到I，为I_g的（R_s+R_g）/R_s倍，即表头指针满偏I_g时，实际流过干路中的电流值为I（此时可在表头刻度盘满偏刻度I_g处刻上电流值I）。

如果确定了扩大后的量程I，由式（1）就可以算出所需要的分流电阻R_s的值。

$$R_S=\frac{I_gR_g}{I-I_g} \qquad (2)$$

令$\dfrac{I}{I_g}=n$表示量程扩大倍数，则：

$$R_S=\frac{1}{n-1}R_g$$

（2）双量程直流电流表的扩程设计

电流表的电流挡一般都有几个电流量程，通常采用图9-3所示的分流电路。图中I_1、I_2为两个不同的电流量程，K是单刀双掷开关。当K与c接通时，电阻R_1、R_2串联后再与表头并联。因此，对量程I_2来说，分流电阻$R_s=R_1+R_2$。当K与b接通时，电阻R_2与表头串联后再与R_1并联，表头支路的电阻由R_g增为（R_g+R_2），分流电阻R_s由（R_1+R_2）减为R_1，这两个变化都有利于扩大安培计的量程，使量程由I_2扩大至I_1。

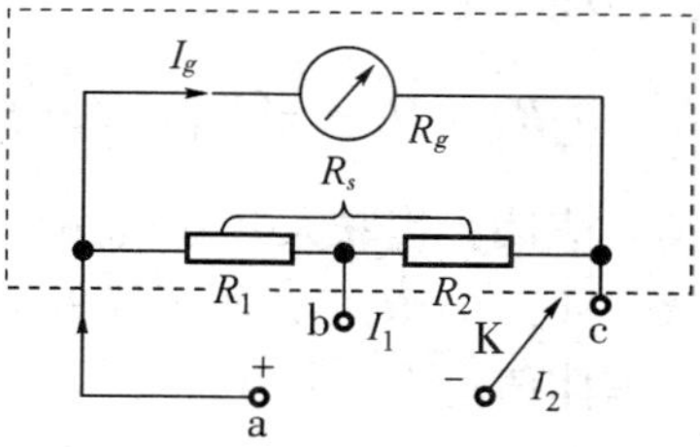

图9-3　双量程直流电流表的扩程设计

对于量程I_2来说，由并联电路各支路电压相等，则有：

$I_gR_g=(I_2-I_g)(R_1+R_2)$ 化简后可得：

$$I_2(R_1+R_2)=I_g(R_S+R_g) \qquad (3)$$

其中$R_s=R_1+R_2$。

对量程I_2来说，同理可得：

$$I_1R_1=I_g(R_1+R_2+R_g)$$

即

$$I_1R_1 = I_g\ (R_S + R_g) \tag{4}$$

由式（3）、（4）可解得：

$$R_1 = \frac{I_gR_g}{I_2 - I_g} \cdot \frac{I_2}{I_1}$$

$$R_2 = \frac{I_1 - I_2}{I_2}R_1 = \frac{I_gR_g}{I_2 - I_1} \cdot \frac{I_1 - I_2}{I_1}$$

$$I_1R_1 = I_2\ (R_1 + R_2)\ = I_g\ (R_S + R_g)\ = V^* \tag{5}$$

式（5）中 $V^* = I_g\ (R_S + R_g)$ 称为直流电流表的测量电压降。由式（5）可见图9－3的分流电路具有这样的特点：各量程与其对应的分流电阻的乘积是个常量，这个常量就是表头量程 I_g 与整个闭合回路的总电阻（$R_S + R_g$）的乘积，即所谓的测量电压降。

2. 直流电压表的扩程

（1）单量程直流电压表的扩程设计　表头是一个量程较小的直流电流表，为什么它既能用来指示电流，又能用来指示电压呢？这是因为表头具有一定的内阻 R_g，当表头流有满偏电流 I_g 时，这也就是意味着表头两端具有相应的电压 $U_g = I_gR_g$。但是单用这一个表头来测量电压是不行的。其一是由于它的电压量程很小，只能测量很低的电压；其二是由于表头本身的内阻不够大，将它并联到被测电路上进行测量时，要对待测电路产生较大的影响。电压挡所要解决的问题是必须扩大表头的电压量程，并减少它对待测电路的影响，这可用串联电阻分压的方法来解决，如图9－4所示。

当电压量程为 $U_g = I_gR_g$ 的表头和一个阻值较大的电阻 R_M 串联后，就设计成一个新的直流电压表，设其量程为 U，则：

$$U = U_g + I_gR_M = U_g + \frac{U_g}{R_g}R_M = U_g\frac{R_g + R_M}{R_g} \tag{6}$$

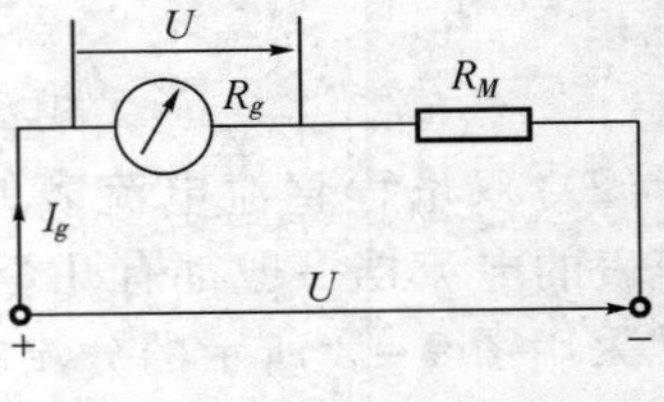

图9－4　单量程直流电压表的扩程设计

式（6）表明：串联大电阻 R_M 后，电压表的量程扩大了（$R_g + R_M$）/ R_g 倍，所串联的大电阻 R_M 称为倍率电阻，其阻值由式（6）可得：

$$R_M = \frac{R_g}{U_g}\ (U - U_g)\ = \frac{R_g}{U_g}U - R_g \tag{7}$$

（2）双量程直流电压表的扩程设计　图9－5是通常采用的双量程直流电压挡的设计电路图。U_1、U_2 是电压挡的两个不同量程，R_3、R_4 是倍率电阻，R_{so} 是量程为 I_{ge} 的电流挡的分流电阻（R_{so} 的作用是将表头的量程由 I_g 扩大至 I_{ge}）。R_{so} 和表头组成新的表头（图9－5虚线框内部分，相当于图9－4中的 U_g、R_g 部分）。

从图9－5看出，要计算倍率电阻 R_3、R_4，应先确定新表头的内阻 R_{ge}（相当于图9－4中的 R_g），内阻 R_{ge} 是分流电阻 R_{so} 与原表头的内阻 R_g 的并联值。

根据式（2）得：

$$R_{S0} = \frac{I_gR_g}{I_{ge} - I_g} \tag{8}$$

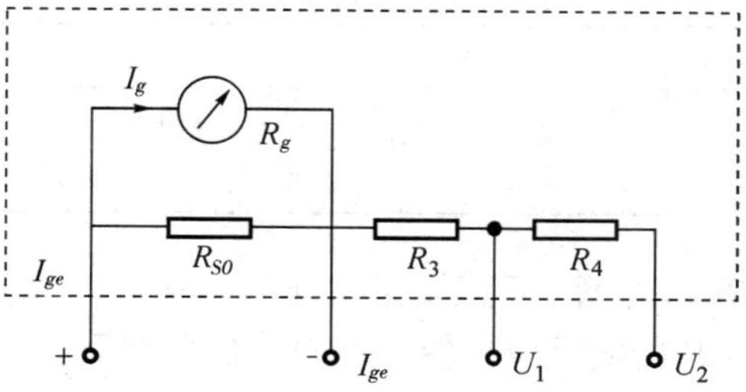

图 9－5　双量程直流电压表的扩程设计

由电阻并联公式得：

$$R_{ge}=\frac{R_{S0}\cdot R_g}{R_{S0}+R_g} \tag{9}$$

然后根据式（7）可确定 R_3、R_4数值：

$$R_3=\frac{R_{ge}}{U_g}U_1-R_{ge}=\frac{1}{I_{ge}}U_1-R_{ge}$$

$$R_4\ \frac{R_{ge}}{U_g}U_2-R_{ge}-R_3=\frac{1}{I_{ge}}U_2-R_{ge}-R_3=\frac{1}{I_{ge}}\left(U_2-U_1\right) \tag{10}$$

【实验内容】

1. 直流电流挡的扩程

（1）如图 9－6，计算 R_1 和 R_2的大小，并表示在相应的电阻箱上。

（2）连接线路，其中 R_1、R_2和 R 是电阻箱。

（3）首先校验 $I_2=5\text{mA}$ 的量程，稳压电源电压 $E=2.0\text{V}$，电阻 R_0 初值取 400Ω，把开关 K 合到右边，通过调节电阻箱 R 的大小，使扩程表读数分别取表 9－1 数值时，观察并记录标准电流表的读数，并填入相应数据表格中。

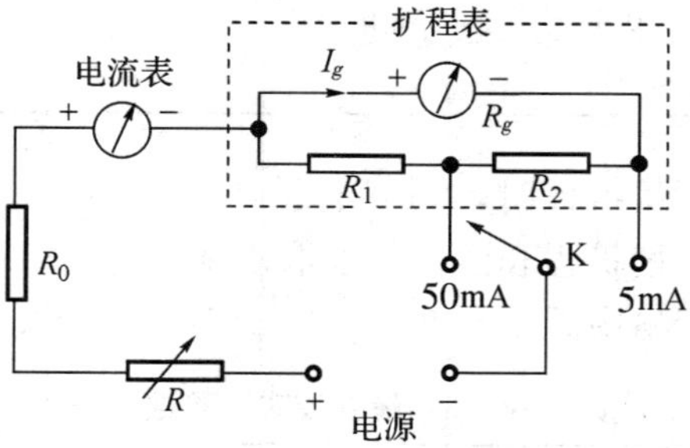

图 9－6　直流电流表的扩程校验电路图

表 9-1　5mA 量程校对实验数据表

电流	电流取值				
$I_{(扩)}$（mA）	1.0	2.0	3.0	4.0	5.0
$I_{(标)}$（mA）					

（4）然后校验 $I_1 = 50$mA 量程，稳压电源电压 $E = 4.0$V，电阻 R 初值取 100Ω，把开关 K 合到左边，通过调节 R 的电阻值，使扩程表读数分别取表 9-2 数值时，观察并记录标准电流表的读数，并填入相应数据表格中。

表 9-2　50mA 量程校对实验数据表

电流	电流取值				
$I_{(扩)}$（mA）	10	20	30	40	50
$I_{(标)}$（mA）					

（注意：实验中标准电流表要选择合适的量程）

2. 直流电压挡的扩程

（1）如图 9-7，计算 R_{so}、R_3 和 R_4 的大小，并表示在相应的电阻箱上，其中 $R_{S0} = R_1 + R_2$。

（2）连接线路，其中 R_{so}、R_3、R_4 是电阻箱，R 是滑线变阻器。

（3）首先校验 $U_1 = 1$V 的量程，稳压电源电压 $E = 1.5$V，闭合 K_2，开关 K 合到左边，移动滑线变阻器触头 d 的位置，使扩程表读数分别取表 9-3 数值时，观察并记录标准电压表的读数，并填入相应数据表格中。

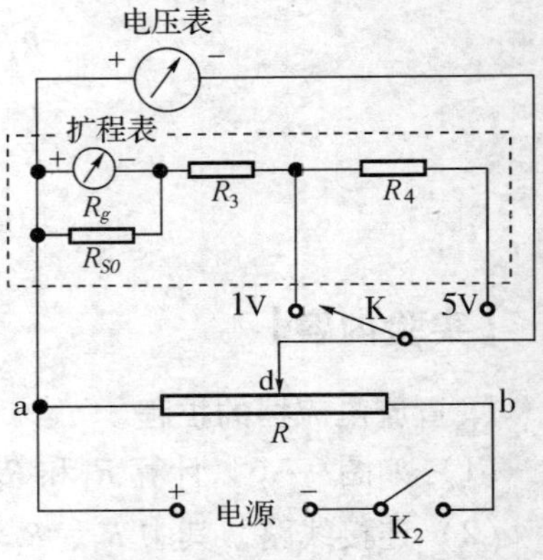

图 9-7　直流电压表的扩程校验电路图

表 9-3　1V 量程校对实验数据表

电压	电压取值				
$U_{(扩)}$（V）	0.2	0.4	0.6	0.8	1.0
$U_{(标)}$（V）					

（4）检验 $U_2 = 5$V 量程伏特计时，调节电源电压 $E = 6.0$V，开关 K 合到右边，移动滑线变阻器触头 d 的位置，使扩程表读数分别取表 9-4 数值时，观察并记录标准电压表的读数，并填入相应数据表格中。

表 9-4　5V 量程校对实验数据表

电压	电压取值				
$U_{(扩)}$（V）	1.0	2.0	3.0	4.0	5.0
$U_{(标)}$（V）					

（注意：实验中标准电压表要选择合适的量程）

【注意事项】

1. 连接电路时要注意电源、电表的正负极；要将仪器布置合理，便于操作和读数。
2. 电路接好后，请老师检查，必须无误后才能打开电源开关。
3. 校验扩程表时，标准表量程必须选择合适。
4. 根据各电阻的大小，选用合适的电阻箱接线柱，并要尽量减小接线电阻。
5. 直流稳压电源输出电压可用电压表测量，确认是否是你所要的电压大小。

【思考题】

1. 图 9－3 中考虑量程 I_1 时，被测电流是否全部流过内阻 R_g？被测电流是怎样由 a 流到 b 点的？计算该量程的测量电压降。

2. 计算 R_{so} 和 R_M 的数值时要已知哪些条件？

3. 为什么校准电表时，必须使电流（或电压）从小到大做一遍，又从大到小再做一遍？如果两次数据不一致说明什么？

4. 实验中是否出现故障？请分析故障原因。

5. 在校验 50mA 电流量程时，电源电压为 4.0V，量程置于 50mA，但最大电流始终低于 50mA，请找出原因？

6. 请在扩程的电压表上增加量程 50V 挡，画出电路图，计算有关电阻值。

7. 由于接线电阻的存在，对实验误差产生较大影响，设各电阻箱的接线电阻都是 1Ω，请讨论接线电阻对哪一个电阻（R_1、R_2、R_3、R_4 和 R_{so}）引入的误差最大？这种误差是否有规律？

实 验 指 导

【预习要求】

1. 实验前必须预习实验讲义，明确自己的实验组号，即第几组仪器，预知所使用表头的内阻 R_g 和满偏电流 I_g。

2. 根据实验原理和实验内容，写好预习实验报告，计算 R_1、R_2、R_{so}、R_3 和 R_4 的大小，写出实验步骤，列出记录数据的表格，交老师检查。

【操作要求】

1. 连接线路要有条理。例如：连接电路图 9－7 时，从电源“＋”极开始，接到滑线变阻器一个固定端 a 点，另一个固定端 b 接开关 K，再回接到电源“－”极，此回路已经接好。然后将分压电路部分：先将表头和 R_{so} 并联部分接好，再依次从 a 串联到表头的“＋”极，表头的“－”串联接到 R_3、R_4，从 R_4 的两端连线接到单刀双掷开关两端接线柱，单刀双掷开关中间接线柱串接到分压电路的“负极”，即滑线变阻器的滑动端 d 上。单刀双掷开关摆好位置，并牢记向哪端接通时是那个量程。

2. 校准用电表的量程一定要选择合适，并注意所有的“＋”“－”极必须对应。

3. 调节电压、电流或电阻大小时，必须要看清楚仪器挡位、量程和电阻箱接线柱是否正确，对调节后应该出现的情况要心中有数，尽量消除接线电阻。

4. 如实记录实验中的各种现象、实验条件和原始数据。

【实验后的思考】

1. 如何测表头内阻

做本实验时需要知道表头的内阻。下面介绍几种常用的测量表头内阻 R_g 的方法。必须指出，无论用哪种方法测量，都要注意通过表头的电流不要超过表头的满偏电流 I_g。

（1）电桥法测表头内阻，请参考实验7。

（2）替代法测表头内阻　测量电路如图 9－8 所示。测量时，闭合开关 K_1，先将开关 K_2 置于 2，调节 R_0，使电流表Ⓐ有一偏转 n，然后把 K_2 倒向 1，调节电阻箱 R_n，使Ⓐ有相同的偏转 n，这时电阻箱的示值 R_n 等于表头的内阻 R_g。

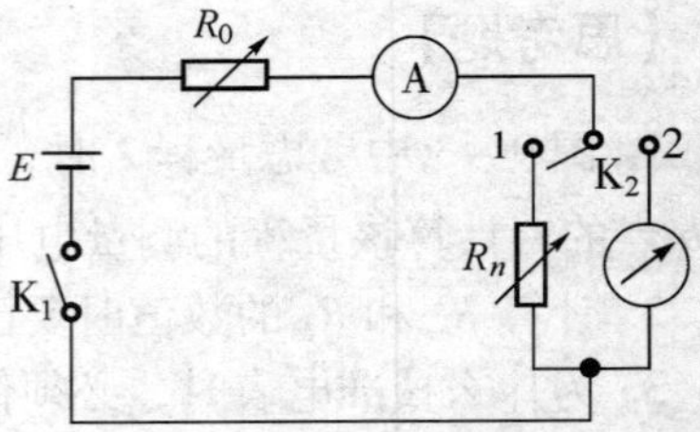

图 9－8　替代法测量表头内阻电路图

（3）半偏法测内阻　半偏法测表头内阻是一种很方便的方法，因为不需要另外的电表，测量电路如图 9－9 所示。

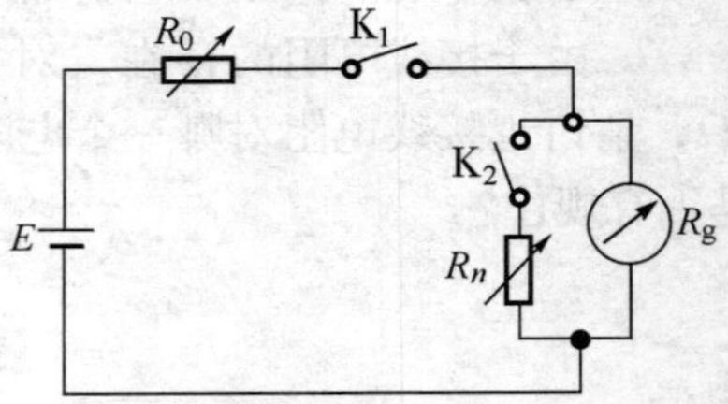

图 9－9　半偏法测量表头内阻电路图

首先合上 K_1，断开 K_2，调节电阻箱 R_0，使表头满偏 n_o，然后合上 K_2，调节电阻箱 R_n，使表头偏转为 $n_o/2$，这时表头内阻 R_g 可用下式求出：

$$R_g = \frac{R_0 R_n}{R_0 - R_n} \qquad (11)$$

下面推导这个公式：当合上 K_1，断开 K_2 时，电路中的总电流为 I，则有：

$$I = \frac{E}{R_0 + R_g} \qquad (12)$$

当合上 K_2 时，电路中的总电流为 I'，则有：

$$I' = \frac{E}{R_0 + \dfrac{R_n R_g}{R_n + R_g}} \qquad (13)$$

调节 R_n，使通过表头的电流为 $I/2$，根据并联电路知识，则有：

$$I' = \frac{I}{2} + \frac{\frac{1}{2} R_g}{R_n} = \frac{R_n + R_g}{R_n} \cdot \frac{I}{2} \qquad (14)$$

将式（13）（14）代入（15），化简后可得：

$$R_n R_0 + R_n R_g = R_0 R_g$$

因此有：

$$R_g = \frac{R_0 R_n}{R_0 - R_n} \qquad (15)$$

最后指出替代法和半偏法测表头内阻时，都要求所使用的电源输出电压比较稳定，否则测量误差很大。

2. 表头满偏电流的测量

（1）替代法测满偏电流　图 9－8 电路还可以同时测量表头的满偏电流 I_g。选一块量程略大于待测表头满偏电流的电流表作为标准表Ⓐ，将 K_2 置于 2，调节 R_0，让待测表头满偏，这时Ⓐ的读数就是表头的满偏电流，当然这时电阻 R_n 是没有用的。

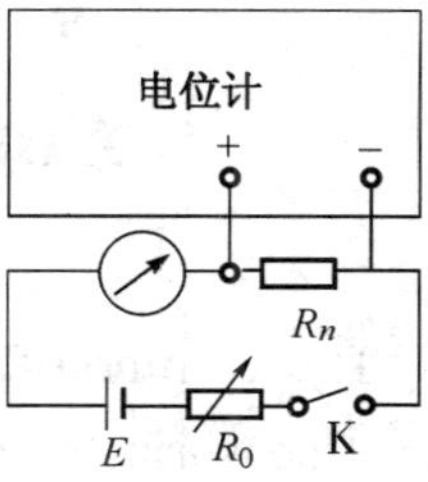

图 9－10　电势差计法测表头满偏电流电路图

（2）电势差计法测表头满偏电流　它是一种常用的方法，此方法准确度高。测量电路如图 9－10 所示。

表头与标准电阻 R_n（可用电阻箱）串联，因而通过 R_n 的电流与通过表头的电流相等。调节 R_0 让表头满偏，用电势差计测出 R_n 两端电压 U_n，则表头满偏电流：

$$I_g = \frac{U_n}{R_n} \tag{16}$$

3. 如何校准电表

扩程以后的电表需要用标准表校准（也叫鉴定）。通过校准，读出扩程表的指示值 I_x（或 U_x）及标准表的指示值 I_s（或 U_s），求出它们的差值 $\Delta I_x = I_s - I_x$（或 $\Delta U_x = U_s - U_x$），并以 I_x（或 U_x）为横坐标，ΔI_x（或 U_s）为纵坐标，两个校准点之间用直线联接的方法（不能画成光滑曲线）画出校准曲线。图 9－11 是电流表的校准曲线，在以后使用这个电表时，根据校准曲线可以修正电表的读数。

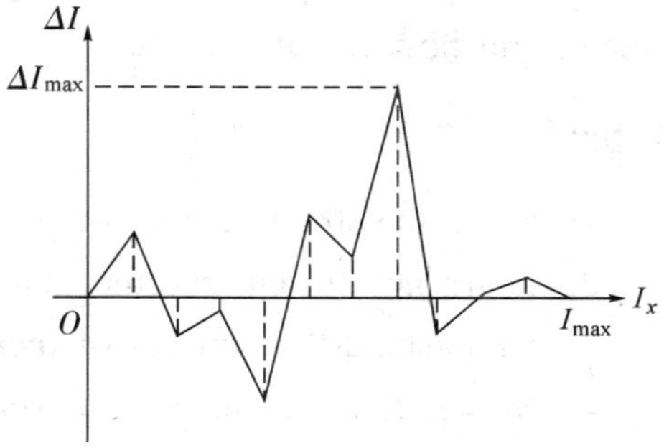

图 9－11　电流表的校准曲线

选取其中最大的绝对误差 ΔI_{max}（或 ΔU_{max}）除以量程，即得该电表的精度等级 K，即：

$$K = \frac{|\Delta I_{max}|}{I_m} \times 100\% \quad \left(或\ K = \frac{|\Delta U_{max}|}{U_m} \times 100\%\right) \tag{17}$$

根据 K 的大小，将电表分为不同的等级（参见实验 5），例如若 $0.5\% < K \leq 1.0\%$，则该电表的等级记为 1.0 级。

校准扩程表须遵循如下步骤。

（1）通电之前调整好扩程表及标准表的零点。

（2）接通电源后，先校准量程。由于给定的 I_g 和 R_g 的数值可能不准确，根据公式（2）算出的 R_s 或根据公式（7）算出的 R_M 值不一定合适，因此当标准表指示到需要的扩程表量程数值时，而扩程表的表头不满偏，则应调整 R_s 或 R_M 值。

（3）校准刻度，使电流（或电压）由大到小校准全部刻度值，然后电流（或电压）由小到大再校准一次。被校表选整刻度，例如 1.00mA（V），2.00mA（V）等读出标准表的相应读数。

9 Enlarging the Ranges of a DC Ampere Meter and a DC Voltmeter

The common DC ampere meter and voltmeter are made up of meter, zeroing resistance, measuring circuit and change – over switch. They are often used to measure the DC current and voltage in the circuit. Usually the indicating part of it is a milliampere or microampere meter with magnetic – electric measurement system. Milliampere or microampere meter can only measure a small current and withstanding a very small voltage. Only if its range has been enlarged it can measure a greater current or voltage. Equipping with rectifier circuit, it also can be used to measure AC parameter, such as AC current, capacity, and inductance and so on. Most of the meters that we used are enlarged, so it's very important to learn the method to enlarge the meter to meet the actual needs. In this experiment, we pay more attention to the method that how to enlarge a meter and how to calibrate it.

Objectives

1. Learn the structure and the principle of a DC ampere meter and a DC voltmeter.
2. Learn how to enlarge the ranges of a DC ampere meter and a DC voltmeter.
3. Understand the concept of measuring potential fall.
4. Master the assembling and verifying methods of a DC ampere meter and a DC voltmeter, and master the usage of meter.

Equipments

Meter (a DC microampere meter or a DC milliampere meter), a DC ampere meter, a DC voltmeter, standard – multimeter, DC voltage source, resistances, slide rheostat, switch, single – pole double – throw (SPDT) switch, and some wires.

Principles

The meter has two important parameters: one is the internal resistance R_g and the other is the full – scale current I_g (also called range), which are all given in this experiment (the ways to measure them are shown in the "Additional Instruction" part). The symbol of a meter is shown in figure 9 – 1.

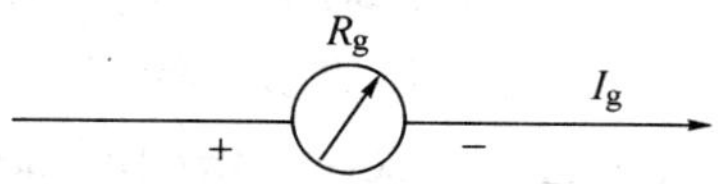

Figure 9 − 1　The symbol of a meter

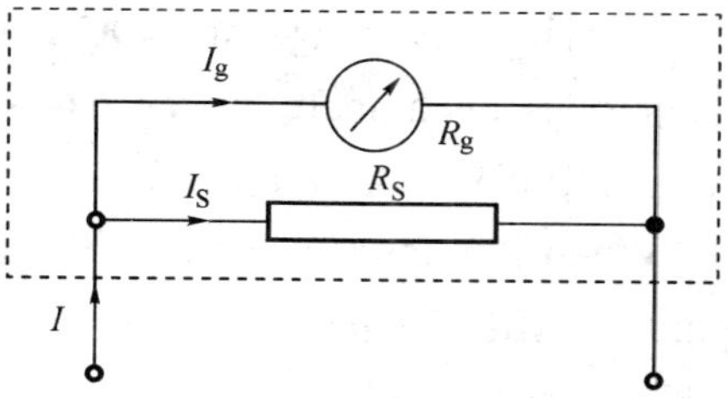

Figure 9 − 2　The circuit of enlarging the range of a DC ampere meter that has single range

1. Enlarge the ranges of a DC ampere meter

(1) Enlarge the range of a DC ampere meter that has single range

In general, the I_g of the meter is very smaller. How to make the meter with the range I_g to measure the current I, which is bigger than I_g? That is, how to expand the range of the meter? It's an important problem that should be solved. We have known that parallel circuit has the shunt effect. According to this, we can install a resistor R_s in parallel with the meter in order to get an ampere meter with the current range I. If the shunt resistor R_s has appropriate value, the meter can be designed into a new range for I (You should know that the range of the meter itself remain unchanged), which is shown in figure 9 − 2.

$$I = I_g + I_S$$

I_s is the current that flow through R_s when the meter gets its full − scale current I_g. Because the voltages of each branch are the same in the parallel circuit, we can get:

$$I_g R_g = I_S R_S$$

$$I_S = \frac{I_g R_g}{R_S}$$

Then, we get:

$$I = I_g + I_S = I_g + \frac{I_g \cdot R_g}{R_S} = I_g \left(\frac{R_S + R_g}{R_S} \right) \qquad (1)$$

We can see that the range of the ampere meter is enlarged to I for $(R_s + R_g) / R_s$ times after installing a shunt resistance R_s in the circuit. That is, the whole current of the circuit is I when the meter gets its full − scale current.

The shunt resistance R_s value can be calculated when I is given.

$$R_S = \frac{I_g R_g}{I - I_g} \qquad (2)$$

If we let $\frac{I}{I_g} = n$, then

$$R_S = \frac{1}{n-1} R_g$$

(2) Enlarge the range of a DC ampere meter that has double ranges

Generally the ampere meter has several different ranges, whose inner circuit is shown in figure 9 – 3. I_1 and I_2 are two different current ranges, and K is a SPDT switch. When K connects with c, the resistors R_1 and R_2 are in series firstly and then they parallel with the meter. Thus, the shunt resistor is $R_s = R_1 + R_2$ for the I_2 range. When K connects with b, the resistor R_2 and the meter are in series firstly and then they parallel with R_1. The resistance of the meter branch is increased from R_g to $(R_g + R_2)$, and the shunt resistor R_s is decreased from $(R_1 + R_2)$ to R_1. These two changes are beneficial to enlarge the range ammeter from I_2 to I_1.

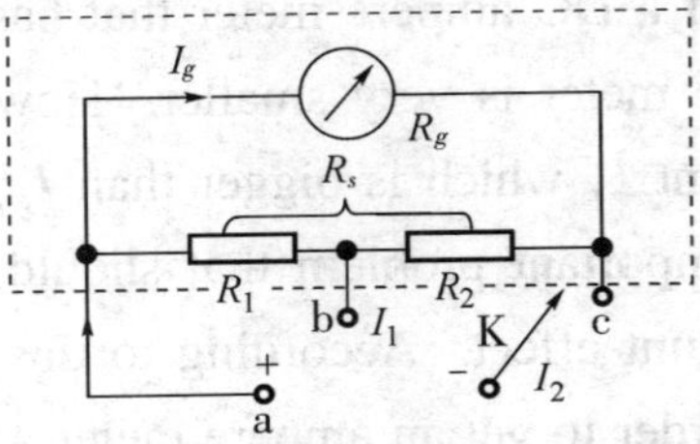

Figure 9 – 3 The circuit of enlarging the range of a DC ampere meter that has double ranges

According to characters of parallel circuit, for the range I_2, we can get:

$I_g R_g = (I_2 - I_g)(R_1 + R_2)$ That is, $I_2 (R_1 + R_2) = I_g (R_S + R_g)$ (3)

And $R_s = R_1 + R_2$

And for the range I_1, we can also get:

$$I_1 R_1 = I_g (R_1 + R_2 + R_g)$$

That is, $I_1 R_1 = I_g (R_S + R_g)$ (4)

From the equation (3) and (4), we can get:

$$R_1 = \frac{I_g R_g}{I_2 - I_g} \cdot \frac{I_2}{I_1}$$

$$R_2 = \frac{I_1 - I_2}{I_2} R_1 = \frac{I_g R_g}{I_2 - I_1} \cdot \frac{I_1 - I_2}{I_1}$$

$$I_1 R_1 = I_2 (R_1 + R_2) = I_g (R_S + R_g) = V^* \quad (5)$$

In the above equations, V^* is called "measuring potential fall" of the ampere meter. From the equation (5) and figure 9 – 3, we can conclude that the product of a range and its corresponding shunt resistance is a constant, and this constant is equal to the product of the I_g and the closed – loop total resistance, which is called "measuring potential fall".

2. Enlarge the ranges of a DC voltmeter

(1) Enlarge the range of a DC voltmeter that has single range

Meter is a DC ampere meter with a small range, but why it can be used to indicate both current and voltage? This is because the meter has a certain internal resistance R_g. When the meter has the current I_g, this means that the corresponding voltage of the meter is $U_g = I_g R_g$. However, it's not acceptable to measure the voltage only by the meter. Firstly the voltage range of the meter is very small, by which only a very low voltage can be determined; secondly the resistance of the meter itself is not big enough, which can bring a big error to the circuit when it is paralleled to the test part of the circuit. The proper method to solve this problem is to enlarge the voltage range of the meter, and at the same time to reduce its impact on the circuit when measuring. The series voltage divider circuit can be used to resolve this problem, which is shown in Figure 9-4.

We can install a resistor R_M in series with the meter to get a voltmeter with voltage range U. From the figure 9-4, we can get:

$$U = U_g + I_g R_M = U_g + \frac{U_g}{R_g} R_M = U_g \frac{R_g + R_M}{R_g} \qquad (6)$$

The voltage range of the meter has been enlarged for $(R_g + R_M)/R_g$ times, and the R_M is called "multiplying resistance".

From the equation (6), we get the resistance of R_M:

$$R_M = \frac{R_g}{U_g}(U - U_g) = \frac{R_g}{U_g}U - R_g \qquad (7)$$

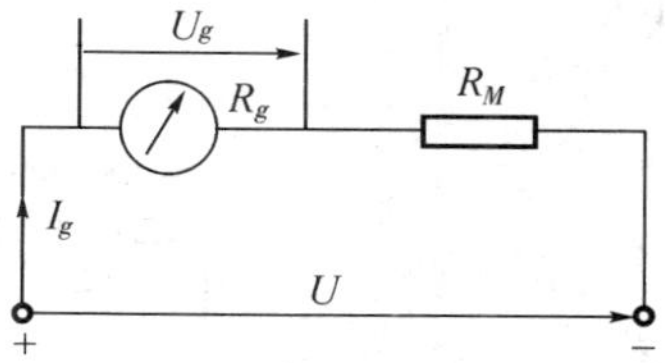

Figure 9-4 The circuit of enlarging the range of a DC voltmeter that has single range

(2) Enlarge the range of a DC voltmeter that has double ranges

In figure 9-5, U_1 and U_2 are two different ranges of the voltmeter. R_3 and R_4 are multiplying resistances. R_{so} is the shunt resistance when the current range is I_{ge}.

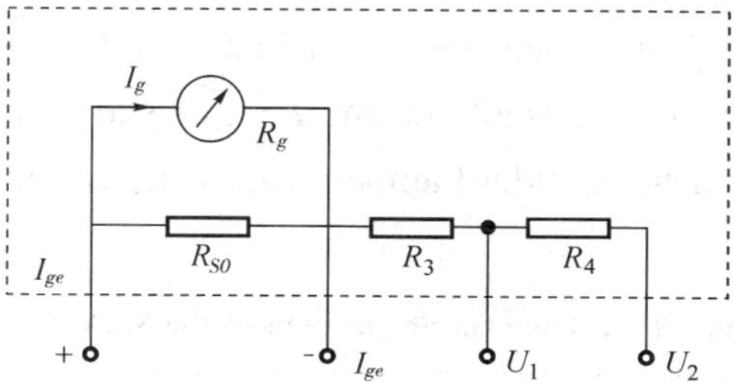

Figure 9-5 The circuit of enlarging the range of a DC voltmeter that has double range

The resistance of the R_{so} is determined by the following equation:

$$R_{S0} = \frac{I_g R_g}{I_{gs} - I_g} \tag{8}$$

And the equivalent resistance of the meter can be written as:

$$R_{ge} = \frac{R_{S0} \cdot R_g}{R_{S0} + R_g} \tag{9}$$

According to the equation (7), we can get:

$$R_3 = \frac{R_{ge}}{U_g} U_1 - R_{ge} = \frac{1}{I_{ge}} U_1 - R_{ge}$$

$$R_4 = \frac{1}{I_{gs}} U_2 - R_{ge} - R_3 = \frac{1}{I_{ge}} (U_2 - U_1) \tag{10}$$

Experiment

1. The steps for enlarging a DC ampere meter

(1) The circuit of the experiment is shown in figure 9 -6. Calculate the values of R_1 and R_2.

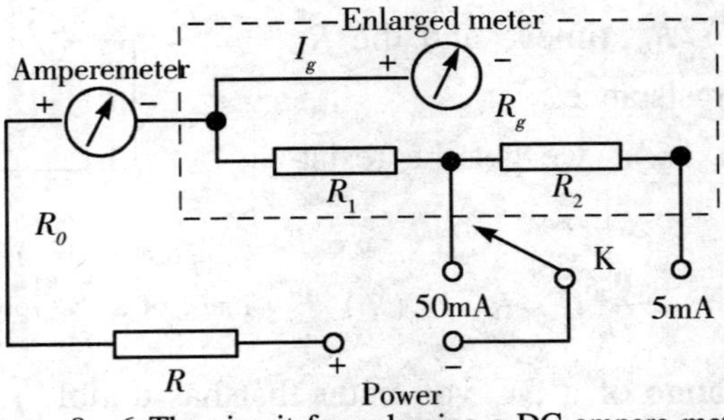

Figure 9 -6 The circuit for enlarging a DC ampere meter

(2) Join the circuit. R_1, R_2 and R are resistance boxes.

(3) Check the current range $I_2 = 5$mA of the ampere meter. The voltage of the regulated power supply is $E = 2.0$V, and the resistance box $R = 400\Omega$. Adjust R to make the enlarged ampere meter get different current $I_{(e)}$, as shown in table 9 -1, and record the corresponding values of the standard ampere meter $I_{(s)}$. Fill the data into the following table.

Table 9 -1 The experiment data of the 5mA range

Current	Current intensity				
$I_{(e)}$ (mA)	1.0	2.0	3.0	4.0	5.0
$I_{(s)}$ (mA)					

(4) Check the current range $I_2 = 50$mA of the ampere meter. The voltage of the regulated power supply is $E = 4.0$V, and the resistance $R = 100\Omega$. Let K connect to the left, and adjust R to make the enlarged ampere meter get different current $I_{(e)}$, as

shown in table 9 -2, and record the corresponding values of the standard ampere meter $I_{(s)}$. Fill the data into the following table.

Table 9 -2 The experiment data of the 50mA range

Current	Current intensity				
$I_{(e)}$ (mA)	10	20	30	40	50
$I_{(s)}$ (mA)					

(Please pay more attention to the range of the standard ampere meter!)

2. The steps for enlarging a DC voltmeter

(1) The circuit of the experiment is shown in figure 9 -7. Calculate the values of R_{so}, R_3 and R_4.

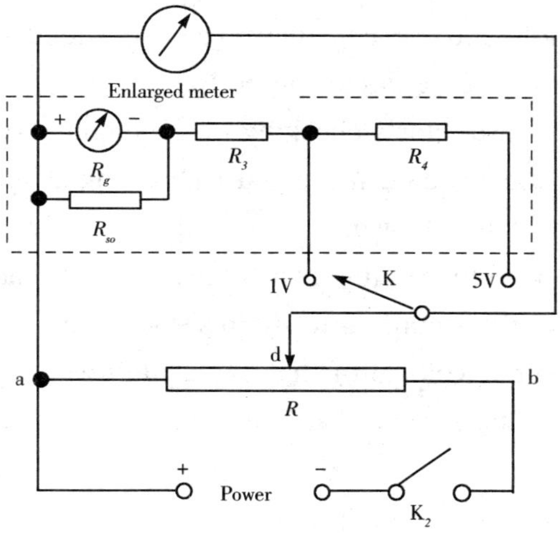

Figure 9 -7 The circuit for enlarging a DC voltmeter

(2) Join the circuit. R_{so}, R_3 and R_4 are resistance boxes, and R is a slide rheostat.

(3) Check the voltage range U_1 =1V of the voltmeter. The voltage of the regulated power supply is E = 1.5V. Adjust the sliding contactor d of the slide rheostat to make the enlarged voltmeter to get different voltage $U_{(e)}$(V), as shown in table 9 -3, and record the corresponding values of the standard voltmeter $U_{(s)}$(V). Fill the data into the following table.

Table 9 -3 The experiment data of the 1V range

Voltage	Value of the Voltage				
$U_{(e)}$ (V)	0.2	0.4	0.6	0.8	1.0
$U_{(s)}$ (V)					

(4) Check the voltage range U_1 =5V of the voltmeter. The voltage of the regula-

ted power supply is $E = 6.0V$. Let K connect to the right, and adjust the sliding contactor d of the slide rheostat to make the enlarged voltmeter to get different voltage $U_{(e)}$ (V), as shown in Table 9 - 4, and record the corresponding values of the standard voltmeter $U_{(s)}$(V). Fill the data into the following table.

Table 9 - 4 The experiment data of the 5V range

Voltage	Value of the Voltage				
$U_{(e)}$ (V)	1.0	2.0	3.0	4.0	5.0
$U_{(s)}$ (V)					

(Please pay more attention to the range of the standard voltmeter!)

Notice

1. The positive pole and the negative pole of the regulated power supply and the meters should be joined into the circuit properly.
2. Only after your instructor's checking of the circuit, you can turn on the power.
3. The ranges of the standard meter must be changed suitably while you checking the ranges of the meter you enlarged.
4. For reducing the connecting resistance, the terminals of the resistance box should be selected properly according to the resistance you need.
5. Measure the output voltage of the regulated power supply with a standard voltmeter to get a correct voltage value that you need.

Problems

1. Does the current flow through the internal resistance R_g fully in the figure 9 - 3 when checking the range I_1? How does the currents get through from a to b? Calculate the measuring potential fall in this Figure.
2. What should we know when calculating the resistance R_{so} and R_M?
3. Why do we make the current (or voltage) increase and decrease when checking the enlarged meter? If the data is not same, what does that indicate?
4. What are the troubles that have happened to this experiment? What are the reasons?
5. When checking the 50mA range (the regulated power supply voltage is 4.0V), the maximum current is always lower than the 50mA, please find out the reasons.
6. If we want to add a 50V range to the meter, please draw the circuit diagram and calculate the resistance value.
7. Because of the connecting resistance, which has a great impact on our experiment, because of the existence of experimental error have the resistance box for the wir-

ing resistance is, please discuss that the connecting resistance of each resistor (R_1, R_2, R_3, R_4 and R_{so}), which does make the error largest in the experiment (If the connecting resistance for each resistor is 1Ω)?

Additional Instruction

Preparation

1. Read the text carefully and get the values of the meter internal resistance R_g and the full – scale current I_g of the meter, according to your experiment group number.

2. Calculate the values of R_1, R_2, R_{so}, R_3 and R_4. Write down the experiment steps. Make the tables for recording the experiment measurements.

Operating key points

1. Join the circuit correctly. Take figure 9 – 7 for example, you'd better connect from the "+" pole of the regulated power supply to the "a" end of the slide rheostat; the "b" end of the slide rheostat should be connected to switch K; and then connect to the "–" pole of the regulated power supply. After that, the first loop has been connected. Then connect from the "a" to the "+" pole of the meter, and "–" pole in series to the R_3 and R_4. At last, the two poles of the R_4 are connected to the outside poles of the SPDT switch; the middle pole of the SPDT switch should be connected to the "–" pole of the regulated power supply. Please keep the right position of the SPDT switch in your mind.

2. The ranges of all standard meters must be chosen suitably, and pay more attention to the "+" "–" pole when connected.

3. When alternating the voltage, current or resistance, it's important to keep the ranges, stalls or connecting poles in their right position. Eliminate the connecting resistance as much as possible.

4. Record the experiment phenomena, experimental conditions and the original data accurately.

Thinking after the experiment

1. How to measure the internal resistance R_g of the meter?

(1) Measure R_g by the Electric bridge method. Please refer to the Experiment 7.

(2) Measure R_g by the substitution method. The circuit is shown in figure 9 – 8. Firstly connect the switch K_1; then connect K_2 to 2, and regulate R_0 to make the ampere meter Ⓐ point to n. Then connect K_2 to 1, and adjust the resistance box R_n to make the

ampere meter Ⓐ have the same reading. The internal resistance R_g of the meter is equal to R_n.

(3) Measure R_g by the semi – partial method. It's convenient to measure R_g by the semi – partial method, which is shown in figure 9 – 9.

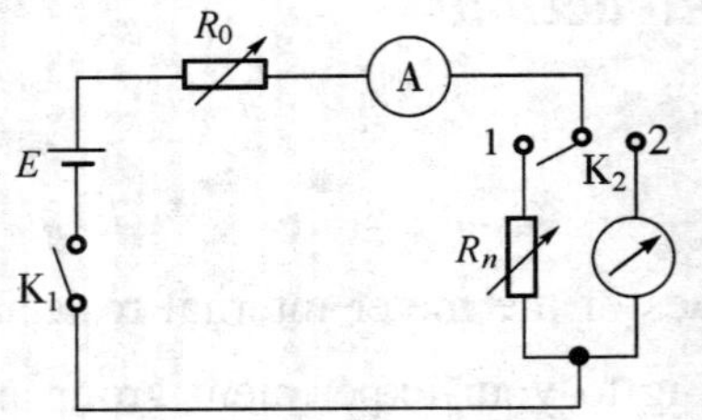

Figure 9 – 8 The circuit of measuring R_g by the substitution method

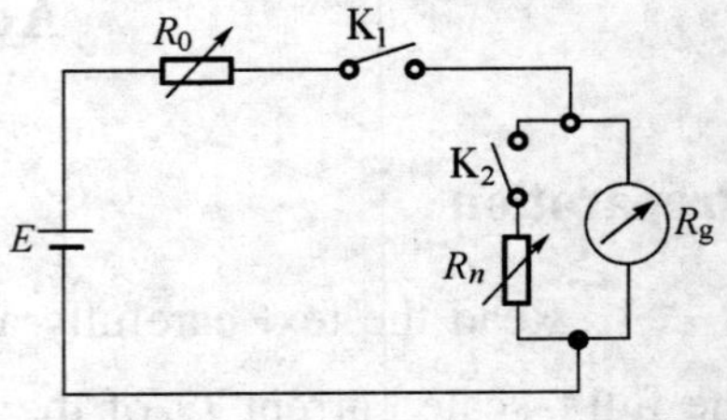

Figure 9 – 9 The circuit of measuring R_g by the semi – partial method

First of all close K_1, and disconnect K_2; then alternate the resistance box R_0 until the meter reach its scale n_o; After that, close K_2, and alternate the R_n to make the meter have the reading of n_o /2. R_g is determined by the following equation.

$$R_g = \frac{R_0 R_n}{R_0 - R_n} \tag{11}$$

Now we deduce this formula as follows:

When close K_1 and disconnect K_2, the whole current of the circuit is I:

$$I = \frac{E}{R_0 + R_g} \tag{12}$$

When close K_2, the whole current of the circuit is I':

$$I' = \frac{E}{R_0 + \dfrac{R_n R_g}{R_n + R_g}} \tag{13}$$

Alternate the R_n to make the meter have the reading of $I/2$. According to the knowledge of parallel circuit, we can get:

$$I' = \frac{I}{2} + \frac{\frac{1}{2}R_g}{R_n} = \frac{R_n + R_g}{R_n} \cdot \frac{I}{2} \tag{14}$$

From the equation (13) (14) (15), we can get the important equation:

$$R_n R_0 + R_n R_g = R_0 R_g$$

Therefore we have:

$$R_g = \frac{R_0 R_n}{R_0 - R_n} \tag{15}$$

Finally we must concluded that if we use the substitution method or the semi – partial method to measure the R_g, the output voltage of the regulated power supply must be relatively stable; otherwise a large measurement error will exist in the experiment.

2. How to measure the full – scale current I_g of the meter

(1) Measure I_g by the substitution method. The circuit is the same as Figure 9 – 8. Select an standard ampere meter Ⓐ whose range is a little larger than that of the meter. Connect K_2 to 2, and regulate the R_0 to make the meter reach its full scale. Obviously the I_g of the meter is equal to the reading of the Ⓐ.

(2) Measure I_g by the potentiometer method. It's a common method with high sensitivity and accuracy. The circuit is shown in figure 9 – 10.

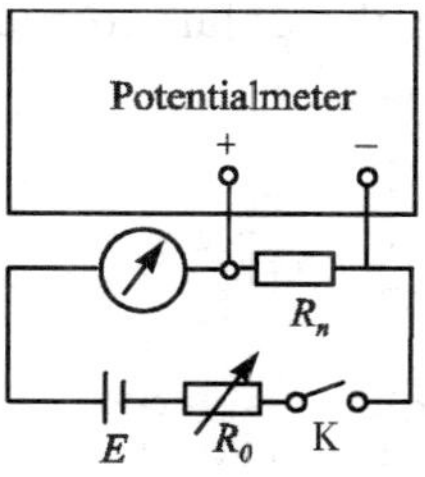

Figure 9 – 10 The circuit of measuring I_g by the potentiometer method

The standard resistance R_n (resistance box) and the meter are in series, so the current flowing through them is the same. Regulate the R_o to make the meter reach its full scale and then measure the voltage U_n of the R_n by the potentiometer. The I_g of the meter is determined by the following equation.

$$I_g = \frac{U_n}{R_n} \tag{16}$$

3. How to calibrate the enlarged meter

The enlarged meter must be calibrated by the standard meter; otherwise it can't be used. Record the reading I_x (or U_x) of the enlarged meter and the standard meter respectively; and then calculate their difference $\Delta I_x = I_s - I_x$ (or $\Delta U_x = U_s - U_x$). Finally draw the calibration curve, as shown in figure 9 – 11. Please notice: I_x (or U_x) for the abscissa coordinate, and ΔI_x (or ΔU_x) for the longitudinal coordinate. The two calibration points are connected with a straight – line not a smooth curve.

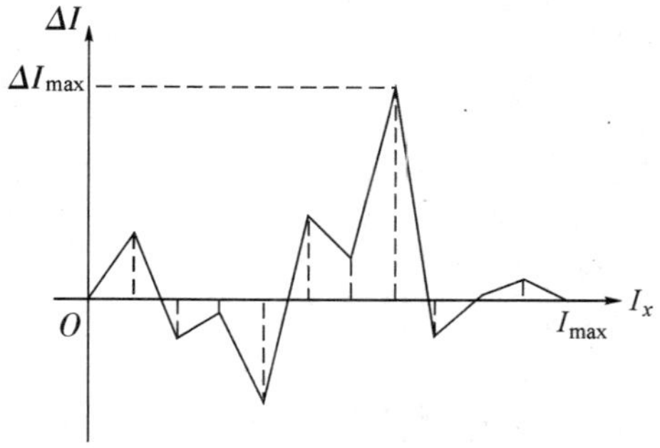

Figure 9 – 11 The circuit of measuring I_g by the potentiometer method

We can get the accuracy of the enlarged meter K by using the largest absolute error ΔI_{max} (or ΔU_{max}) divided by the range.

$$K = \frac{|\Delta I_{\max}|}{I_m} \times 100\% \quad \left(\text{or } K = \frac{|\Delta U_{\max}|}{U_m} \times 100\%\right) \tag{17}$$

According to the K, the meter is divided into different grades (see Experiment 5).

When calibrating the enlarged meter, we should follow the following steps:

(1) Adjust the zero point of the enlarged meter and the standard meter well before the experiment.

(2) Calibrate the range firstly. According to the formula (2) and (7), the calcu-

lated R_s and R_M may be not suitable because the given I_g and R_g values may be inaccurate. You should adjust the R_s and R_M when the standard meter reach the range but the enlarged meter not.

(3) Calibrate the scale. Check the scales twice (from the small scale to the large one and from the large scale to the small one). Record the readings of the standard meter when the enlarged meter reach its scale, such as 1.00mA (V), 2.00mA (V) and so on.

实验十　欧姆表的改装

导体的电阻一般采用“伏安法”来测定，即使用电压表和电流表分别测量出导体两端的电压和通过导体中的电流，利用欧姆定律（导体的电阻等于其两端的电压 U 与流过导体的电流 I 的比值）就可以得出该导体的电阻值 R，即 $R=U/I$。伏安法测量电阻需要两只电表（一只电压表和一只电流表），并且最终还要对测量数据进行计算处理才能得到导体的电阻值，这种方法对一般测量很不方便。如果能把导体两端的测量电压固定不变，那么就可以省掉电压表，而直接用电流表的读数来反映待测电阻的大小，欧姆表就是依据这个思想设计的。本实验主要是将表头改装成为欧姆表并对其刻度进行校验。

【实验目的】

1. 学习欧姆表的设计方法；
2. 掌握欧姆表的中值电阻的概念。

【实验仪器】

表头（直流微安计或毫安计）、稳压电源、电阻箱、滑线变阻器、开关、导线若干。

【实验原理】

1. 欧姆表的设计方法与特点

图 10－1 是欧姆表的电路图。E 为干电池，其电压 U 是一定值，R_g 为表头内阻，R_0 为限流电阻，均为已知。R_x 是待测电阻，测量时将它接在图中 a、b 两点之间，由欧姆定律可知，接入 R_x 后，表头所指示的电流 I_x 为：

$$I_x=\frac{U}{R_0+R_g+R_x}$$

化简可得：$R_x=\dfrac{U}{I_x}-(R_0+R_g)$　　　(1)

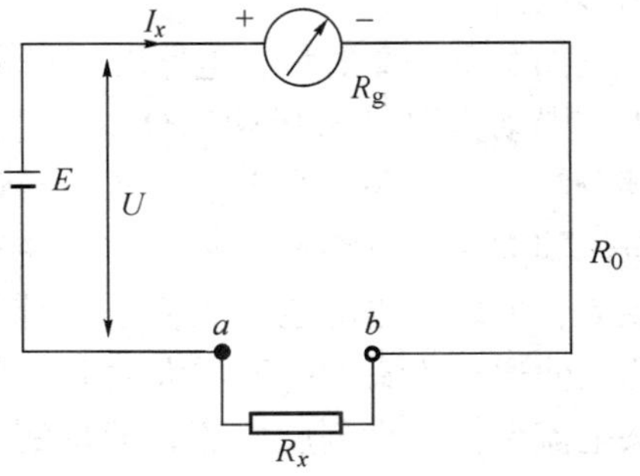

图 10－1　欧姆表电路图

根据式（1），可以从表头的指示值 I_x 得出待测电阻 R_x 的值。如果直接在表头的表盘上刻上与 I_x 所对应的电阻值 R_x，这样表头就成了一只欧姆表。

在图 10－1 线路中，我们适当选择 R_0 的大小，使待测电阻 $R_x=0$（即 a、b 之间短路）时表头指针满偏，即 R_0 满足：

$$R_0=\frac{U}{I_g}-R_g$$

这样，接有待测电阻 R_x 时的电流 I_x 与 I_g 的比值为

$$\frac{I_x}{I_g}=\frac{R_0+R_g}{R_0+R_g+R_x}=\frac{1}{1+\frac{R_x}{R_0+R_g}} \tag{2}$$

根据式（2），可以讨论欧姆表的一些特点：

（1）当待测电阻 $R_x=0$ 时，$I_x/I_g=1$，即 $I_x=I_g$，此时流过表头的电流等于表头满偏电流 I_g，表头指针满偏。因此，欧姆表的零点应在表盘刻度的右端，相当于电流挡的满刻度。

（2）当 $R_x=R_0+R_g$ 时，$I_x/I_g=1/2$，即 $I_x=I_g/2$，表头指针在表盘刻度中央，将表盘中心刻上 R_0+R_g 的值，就成为了一个中心值为 $R_T=R_0+R_g$ 的欧姆表。一般把 R_T 称为中值电阻。

中值电阻 R_T 的数位就是该电阻挡的倍率。例如 $R_T=600\Omega$ 为三位数，则该电阻挡倍率为 100。当 R_x 和 R_T 相接近时，测量就比较准确。

（3）当 $R_x=\infty$ 时，即被测电路开路，$I_x=0$，表头指针不偏转，因此欧姆表左端刻度线表示的电阻值为无穷大。欧姆表的刻度为自右至左电阻值逐渐增大，而且刻度不均匀。因此读数时必须注意。

2. 零点漂移

图 10－1 电路中，如果干电池的电压 U 不等于设计所确定的电压，那么将表棒短路时指针就不会刚好在零欧姆处，即电流 $I_x\neq I_g$，这一现象叫零点漂移，它给测量带来很大的误差。为了克服零点漂移，最简单的办法是将限流电阻 R_0 换成可变电阻。当电池电压变化时，只需调整 R_0，引起中值电阻的改变，就可使表头指针仍指零点。这样虽然能补偿零点漂移，但引起的测量误差较大，原因是 R_0 的改变引起了中值电阻的改变。因此应该选用对表头电流影响较大，而对中值电阻影响小的电路来补偿零点漂移，一般采用并联调节电路。

如图 10－2 所示，电路的特点是把对零点漂移起补偿作用的电阻 R_6 安置在表头回路里。滑动头把 R_6 分成两部分，一部分与表头串联，其余部分与表头并联。当电池电压高于设计值以致电路中总电流偏大时，可将滑动头下移，使通过表头的电流减少到 I_g；反之，当电池电压低于设计值时，可将滑动头上移，使通过表头的电流增大到 I_g。总之，调节 R_6 可使表头电流保持为满刻度电流，即待测电阻为零时，表头指针为 0Ω。

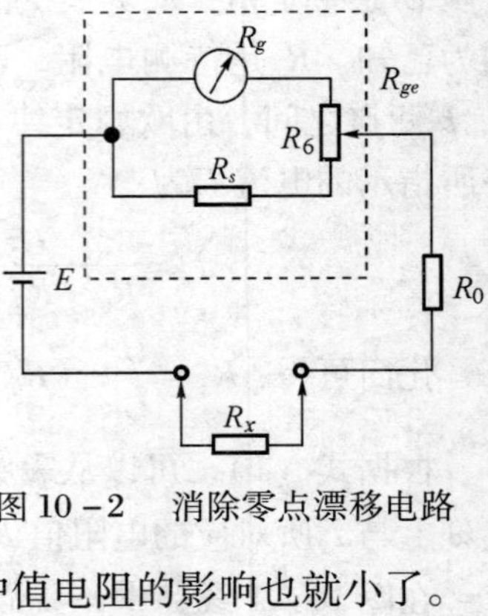

图 10－2　消除零点漂移电路

当改变调零电位器 R_6 的滑动头时，表头回路的等效电阻 R_{ge} 要随之改变，因而中值电阻 $R_T=R_{ge}+R_0$ 也要发生变化，但是我们可以把限流电阻 R_0 取得大些，这样 R_{ge} 的变化，对于整个中心电阻 R_T 来讲影响就较小，因而对中值电阻的影响也就小了。

电阻挡的线路如图 10－3 所示。

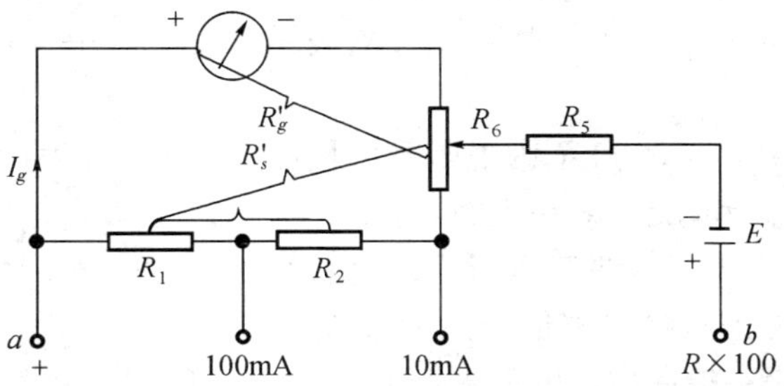

图 10－3　欧姆表的电阻挡线路图

设 $I_g = 157 \times 10^{-6}$ A，$R_g + R_6 = 1200\Omega$ 均为已知，表头分流电阻 R_s 仍取 $R_1 + R_2 = 19.1\Omega$，电池的标称电压 $U = 1.5$V。如果将该电流计改装成中值电阻为 $R_T = 600\Omega$ 的欧姆计，则表头等效电阻 R_{ge} 和限流电阻 R_0（$= R_5$）的计算方法如下：

图中 R_g' 为表头内阻 R_g 和 R_6 的滑动端上部分电阻串联值，R_s' 为 R_s 和 R_6 滑动端以下部分电阻的串联值。当 $R_x = 0$ 时，从 R_6 滑动端上流出的电流为工作电流

$$I = \frac{U}{R_T} = \frac{1.5\text{V}}{600\Omega} = 2.5 \times 10^{-3}\text{A}$$

R_s' 和 R_g' 并联，两端电压相等，则：

$$I_g R_g' = (I - I_g) R_S'$$

即：

$$I_g R_g' = (I - I_g)(R_S + R_g + R_6 - R_g')$$

其中

$$R_S' = R_S + R_g + R_6 - R_g'$$

$$= R_S + R_g + R_6 - \frac{(I - I_g)(R_S + R_g + R_6)}{I}$$

$$= R_S + R_g + R_6 - \left(I - \frac{I_g}{I}\right)(R_S + R_g + R_6)$$

$$= \frac{R_S + R_g + R_6}{I} I_g$$

代入实验数据可得：$R_s' = \dfrac{R_S + R_g + R_6}{I} I_g = \dfrac{19.1 + 1200}{2.5 \times 10^{-3}} \times 157 \times 10^{-6} = 76.6\Omega$

则：$R_g' = R_g + R_6 + R_s - R_s' = 19.1 + 1200 - 76.6 = 1142.5\Omega$

因此表头等效电阻：

$$R_{ge} = \frac{R_S' R_g'}{R'_S + R'_g} = \frac{76.6 \times 1142.5}{76.6 + 1142.5} = 71.8\Omega$$

故 $R_0 = R_T - R_{ge} = 600 - 71.8 = 528.2\Omega$

图 10－3 中在 a、b 端之间接上待测电阻 R_x 时，就可通过表头直接测出其电阻的大小。但需注意：本量程是中心电阻为 600Ω，具有三位数位，因此表头读数乘以 100 才是待测电阻的阻值。

【实验内容】

1. 将表头改装成中值电阻 $R_T = 600\Omega$ 的欧姆计，如图 10－4，表头内阻 $R_g = R_{go} + R_6$

已给出，$R_s=R_1+R_2$，R_6为调零电阻，电池标称电压$E=1.5V$，计算R_5的大小，并分别将R_s、R_5表示在电阻箱上。

2. 连接线路，其中R_s，R_5和R_x为电阻箱，R_6已接在表头上，其滑动端调节旋钮P在正面。

3. 欧姆表调零：闭合K，调节$R_x=0$，使欧姆表短路，观察表头是否指示电阻零值，并通过R_6上的P点旋钮调节，使表头指针指示电阻零点。

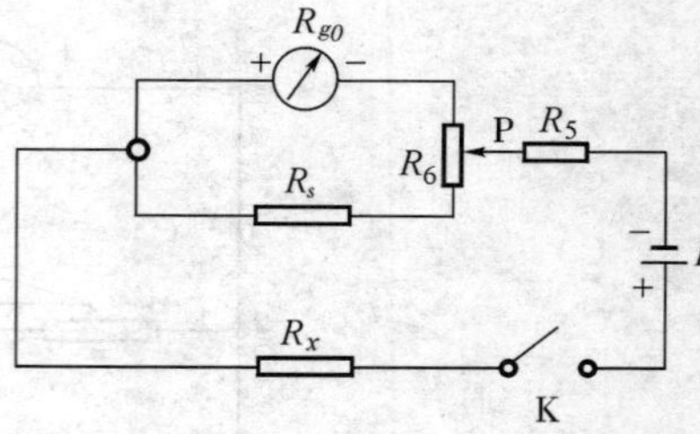

图 10－4　欧姆表改装实验电路图

4. 分别调节R_x为表10－1所列数据，观察并记录改装欧姆表的示数（注意读数乘以倍率100）。

表10－1　欧姆表改装实验数据表格

电阻	电阻取值				
R_x（Ω）	200	400	600	1000	2000
$R_{改}$（Ω）					

5. 分别改变电源电压为$E=1.2V$和$E=1.6V$，调零后按表10－1重新校验。

【注意事项】

1. 同实验九。
2. 注意电阻挡调零电位器R_6的位置及内部的线路。

【思考题】

1. 图10－2中，为什么要用固定电阻R_0？
2. 欧姆表为什么要调零？为什么用并联调零电路比较好？
3. 欧姆表的中心电阻值只决定于E和I_g，若$I_g=50\mu A$，要求设计中心阻值为250kΩ，问应该用多大的电源电压？

实　验　指　导

【预习要求】

同实验九，并要求预先计算出R_5、R_s的大小。

【操作要点】

同实验九。

【实验后的思考】

1. 设中心电阻$R_T=600\Omega$，则当外测电阻为多少时，表头电流最大？电流多大？同时改变哪一电阻才能使中心电阻 R_T 改变较大？

2. 表头上电流刻度是均匀变化的，而欧姆数刻度是不均匀的，其内在原因是什么？能不能设计出一种电阻测量表，其欧姆数刻度是均匀变化的呢？

3. 不改变 R_s 和 R_6 的阻值，怎样才能扩展中心电阻，使改装多用表能准确测量 6000Ω 附近的电阻？

附录：多用电表

多用电表（也称万用电表）是一种集电流表、电压表和欧姆表等功能的最常用的电学仪表，主要有指针式（如：MF47）和数字式（如：DT－830）等种类。它可以测量交流和直流的电压、电流、电阻、电容、电感及二极管反向截止电流、放大倍数等，其详细使用方法可参考实验5或相应说明书，在此不作详细介绍。下面主要介绍多用电表对测量电路的影响和使用多用电表检查电路故障方法：

1. 多用电表接入误差的成因及修正方法

多用电表在使用时往往不是固定连接在待测电路上，而是在测量时接上，读数后即拆离，所以接入误差成为经常要考虑的问题。

（1）直流电压挡　多用电表的电压挡测量被测电路电压时，必须将其并联于被测电路两端，如图 10－5 所示。

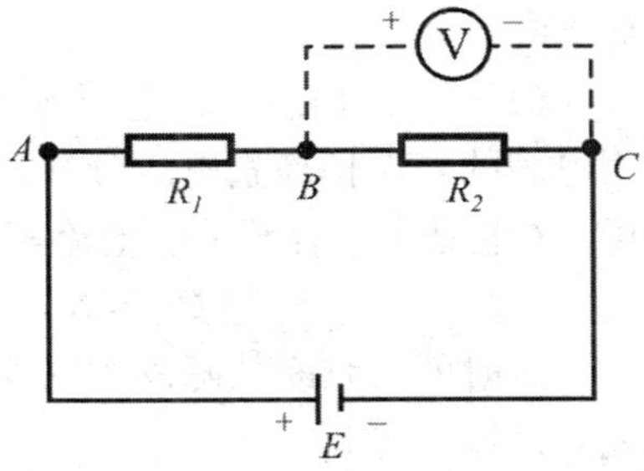

图 10－5　多用电表测量电压示意图

BC 间的电压 $U_{BC}=[R_2/(R_1+R_2)]E$，如果多用电表接在 B、C 两点，测出的电压是否等于 U_{BC} 呢？显然不是的。由于多用电表有一定的内阻 R_V，将其接入电路后，电路的电压分配就会发生改变，BC 间电压变为 U'_{BC}。我们想要知道的是多用电表未接入前的电压 U_{BC}，这两者之差称为接入误差 ΔU，定义为：

$$\Delta U=U_{BC}-U'_{BC}$$

则有：

$$U'_{BC}=\frac{\dfrac{R_VR_2}{R_V+R_2}}{R_1+\dfrac{R_VR_2}{R_V+R_2}}\cdot E=\frac{1}{1+\dfrac{R_1(R_V+R_2)}{R_VR_2}}\cdot E$$

$$\frac{\Delta U}{U'_{BC}}=\frac{U_{BC}-U'_{BC}}{U'_{BC}}=\frac{U_{BC}}{U'_{BC}}-1=\frac{R_1}{R_1+R_2}\left[1+\frac{R_1(R_V+R_2)}{R_VR_2}\right]-1=\frac{R_1R_2}{R_V(R_1+R_2)}$$

观察图 10－5 的电路可知，$R_1R_2/(R_1+R_2)$ 正是以多用电表接入点 BC 为考察点的等效电阻 $R_{等效}$，故得 $\Delta U=(R_{等效}/R_V)U'_{BC}$。根据上式很容易知道接入误差的大小，在必要时可用以对测量值进行修正。

（2）直流电流挡　多用电表电流挡测量电流时，须将电表串联于被测电路中，如图 10－6 所示。

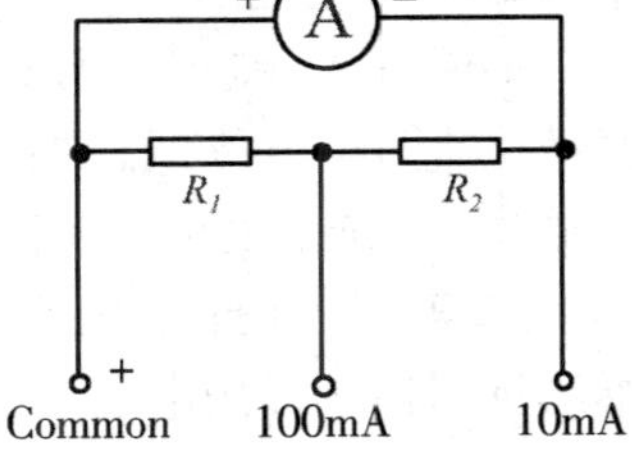

图 10－6　多用电表的电流挡

多用电表具有一定的内阻，接入电路时也有接入误差。若电表内阻为 R_A，以电表接入点为考察点时，电路的电阻

为 $R_{等效}$。

同理可推出接入误差为：

$$\Delta I = \frac{R_A}{R_{等效}} I'$$

式中 I' 即从多用电表读出的电流值。

（3）电阻挡　我们知道欧姆计的刻度是非线性的，当 $R_X \ll R_T$时，近似地有 $I_X = E/R_T = I_g$，此时偏转接近满度，I_X随 R_X的变化未有明显变化，因而测量误差很大；当 $R_X \gg R_T$，则 $I_X \approx 0$，因而测量误差也很大。所以在实用上，通常只用欧姆计刻度线上中间的一段来测量，例如，$1/5 \sim 5R_T$这段电阻范围。实际上欧姆计都有几个量程，每个量程的 R_T都不同，但每个量程中可用的范围都是 $1/5 \sim 5R_T$，如 $R_T = 600\Omega$，则测量范围为 $120 \sim 3000\Omega$；$R_T = 6000\Omega$，则测量范围为 $1200 \sim 30\,000\Omega$。

2. 多用电表检查电路的方法

多用电表常用于检查电路，排除故障。实验中往往遇到这样情况：电路经仔细检查，连接没有错误，但合上开关后不能正常工作，说明电路出现故障。故障原因大致为导线内部断线、开关或接线柱接触不良或电表及元件内部损坏等。常用多用电表检查，其方法如下。

（1）电压挡法　在电源接通的情况下从电源的两端开始，按对接点依电流顺序用多用电表的电压挡检查电压分布，若出现电压分布反常点，就是产生故障之处。例如，在图 10－7 电路中，合开关后电流表无指示，用电压挡法测出：

$$U_{ab} = U_{a'b'} = U_{cd} = U_{c'd'} = E，\ U_{c'd'} = U_{e'd''}，\ U_{fd''} \neq 0$$

这时就可判断故障发生在 ff'之间，可能是 f 或 f' 处接触不良，也可能导线 ff' 断了，或毫安计损坏。

采用伏特计法不必拆电路，检查方便，可较快发现故障的位置，但不适合于检查电压太小的部位。

（2）欧姆计法　将电路逐段拆开，特别要注意将电源和电表断开，而且应使待测部分无其他支路；再用多用电表的欧姆挡检查无源部分的电阻分布，特别要检查导线和接触点通不通。如图 10－7 电路中，将 K 断开，电源开关关掉（或取下），测得 aa'，bb'，cc'，dd'，ee' 都是通的，即：$R_x = 0$，而 ff' 不通，即 $R_x = \infty$，则故障必在 ff'之间。

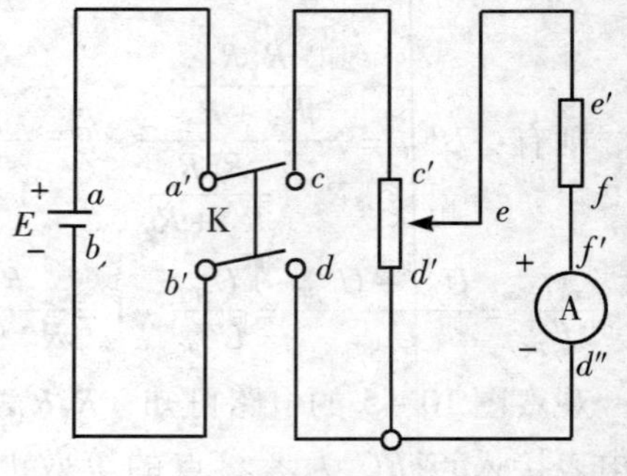

图 10－7　多用电表检查电路故障

如断开 f，测得 ff' 导线不通，则应换上用欧姆挡检查过的好导线，如检查 ff' 导线也是通的，则故障为 f 或 f' 处接触不良。

迅速查清并排除电路故障是电学实验的基本训练内容之一，在今后做电学实验时，应继续练习使用多用电表检查电路，以培养分析问题、解决问题的独立工作能力。

10 Assembling an Ohmer

The resistance of a conductor is usually measured by the volt – ampere method. The voltage and the current can be measured by voltmeter and ampere meter respectively. According to the Ohm's law, the resistance R of the conductor is equal to the ratio of voltage U and the current I, that is $R = U/I$. The volt – ampere method must have two meters (a voltmeter and an ampere meter), and the data of the measurement should be calculated in order to get the resistance of the conductor. Therefore, it's very inconvenient for the general measurement. If the applied voltage on the conductor is fixed, the voltmeter can be omitted, and the resistance can be determined by the reading of the ampere meter. An ohmer is designed based on this method. In this experiment, the meter is assembled as an ohmer and its scale is calibrated.

Objectives

1. Learn how to design an ohmer.
2. Understand the concept of central resistance.

Equipments

Meter (milliamperemeter or microampere meter), DC regulated power supply, resistance box, slide rheostat, switch, and some wires.

Principle

1. Design an ohmer and its characters

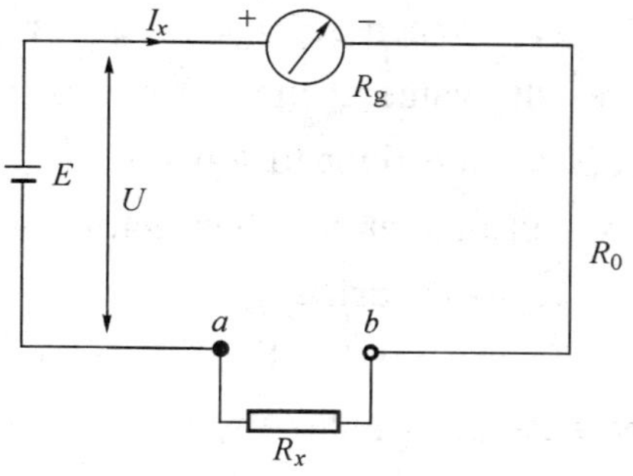

Figure 10 – 1 The inner circuit of an ohmer

E is a battery, whose voltage U is a constant. The internal resistance of the meter is R_g. And R_0 is a current – limiting resistance. R_g and R_0 are given. R_x is a resistance to be measured. When R_x is connected to the circuit, the current of the meter is

$$I_x = \frac{U}{R_0 + R_g + R_x}$$

So we can get:

$$R_x = \frac{U}{I} - (R_0 + R_g) \qquad (1)$$

Therefore, the resistance R_x to be measured can be calculated by the current of the meter, which is shown in (1). An Ohmer can be designed if the corresponding resistance value is caved on the meter panel.

In figure 10 – 1, we can choose proper R_0 to make the meter reach its full – scale when $R_x = 0$, that is, there is no resistance to be measured. Now (1) can be written as:

$$R_0 = \frac{U}{I_g} - R_g$$

And we can get the ratio of I_x and I_g will be:

$$\frac{I_x}{I_g} = \frac{R_0 + R_g}{R_0 + R_g + R_x} = \frac{1}{1 + \frac{R_x}{R_0 + R_g}} \tag{2}$$

Let's discuss the characters of an ohmer for measuring resistance.

(1) When $R_x = 0$, we can get $I_x = I_g$. The current of the meter equals I_g. Therefore, zero – point of the ohmer is at the right end of its panel.

(2) When $R_x = R_0 + R_g$, we can get $I_x / I_g = 1/2$. The pointer of the meter will be at the middle of its panel. And $R_0 + R_g$ is called central resistance, which is represented by R_T.

The digits of the central resistance are called the magnification of this range. For instance, if $R_T = 600\Omega$, which is a number of three digits, the magnification will be 100. The result will be more accurate if the resistance R_x is near the central resistance R_T.

(3) When $R_x = \infty$, we can get $I_x = 0$. There is no current in the circuit. Therefore, the value at the left end of the panel should be infinite. The resistance value increases from right to left on the panel, and the numbers are not uniform. You should pay more attention when getting a reading from an ohmer.

2. Zero drifting

In Figure 10 – 1, if the voltage of the battery changes, that is it does not equal to the designed value, the hand of the meter will not point to zero when $R_x = 0$. This phenomenon is called zero drift. It will produce a big error in the measurement. To avoid the zero drift, the simplest way is to adjust the current – limiting resistance R_0. Choose a proper R_0 to make the hand point to zero again. But at the same time the R_T may be also change. For a good way to solve this problem, we choose a parallel circuit to avoid great change that happens to the central resistance R_T, which is shown in figure 10 – 2.

R_6 is put into the loop of the meter. R_6 is divided into two parts. One part is in series with the meter. The other is parallel with the meter. When the voltage of the battery is higher than its designed value, we move down the sliding contactor to decrease the current that flows through the meter until the hand points to the full scale. On the other hand, when it is lower than the designed value, we move up the sliding contactor to increase the current. In general, adjust R_6 to make the meter get its full scale I_g. That is to say, when the resistance to be measured is zero ($R_x = 0$), the hand of the meter will point to its full scale I_g at any condition.

We should know that the equivalent resistance R_{ge} of the meter will change because of the moving of the sliding contactor of R_6, and the central resistance $R_T = R_{ge} + R_0$ will also change. If the current – limiting resistance R_0 is big enough, the influence of R_{ge} to the central resistance R_T will be relatively small.

The circuit of an ohmer can be shown in figure 10 – 3. $U = 1.5\text{V}$, $I_g = 157 \times 10^{-6}$ A, $R_g + R_6 = 1200\Omega$ and $R_s = R_1 + R_2 = 19.1\Omega$ are given. For assembling an ohmer with the central resistance of $R_T = 600\Omega$, the equivalent resistance of the meter R_{ge} and the current – limiting resistance $R_0 (= R_5)$ can be calculated as following:

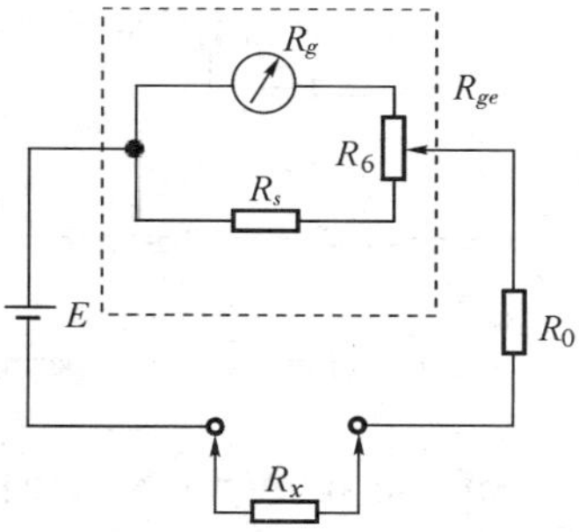

Figure 10 – 2 The circuit of avoiding zero drift

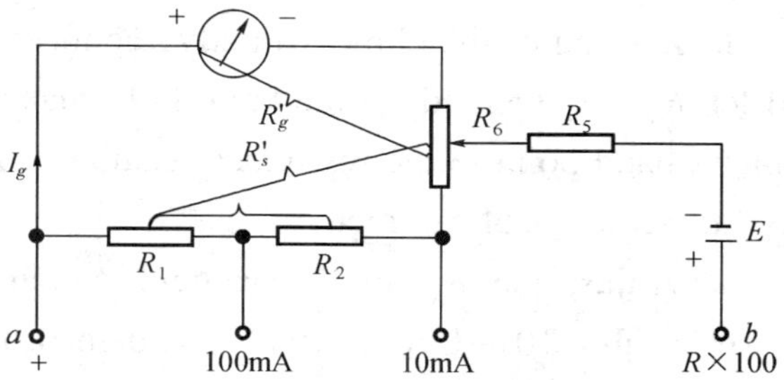

Figure 10 – 3 The inner circuit of an ohmer

R_g' is the sum of the meter's internal resistance R_g and the upper part of R_6; R_s' is the sum of R_s and the lower part of R_6. When $R_x = 0$, the current flowing out of the sliding contactor of R_6 is:

$$I = \frac{U}{R_T} = \frac{1.5\text{V}}{600\Omega} = 2.5 \times 10^{-3}\text{A}$$

R_g' is parallel with R_s', so we can get:

$$I_g R_g' = (I - I_g) R_S'$$

That is,

$$I_g R_g' = (I - I_g)(R_S + R_g + R_6 - R_g')$$

$$R_S' = R_S + R_g + R_6 - R_g'$$

$$= R_S + R_g + R_6 - \frac{(I - I_g)(R_S + R_g + R_6)}{I}$$

$$= R_S + R_g + R_6 - \left(1 - \frac{I_g}{I}\right)(R_S + R_g + R_6)$$

$$= \frac{R_S + R_g + R_6}{I} I_g$$

Put the experiment data to the equation, we can get:

$$R_s{}' = \frac{R_S + R_g + R_6}{I} I_g = \frac{19.1 + 1200}{2.5 \times 10^{-3}} \times 157 \times 10^{-6} = 76.6\Omega$$

So we can get : $R_g{}' = R_g + R_6 + R_s - R_s{}' = 19.1 + 1200 - 76.6\Omega = 1142.5\Omega$

$$R_{ge} = \frac{R_S{}' R_g{}'}{R'_S + R'_g} = \frac{76.6 \times 1142.5}{76.6 + 1142.5} = 71.8\Omega$$

So $R_0 = R_T - R_{ge} = 600 - 71.8\Omega = 528.2\Omega$

Experiment

1. The experiment circuit is shown in figure 10 – 4. $E = 1.5V$, $R_T = 600\Omega$. Firstly, calculate the value of R_5.

2. Join the circuit. R_6 has been already installed into the meter and the knob P can be used to adjust its resistance.

3. Zeroing of the ohmer. Firstly, connect K, and let $R_x = 0$; then adjust the knob P to make the ohmer's hand point to the resistance reading of zero (its full scale I_g of the meter).

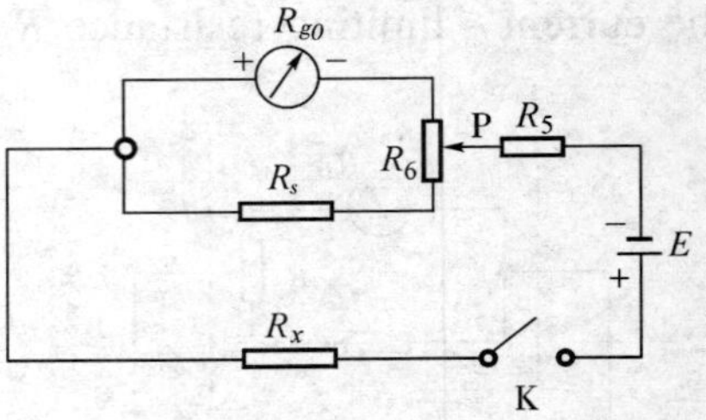

Figure 10 – 4 The circuit of the experiment

4. Adjust the R_x to the values, which are shown in table 10 – 1 respectively, and record the reading of the ohmer (Notice: the values should be multiplied by 100 for magnification).

Table 10 – 1 The data of assembling an ohmer

Resistance	The value of the resistor				
R_x (Ω)	200	400	600	1000	2000
R_a (Ω)					

5. Finally, change the voltage of the regulated power supply source to $E = 1.2V$, and $E = 1.6V$, then check the ohmer again.

Notice

1. Notice the same items as in Exp. 9.
2. Please notice to the right position of the zero adjusting resistance R_6.

Problems

1. What is the reason for setting a fixed resistance R_0 in the circuit of figure10 – 2?

2. Why should we make the ohmer point to zero? What are the advantages of the parallel zero adjusting circuit?

3. The central resistance of an ohmer is determined by the E and I_g. If $I_g = 50\mu A$,

and the central resistance is designed as 250kΩ, then how much should the power supply voltage E be?

Additional Instruction

Preparation Requirement

The requirement is the same as in Exp. 9.

Calculate the values of R_5 and R_s.

Problems after the experiment

1. If the $R_T = 600\Omega$, what' s the value of R_x when the meter reach its full scale I_g? And how much is the I_g? Which resistor can make the R_T change very much?

2. The marks for current values on the meter' s panel are uniform, but the marks for an ohmer are not, why? Can you assemble an ohmer with uniform marks?

3. How to enlarge the central resistance to measure a resistance about 6000Ω without changing the values of R_s and R_6?

Appendix: Mutimeter

Multimeter is a common electric meter which has the function of ampere meter, voltmeter and ohmmeter ans so on. It has many kinds, such as pointer multimeter (MF47) and digital multimeter (DT-830). It can measure AC and DC voltage, current, resistance, capacitance, inductance and diode reverse cut-off current, diode magnification, etc. The details of operating it can refer to the experiment 5 or the corresponding instructions, and we don't intend to pay more attention to them. In the following part, we discuss the measuring error of the multimeter and its usage of checking the circuit.

1. The joint error and its revision

(1) The step of voltage

The multimeter is often connected to the circuit when testing. But when connected to the measurement, it can bring impact to the circuit. Therefore, the result should be revised after the measuring. The Multimeter is paralleled to the equipment when measuring, as is shown in figure 10-5.

So the true value of the voltage between BC is U_{BC}, we can get:

$$U_{BC} = \frac{R_2}{R_1 + R_2} \cdot E$$

But when it is measured by a multimeter (voltage step) with a internal resistance of R_V, the measuring result will be U'_{BC}, and the difference between U_{BC} and U'_{BC} is ΔU, which is called joint error, that is:

$$\Delta U = U_{BC} - U'_{BC}$$

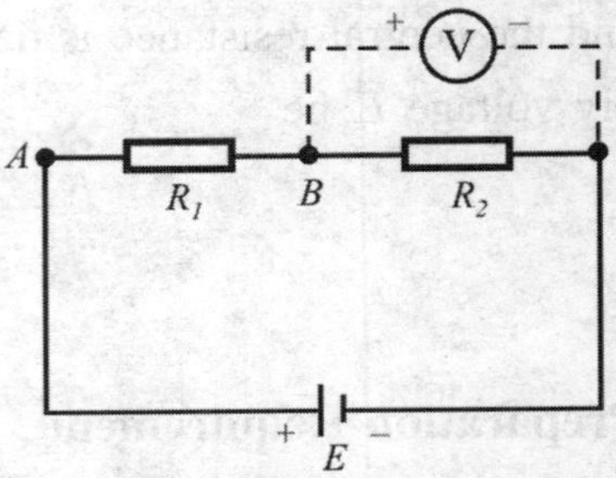

Figure 10 – 5 The circuit of measuring the voltage by a multimeter.

So we can get:

$$U'_{BC} = \frac{\frac{R_V R_2}{R_V + R_2}}{R_1 + \frac{R_V R_2}{R_V + R_2}} \cdot E = \frac{1}{1 + \frac{R_1 (R_V + R_2)}{R_V R_2}} \cdot E$$

And the relative error is:

$$\frac{\Delta U}{U'_{BC}} = \frac{U_{BC} - U'_{BC}}{U'_{BC}} = \frac{U_{BC}}{U'_{BC}} - 1 = \frac{R_1}{R_1 + R_2} \left[1 + \frac{R_1 (R_V + R_2)}{R_V R_2}\right] - 1 = \frac{R_1 R_2}{R_V (R_1 + R_2)}$$

We know that the equivalent resistance R_e between BC is:

$$R_e = R_1 R_2 + R_1 + R_2$$

And now we can get the correction value of the joint error of the voltage will be:

$$\Delta U = \frac{R_e}{R_V} U'_{BC}$$

Based on the upper equation, the error is easy to know and if necessary, it can be used to carry out amendments to the measured value.

(2) Step of electricity

The multimeter is connected in series for measuring the current in the circuit. There is a different internal resistance R_A for every step, as is shown in figure 10 – 6.

Let R_e for the equivalent resistance between the joint points, the value of joint error will be:

$$\Delta I = \frac{R_A}{R_e} I'$$

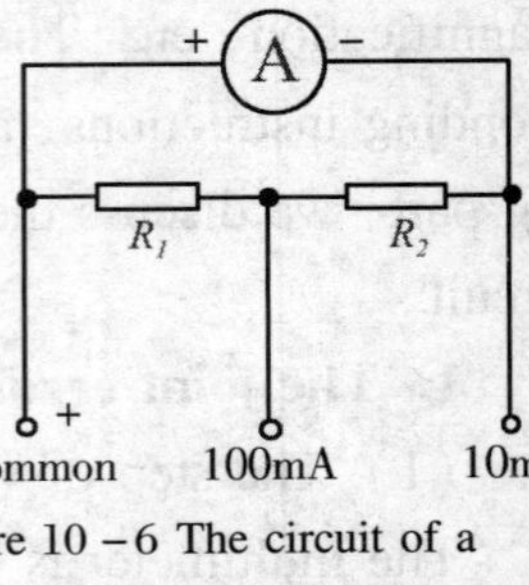

Figure 10 – 6 The circuit of a multimeter – step of electricity

I' is the reading of the multimeter.

(3) Step of Ohmer

We have known that the scale of an ohmer isn't uniform, and for a resistance $R_X << R_T$ or $R_X >> R_T$, the measuring error will be great. Therefore, only the middle section of the scale, about 1/5 ~ $5R_T$, can measure resistance with a relative small error. For example, if $R_T = 600\Omega$, the relative correct range will be 120 ~ 3000Ω, and if $R_T = 6000\Omega$, the range will be 1200 ~ 30000Ω.

2. Application of multimeter for checking the circuit

This situation may sometimes happen that a circuit has no joint error but it don't work, so there should have some troubles with the circuit, such as broken wires, bad contact and so on. It is usually checked by a multimeter.

(1) Voltmeter checking

Under the condition of turning on the power, check the equipments or wires in the order of the current. When there is an abnormal voltage, there maybe have troubles. For example, in the circuit of figure 10 – 7, the switch K is on, but there is no reading on the ampere meter, and then we check the circuit and find:

$$U_{ab} = U_{a'b'} = U_{cd} = U_{c'd'} = E$$

It means that the power source, switch and the slide rheostat works well. And go on checking, and we find that:

$$U_{c'd'} = U_{e'd''}, \quad U_{fd''} \neq 0$$

Now we can conclude that there may be a trouble between ff', such as bad contact or the ampere meter does not work.

The convenience of checking circuit by a multimeter is that it can be done without taking apart the circuit, but it is unpractical for a small voltage.

(2) Ohmer checking

Check a circuit by an ohmer is to measure the resistance of an equipment or a wire. Be sure to disconnect the circuit, especially the power source. For example, the circuit in figure 10 – 7, check the wires and find that aa', bb', cc', dd' and ee' are all good, that is $R_x = 0$, but ff' not, $R_x = \infty$, so the wire ff' isn't good.

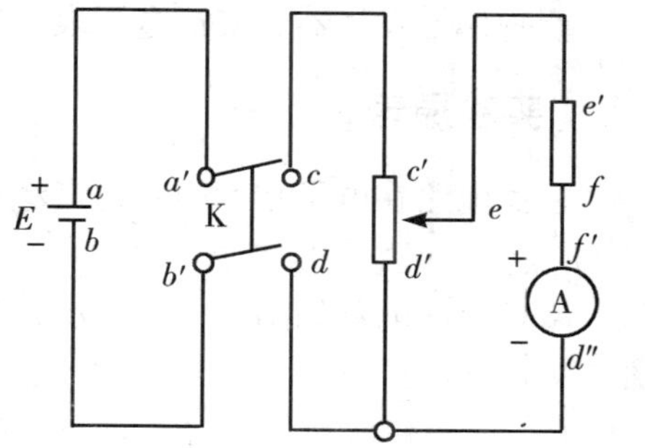

Figure 10 – 7 Check the circuit by a multimeter

Check and eliminate the electrical circuit fault quickly is one of the important skills to be trained. You should continue to practice the use of multimeter to check the circuit, and it can improve your analysis, problem – solving ability to work independently.

实验十一　自感系数的测定

谐振现象是正弦稳态电路的一种特定工作状态，它在无线电和电工技术中得到了广泛的应用，但另一方面谐振有时会破坏系统的正常工作，所以，对谐振现象的研究，有重要的实际意义。本实验采用的是 *RLC* 串联电路，通过测定共振曲线，可以计算线路的自感系数和品质因数。

【目的要求】

1. 用串联共振法测定自感系数。
2. 研究 *RLC* 串联电路的幅频特性。
3. 理解回路 Q 值的物理意义。

【仪器器材】

音频信号发生器、交流电压表（VC890D 数字万用电表）、镇流器、电容器。

【实验原理】

共振电路在电子技术中获得广泛应用，*RLC* 电路有串联共振与并联共振两种。

RLC 串联电路如图 11－1 所示，其电路方程为

$$\dot{U}=\dot{I}\left[R+j\left(\omega L-\frac{1}{\omega C}\right)\right]$$

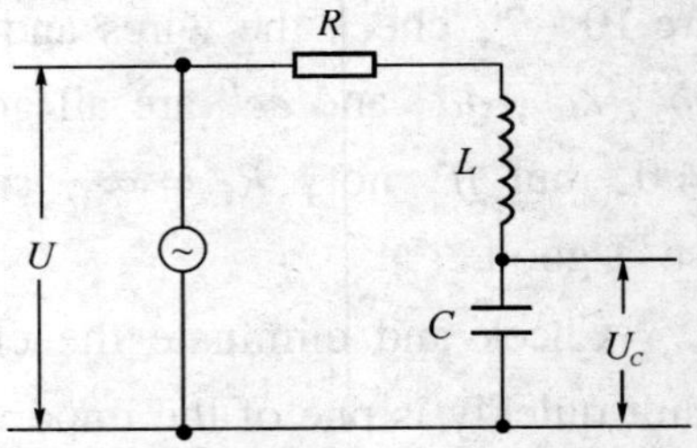

图 11－1 *RLC* 电路图

由上式可得

$$\dot{I}=\frac{\dot{U}}{R+j\left(\omega L-\frac{1}{\omega C}\right)}$$

通过电路的交流电流有效值为

$$I=\frac{U}{\sqrt{R^2+\left(\omega L-\frac{1}{\omega C}\right)^2}} \tag{1}$$

式中 U 为电压的有效值

$$Z=\sqrt{R^2+\left(\omega L-\frac{1}{\omega C}\right)^2} \tag{2}$$

Z 称为回路的阻抗。

改变 *RLC* 串联电路的信号源频率，而保持信号源电压不变，由式（1）可知，回路电

流也将变化。当回路电流出现最大值时，即称电路共振。由式（1）可知产生共振条件为

$$\omega L = \frac{1}{\omega C} \tag{3}$$

此时的频率称为共振频率，记为f_0。

$$f_0 = \frac{\omega_0}{2\pi} = \frac{1}{2\pi\sqrt{LC}} \tag{4}$$

从以上讨论可得出如下结论：

（1）共振时，回路阻抗为纯电阻，且取最小值$Z=R$。

（2）在保持电源电压U恒定的条件下，$f=f_0$，电流有极大值，且$I=\frac{U}{R}$。当f偏离f_0，电流减小，偏离越多，电流越小。

（3）Q值标志共振电路性能的好坏，称为电路的品质因数，它定义为回路的特性阻抗与回路电阻的比值，即

$$Q = \frac{\omega_0 L}{R} = \frac{\frac{1}{\omega_0 C}}{R} = \frac{1}{R\omega_0 C} = \frac{1}{R}\sqrt{\frac{L}{C}} \tag{5}$$

（4）共振时，电容和电感上的电压均是电源电压的Q倍，即

$$U_C = U_L = QU \tag{6}$$

通常因$Q>>1$，所以共振时，U_C（U_L）比U大很多倍，故又称串联共振为电压共振。

（5）由式（4）知共振频率只与电路中感抗及容抗有关，与电路中阻值R无关，阻值的大小仅影响Q值，也就是影响共振曲线的尖锐程度。

由式（4）有

$$L = \frac{1}{4\pi^2 f_0^2 C} \tag{7}$$

若由实验测出共振频率f_0，并已知C，则由式（7）可算出L的大小。

实验电路如图11－2所示，以音频信号发生器为信号源。通常情况下，$Q>>1$，当电路发生共振时，电路中的回路电流达到最大，同时电容器两端的电压U_C也几乎达到最大值。实验时，改变音频信号发生器的频率，但保持输出电压不变，当电容器两端电压为最大时，音频信号发生器所示频率，即可近似看作共振频率f_0。

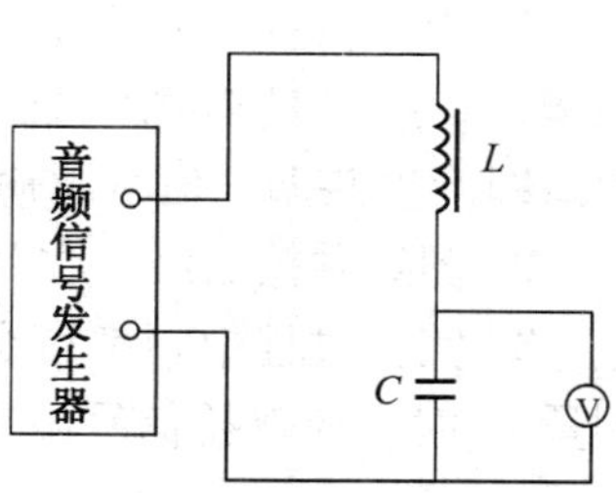

图11－2　实验电路图

【实验内容】

1. 按图11－2连接电路。交流伏特计可用多用电表20V交流电压挡代替。

2. 经教师检查无误后，开启音频信号发生器。调节输出旋钮使输出电压为2V（用数字表测量）。“频率范围”置于1kHz挡（调节小频率时置于100Hz挡）。

3. 连续改变频率f，同时注意观察电容两端的电压变化，找出对应于最大U_C的频率

f_0，然后在大于f_0和小于f_0的附近各取10个f数据，同时记下相应的U_C值，在共振点附近，频率间隔应取得小些。在整个测量过程中应始终保持音频信号发生器的输出电压（用数字表测量）为某一恒值（例如2V）。

4. 以U_C为纵坐标，频率f为横坐标，作为共振曲线，并在曲线上标出共振频率f_0。

5. 将已知电容值C和测得的f_0值代入式（7）中，求出自感系数L。

6. 由式（6）计算出该回路的品质因素Q值。

【注意事项】

实验完毕，应将音频信号发生器的输出电压调至零后再关闭电源。

【思考题】

1. 在用共振法测量电感的电路（图11－3）中，接入三只交流伏特计，它们的读数是否符合$U=U_1+U_2$，为什么？

2. 如用并联共振，能否测定自感系数？

3. 能否用串联共振测定电容器的电容？

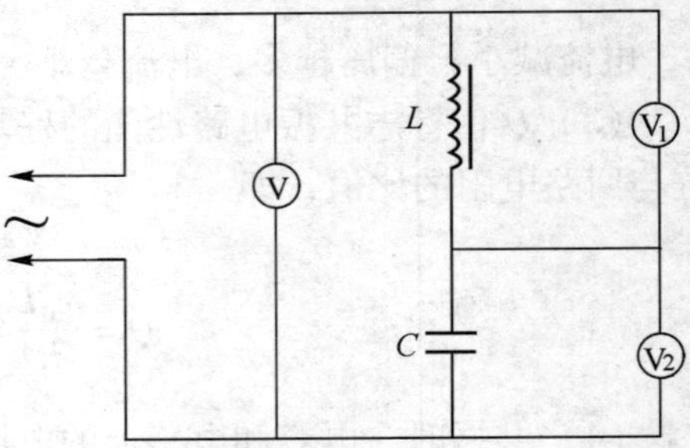

图11－3　思考题1

实　验　指　导

【预习要求】

1. 什么是串联共振？共振条件是什么？

2. 怎样测绘串联共振曲线？

3. 测绘串联共振曲线时，为什么要维持回路电压不变，怎样满足这一要求？

【操作要点】

1. 首先要找到共振点，即在保持信号源输出电压一定的情况下增加（或减小）输出频率，观察到U_C逐渐增加，当频率继续增加（或减小）时，U_C反而减小。此时，U_C达极大值即产生共振，信号源输出频率为共振频率f_0。

2. 测绘共振曲线，在共振点两侧各取10个点，共振点附近曲线陡直，应多测些点。

3. 实验过程中频率变化，外电路阻抗因而改变，此时信号源输出电压也随之变化。测量U_C时，应保持信号源输出电压不变。

【实验后的思考】

1. 在实验过程中，为什么音频信号发生器的输出电压经常变化？

2. 实验线路图11－2中，能否用交流伏特计与镇流器并联来测量U_L，为什么？

3. 怎样用示波器判断串联共振？绘出电路图，说明主要步骤。

【品质因数较小时对实验结果的修正】

在 RLC 串联电路中，根据式（1），可知电容上电压的有效值为

$$U_C=\frac{I}{\omega C}=\frac{1}{\omega C}\frac{U}{\sqrt{R^2+(\omega L-\frac{1}{\omega C})^2}} \tag{8}$$

保持电源电压 U 不变，当 U_C 取极大值时，$(\frac{U}{U_C})^2$ 取极小值，令此时的角频率为 ω_0'，则

$$\frac{\mathrm{d}}{\mathrm{d}\omega}\left((\frac{U}{U_C})^2\right)\bigg|_{\omega=\omega_0'}=\frac{\mathrm{d}}{\mathrm{d}\omega}\left(\omega^2C^2\left[R^2+(\omega L-\frac{1}{\omega C})^2\right]\right)\bigg|_{\omega=\omega_0'}$$
$$=2\omega_0'C^2\left[R^2+2\ (\omega_0'L)^2-2\frac{L}{C}\right]=0$$

可以解出 ω_0'为

$$\omega_0'=\sqrt{\frac{1}{LC}-\frac{R^2}{2L^2}} \tag{9}$$

显见 ω_0'与式（3）中的共振频率 ω_0 不同。电容上的电压最大时，令 U_{Cm}为最大电压值，在 $U_C-\omega$ 图线上求出的品质因子 $Q'=\frac{U_{Cm}}{U}$，可以解得

$$Q'=\frac{1}{\omega_0'C\sqrt{R^2+\ (\omega_0'L-\frac{1}{\omega_0'C})^2}}=\frac{1}{RC\sqrt{\frac{1}{LC}-\frac{R^2}{4L^2}}} \tag{10}$$

根据式（3）和式（5）中，共振角频率 ω_0 和品质因子 Q 的定义，

$$\omega_0=\sqrt{\frac{1}{LC}}$$

$$Q=\frac{\omega_0 L}{R}=\frac{1}{\omega_0 RC}=\frac{1}{R}\sqrt{\frac{L}{C}}$$

在式（9）和式（10）中消去 R、L、C 后可得，

$$\omega_0'=\omega_0\sqrt{1-\frac{1}{2Q^2}} \tag{11}$$

$$Q'=\frac{Q}{\sqrt{1-\frac{1}{4Q^2}}} \tag{12}$$

从式（11）和式（12）中可知，U_C 存在极大值就必须要求 $Q>\frac{\sqrt{2}}{2}$和 $Q'>1$。

当 Q 值分别为 1、2、5 时，$\frac{U_C}{U}-\frac{\omega}{\omega_0'}$图线如图 11 –4 所示。

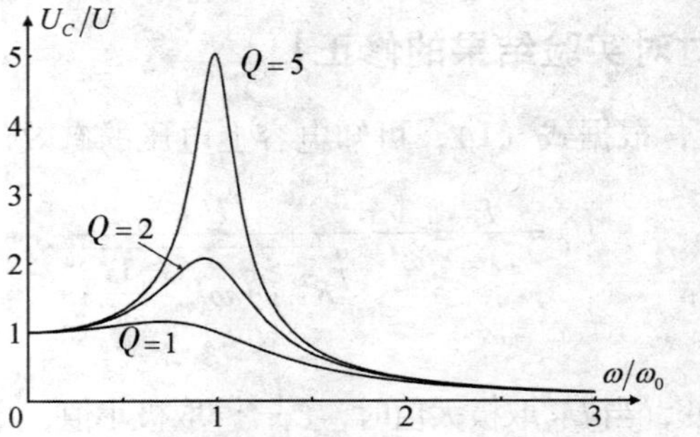

图 11－4　$\dfrac{U_C}{U}-\dfrac{\omega}{\omega_0}$图线

在实验中，当 Q 值较小时必须对式（6）和式（7）做相应的修正。联立式（11）和式（12），可以解出 Q 和 ω_0（其中舍去 $Q<Q'/\sqrt{2}$的增根），

$$Q=Q'\sqrt{\frac{1+\sqrt{1-\dfrac{1}{(Q')^2}}}{2}} \tag{13}$$

$$\omega_0=\frac{\omega_0{}'}{\sqrt{1-\dfrac{1}{(Q')^2+\sqrt{(Q')^4-(Q')^2}}}} \tag{14}$$

设实验中共振曲线共振峰对应的频率为$f_0{}'=\dfrac{\omega_0{}'}{2\pi}$，共振峰值电压与电源电压的比值记为 $Q'=\dfrac{U_{Cm}}{U}$，根据式（7）和式（14）可得电感 L 为

$$L=\frac{K(Q')}{4\pi^2(f_0{}')^2C} \tag{15}$$

其中 $K(Q')=1-\dfrac{1}{(Q')^2+\sqrt{(Q')^4-(Q')^2}}$

品质因数较小时，式（13）和式（15）即为修正后的根据电压共振曲线计算品质因数和电感的公式。

由式（15）和式（7）的比较中可以看出，在 Q 值较低时式（7）的实验系统误差较大（L 的相对误差约为$\dfrac{1}{2Q^2}$），只有品质因数相当大时，才能按照式（7）计算电感 L（例如品质因数 $Q>5$ 时，电感 L 的相对误差将小于 2%）。

11 Self – inductance Measurement

The resonance phenomenon is a certain working state in sine steady circuit. It is widely used in the field of wireless and electronics technology. On the other hand resonance can destroy the normal work in the system sometimes. There are two kinds of *RLC* resonance circuits: in parallel and in series. So it is important to study the resonance phenomenon practically. In this experiment a *RLC* resonance circuits in series is studied. We will calculate the self – inductance L and the quantity Q of the circuits by measuring the resonance curve.

Objectives

Measure self – induction coefficient by the method of series – resonance.

Study the resonance characters of the *RLC* circuit.

Understand the meaning of the quantity Q.

Equipments

Function generator, AC voltmeter (VC890D signal multimeter), inductor, capacitor.

Principle

The resonance circuit is useful in the field of electronics technology. There are two kinds of *RLC* resonance circuits: in parallel and in series.

The *RLC* series circuit is shown in figure 11 –1. Its circuit equation is:

$$\dot{U} = \dot{I}\left[R + \dot{J}\left(\omega L - \frac{1}{\omega C}\right)\right]$$

The electricity is:

$$\dot{I} = \frac{\dot{U}}{R + \dot{J}\left(\omega L - \frac{1}{\omega C}\right)}$$

The effective value of the electricity is:

$$I = \frac{U}{\sqrt{R^2 + \left(\omega L - \frac{1}{\omega C}\right)^2}} \qquad (1)$$

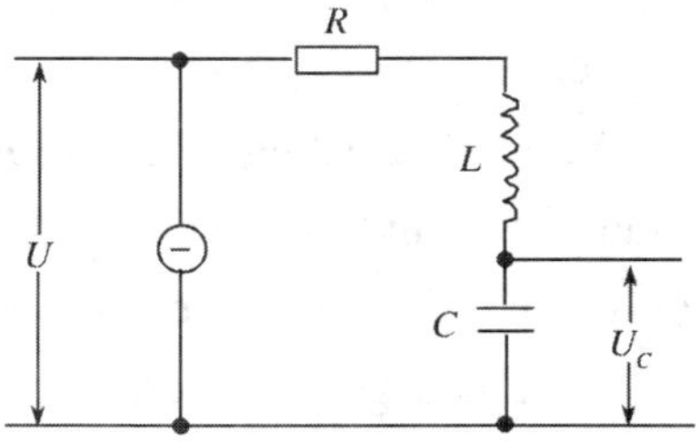

Figure 11 –1 *RLC* Series circuit

Where the U is also a effective value.

The impedance of the circuit is:

$$Z=\sqrt{R^2+\left(\omega L-\frac{1}{\omega C}\right)^2} \tag{2}$$

According to equation (1), when the voltage U of the AC source is kept unchanging, the electricity will be changed as the frequency ω changing. It is called resonance when the electricity reaches is maximum value. And we can get:

$$\omega L=\frac{1}{\omega C} \tag{3}$$

as the condition of resonance. The resonance frequency, denoted as f_0, is:

$$f_0=\omega_0/2\pi=\frac{1}{2\pi\sqrt{LC}} \tag{4}$$

We can also get some useful conclusions from the resonance of RLC series circuit:

(1) When it is resonance, the impedance of the circuit is pure resistance and gets the minimum value at the same time $Z=R$.

(2) When it is resonance, the frequency is $f=f_0$, and the electricity will be the maximum value $I=\frac{U}{R}$. If f deviates f_0, the electricity will be decreased, the more of the deviation the more I decreases.

(3) The ratio of inductive reactance or capacitive reactance with the resistance of the circuit is called the quality factor Q, which can be used for evaluation of a certain circuit:

$$Q=\frac{\omega_0 L}{R}=\frac{\frac{1}{\omega_0 C}}{R}=\frac{1}{R\omega_0 C}=\frac{1}{R}\sqrt{\frac{L}{C}} \tag{5}$$

(4) The resonating voltage of the capacitor or the inductance is Q times of the voltage of source:

$$U_C=U_L=QU \tag{6}$$

Usually $Q>>1$, thus, the resonance of series circuit is also called as voltage resonance.

(5) According to equation (4), the resonance frequency is only determined by the inductance and the capacitance of the circuit, and the resistance is related with the Q value. The resistance of the circuit can affect sharpness of the resonance curve.

We can get from equation (4) that:

$$L=\frac{1}{4\pi^2 f_0^2 C} \tag{7}$$

When the resonance frequency f_0 is measured, we can calculate the L.

The experiment circuit is shown in figure 11 - 2. The AC source is the signal generator and there is no resistance in series because it can be included in the inductance. Generally $Q >> 1$, the voltage of the capacitor nearly reaches its maximum value while the current reaches its maximum value and the circuit resonates. In this experiment, keep the output of the generator unchanging, and when the maximum voltage of the capacitor appears, the corresponding frequency of the generator is similar the same with the resonance frequency f_0.

Experiment

Join the circuit in figure 11 -2. The AC voltmeter is the 20V step of a multimeter.

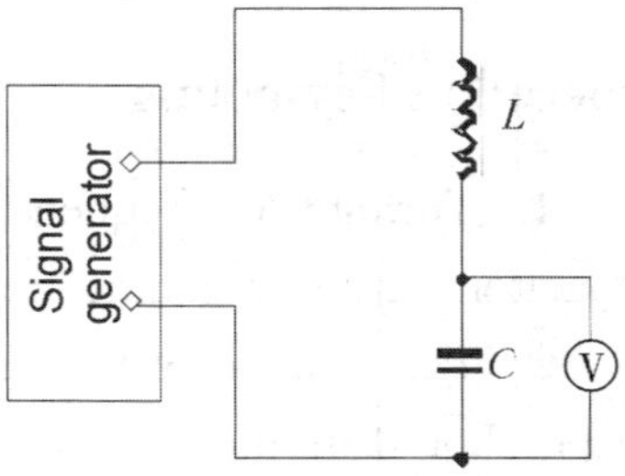

Figure 11 -2 Experiment circuit

Power on the signal generator; adjust the "frequency range" to 1 kHz (it could be switched to 100Hz if necessary).

Pay attention to the voltage changing of the capacitor when the frequency is changing, and find the frequency f_0 corresponding to the maximum voltage U_{C0}. Then, record 10 groups' data of frequency and voltage with the frequency larger and smaller than the f_0 either. For the place of nearing the resonance point, the frequency interval should be small. It is suggested that keeping the voltage output (measured by multimeter) to a certain value (such as 2V).

Draw the resonance curve with the horizontal axis of U_C and the vertical axis of f, and mark the resonance frequency f_0 on the curve.

Calculate the self - inductance L for known value of C and the measurement of f_0 according to equation (7).

Calculate the quality factor Q according to equation (6).

Notice

The signal generator cannot be powered off before the output having been turned to zero.

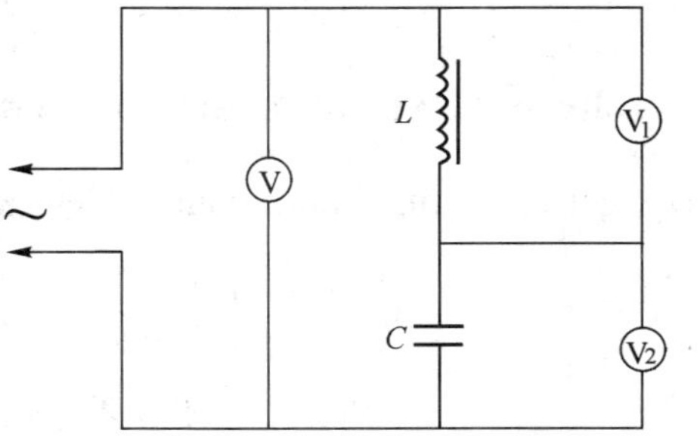

Figure 11 -3 Circuit of Problem 1

Problems

1. There are three voltmeters in the circuit of figure 11 -3, is it true for the following equation $U = U_1 + U_2$? And why?

2. How can we determine the self - inductance by a parallel circuit?

3. Can we determine the capacitance by the series circuit?

Experiment Instruction

Preparation requirements

1. What is series resonance, and what is the resonance condition?
2. How to get a resonance curve?

Operation key points

1. Turning the frequency carefully to find the maximum value of the voltage of the capacitor, in the meaning time, the voltage of signal generator should be the same.
2. The voltage of the measuring points near the resonance frequency changes fast, so, it should be measured carefully.

Problems after experiment

1. What are the reasons for the changing of the voltage output of the signal generator?
2. Can we measure the voltage of the inductance with a parallel voltmeter?
3. How to judge the resonance of the series circuit? Try to arrange the experiment circuit diagram and steps.

Correction the result if Q is small

In the *RLC* series circuit, according to equation (1), the effective value of voltage of the capacitor is shown as,

$$U_C = \frac{I}{\omega C} = \frac{1}{\omega C} \frac{U}{\sqrt{R^2 + (\omega L - \frac{1}{\omega C})^2}} \tag{8}$$

Keep the voltage output to a certain value, $(\frac{U}{U_C})^2$ will be its minimum value when U_C get to its maximum value, let ω_0' be the angular frequency at that time. Then

$$\frac{\mathrm{d}}{\mathrm{d}\omega}\left((\frac{U}{U_C})^2 \right)\Bigg|_{\omega=\omega_0'} = \frac{\mathrm{d}}{\mathrm{d}\omega}\left(\omega^2 C^2 \left[R^2 + \ (\omega L - \frac{1}{\omega C})^2 \right] \right)\Bigg|_{\omega=\omega_0'}$$

$$= 2\omega_0' C^2 \left[R^2 + 2\ (\omega_0' L)^2 - 2\frac{L}{C} \right] = 0$$

We can get,

$$\omega_0' = \sqrt{\frac{1}{LC} - \frac{R^2}{2L^2}} \tag{9}$$

Obviously ω_0' is different with ω_0 in equation (4). When U_C get to its maximum value U_{Cm}, we can get the quality factor $Q' = \frac{U_{Cm}}{U}$ according to the curve $U_C - \omega$,

$$Q' = \frac{1}{\omega_0' C \sqrt{R^2 + (\omega_0' L - \frac{1}{\omega_0' C})^2}} = \frac{1}{RC\sqrt{\frac{1}{LC} - \frac{R^2}{4L^2}}} \tag{10}$$

According to the equation (4) and equation (5), we know that ,

$$\omega_0 = \sqrt{\frac{1}{LC}}$$

$$Q = \frac{\omega_0 L}{R} = \frac{1}{\omega_0 RC} = \frac{1}{R}\sqrt{\frac{L}{C}}$$

According to the equation (9) and equation (10), we know that ,

$$\omega_0' = \omega_0 \sqrt{1 - \frac{1}{2Q^2}} \tag{11}$$

$$Q' = \frac{Q}{\sqrt{1 - \frac{1}{4Q^2}}} \tag{12}$$

From the equation (11) and equation (12) we know that, both $Q > \frac{\sqrt{2}}{2}$ and $Q' > 1$ must be satisfied if U_C can get its maximum value.

The relation of $\frac{U_C}{U} - \frac{\omega}{\omega_0}$ is shown in figure 11-4 when Q is 1, 2, 5.

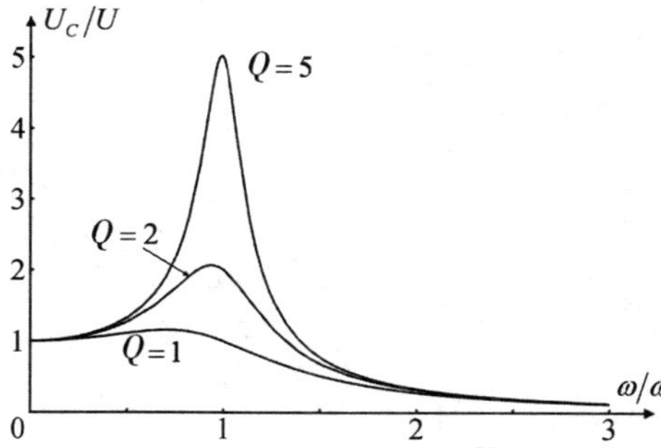

Figure 11-4 The relation of $\frac{U_C}{U} - \frac{\omega}{\omega_0}$

In this experiment, we must correct the equation (6) and equation (7) if Q is small. From the equation (11) and equation (12) we know that (the additional root of Q is given up when ($Q < Q'/\sqrt{2}$),

$$Q = Q'\sqrt{\frac{1 + \sqrt{1 - \frac{1}{(Q')^2}}}{2}} \tag{13}$$

$$\omega_0 = \frac{\omega_0'}{\sqrt{1 - \frac{1}{(Q')^2 + \sqrt{(Q')^4 - (Q')^2}}}} \tag{14}$$

In this experiment, let the resonance frequency $f_0' = \frac{\omega_0'}{2\pi}$ in the resonance curve, the ratio of the resonating voltage of the capacitor to the voltage output of the signal generator $Q' = \frac{U_{Cm}}{U}$. According to the equation (7) and equation (14), we know that,

$$L = \frac{K\ (Q')}{4\pi^2\ (f_0{}')^2 C} \tag{15}$$

Where, $K\ (Q')\ = 1 - \frac{1}{(Q')^2 + \sqrt{(Q')^4 - (Q')^2}}$. If Q is small, the equation (13) and equation (15) should be equations for Q and L according to resonance curve.

Comparing the equation (7) and equation (15), we can see that the experiment error should be big if Q is small (the relative error of L is about $\frac{1}{2Q^2}$). We can calculate L according to the equation (7) only if Q is big enough (the relative error of L is less than 2% when $Q > 5$).

实验十二　用模拟法测绘静电场

模拟法是物理学、工程学以及其他许多学科领域中常用的一种方法，它可以对那些在实验中难以进行直接测量或无法测量的物理量进行系统研究。用实验手段直接测量静电场有一定难度，本实验用稳恒电流场来模拟静电场，通过描绘稳恒电流场等位线，模拟静电场的等势线，再绘制静电场的电场线。

【目的要求】

1. 学习模拟法描述和研究电场的方法。
2. 加深对电场强度和电势概念及其相互关系的理解。

【仪器器材】

HLD－DZ－Ⅳ型静电场描绘实验仪，导线若干，圆规，三角板等。

描迹仪由电极架，同步探针，插件式电极和专用稳压电源（包括指示仪表）组成。

1. 描迹仪电极架

描迹仪电极架分上下两层：上层是载纸板，翻起活动翻板压入坐标纸，供模拟描迹时同步找点描迹。下层是载极板和探针座移动板，供各类电极更换时插入和同步探针座水平移动，侧面有输入电压的两个接线孔。

2. 同步探针

同步探针由装在探针座上的两根长短相同的弹性簧片和两根细而圆滑镀铬探针组成，同步探针座可沿载极板的右侧水平移动。下探针紧贴导电纸，用来探测导电纸中电流场所形成的各处电势数值，由于上下两探针始终保持在同一直线上，其运动轨迹一样，当下探针测出电势后，可用上探针在坐标纸上同步地打出相应等势点。同步探针竖杆上有一接线柱。

3. 描迹仪电源

描迹仪电源可提供13V连续可调的稳定电压。测量时，将“电表指示”开关拨向“内侧”，伏特计显示读数是稳压电源的输出电压值；将“电表指示”开关拨向“外侧”，伏特计显示读数是探针所测得的导电纸中电流场所形成的电压值。图12－1为HLD－DZ－Ⅳ型静电场描绘实验仪正面板示意图。

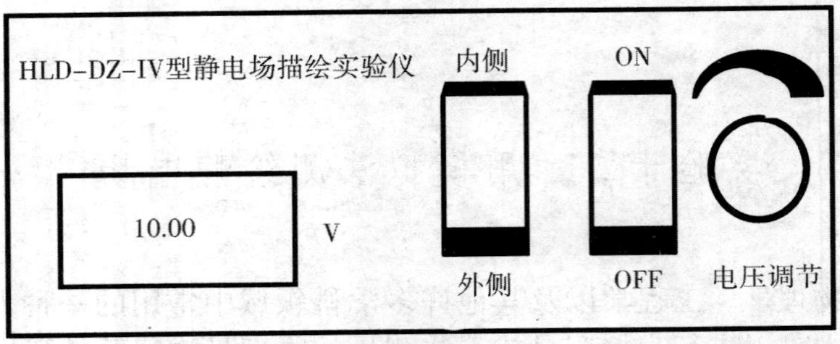

图 12 - 1 HLD - DZ - IV 型静电场描绘实验仪正面板示意图

4. 描迹仪电极板

描迹仪配有三种极板：

（1）同轴电缆电极。

（2）示波管聚焦电极。

（3）一对圆柱形正负电极。

【实验原理】

模拟可分为物理模拟，计算机模拟及半物理半计算机模拟等。所谓物理模拟就是用容易测量，观察以及修改参数方便的物理系统，来研究和它相似的各种系统。本实验所用模拟法属物理模拟。

在实验中，静电场是难以测量的物理量。但在一定条件下电解质中的稳恒电流场与静电场服从相类似的规律，且易于测量。因此，可用稳恒电流场来模拟静电场。

由电磁学理论可知，当空间不存在体分布自由电荷时，各向同性介质中电位移矢量 D 和电场强度 E 有如下关系；

$$D = \varepsilon E = \varepsilon_r \varepsilon_0 E$$

其中 ε_r 为相对电容率，$\varepsilon_0 = 8.85 \times 10^{-12} C^2 \cdot N^{-1} \cdot m^{-2}$。而在各向同性导电媒质中，稳恒电流场的电流密度 δ 和电场强度 E 有如下关系：

$$\delta = \sigma E$$

可见 D、E、ε 与 δ、E、σ 之间有一一对应关系，因此，可以用稳恒电流场来模拟静电场。由于静电场中导体是等势体，导体内部电场为零，这要求电极的电导率远大于电解质的电导率，这样才能使电流场中的电极是等势的。本实验中，用金属做电极，电解质选用导电纸。

1. 同轴圆柱形电极电场的模拟

一根同轴圆柱面，横断面如图 12 - 2，a 为中心电极，b 为同轴外电极，将它们放于导电纸上，在 a、b 电极之间加电压 U_0（内电极 a 接正，外电极 b 接负）。由于电极是对称的，电流均匀地沿径向从内电极流向外电极。

两个电极之间的电流场所形成的同心圆等势面，可以模拟一个无限长均匀带电同轴圆柱面（如同轴电缆）所形成的等势面，如图 12－3。

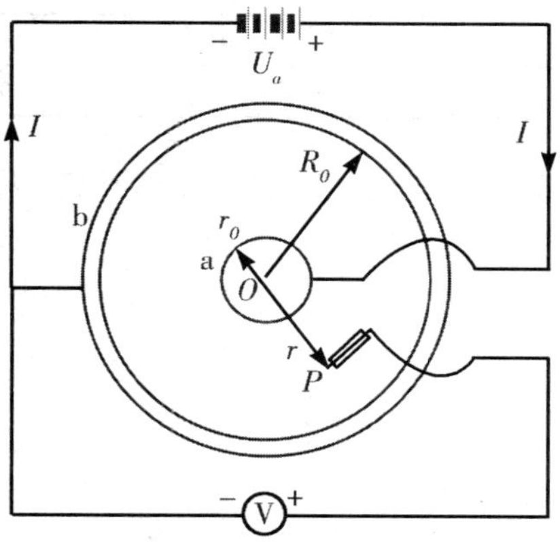

图 12－2　同轴电极的电场模拟电路

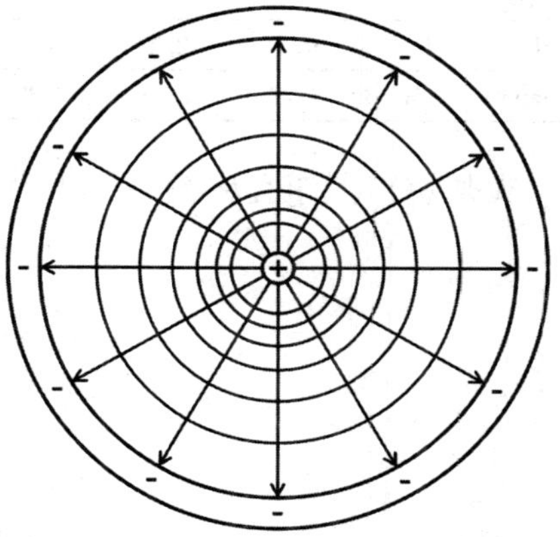

图 12－3　同轴电极的等势面

我们知道，无限长均匀带电的同轴导体之间的电场强度为：

$$E=\frac{\mathrm{d}U}{\mathrm{d}r}=-\frac{\lambda}{2\pi\varepsilon r} \tag{1}$$

式中　λ——导体上电荷的线密度；

ε——电介质的电容率；

r——两导体之间任意一点的半径。

设内电极半径为 r_0，电势为 U_a，外电极半径为 R_0，电势为 U_b，则距轴心为 r 的 P 点的电势为：

$$U=\int_r^{r_0}E\mathrm{d}r+U_a=U_a-\int_{r_0}^{r}E\mathrm{d}r$$

$$U=U_a-\int_{r_0}^{r}\frac{\lambda}{2\pi\varepsilon r}\mathrm{d}r$$

$$U=U_a-\frac{\lambda}{2\pi\varepsilon}\ln\left(\frac{r}{r_0}\right) \tag{2}$$

在 $r=R_0$处应有：

$$U_b=U_a-\frac{\lambda}{2\pi\varepsilon}\ln\left(\frac{R_0}{r_0}\right) \tag{3}$$

如取 $U_b=0$，$U_a=U_0$，将式（3）代入式（2），得：

$$U=U_a\frac{\ln\left(\frac{R_0}{r}\right)}{\ln\left(\frac{R_0}{r_0}\right)} \tag{4}$$

由于 R_0、r_0、U_0均为已知，只要知道 P 点的 r 值，即可从上式求得该点的电势 U_0。将式（4）作适当变换，即得：

$$r=R_0\left(\frac{r_0}{R_0}\right)^{U/U_0} \tag{5}$$

若取 $R_0=7.50\text{cm}$，$r_0=0.50\text{cm}$，r 值随 U/U_0 的变化如表 12－1。

表 12－1　不同的 U/U_0 值对应的 r 值

U/U_0	0	0.1	0.2	0.3	0.4	0.5	0.6	0.7	0.8	0.9	1.0
r（cm）	7.50	5.72	4.36	3.33	2.54	1.94	1.48	1.13	0.86	0.66	0.50

从式（5）我们可以很容易找出两极间具有某一电势值 U 的等势面所在的半径（图 12－4）。

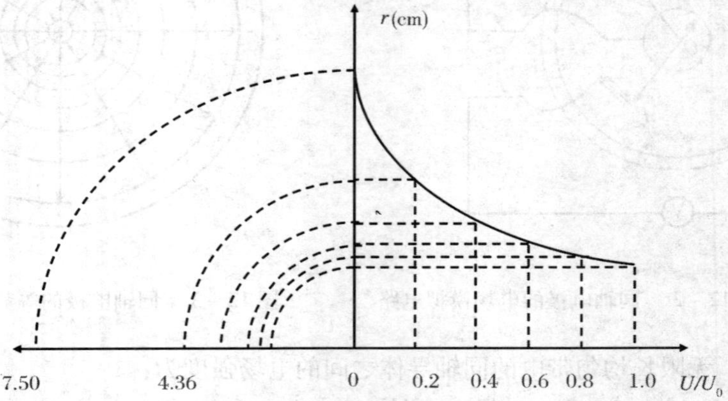

图 12－4　$r-U/U_0$ 关系曲线

2．示波管的聚焦电场

从示波管中，从阴极发射出来的电子经过栅极和第一阳极内中心孔变成电子束射向荧光屏。由于从小孔出来的电子方向各不相同，到达荧光屏时会散成一个较大的斑点。为了使电子束能在荧光屏上形成清晰的亮点，必须将电子流聚焦成细束。聚焦的办法有静电式和磁式两种。

静电聚焦的原理如图 12－5 如示。在电子束的前进途中，放置两个相邻的同轴圆筒，在两筒上加上不同电势，使之形成非均匀电场。当电子束进入非均匀电场时，因受电场力作用而使电子的运动轨道发生偏转，以致散开的电子可能会聚成一个细小的电子束，打到荧光屏上形成细小清晰的亮点，恰如光线穿过透镜以后的聚焦效果。因此，我们把能对电子束有聚焦作用的非均匀静电场称为聚焦电场或静电透镜。

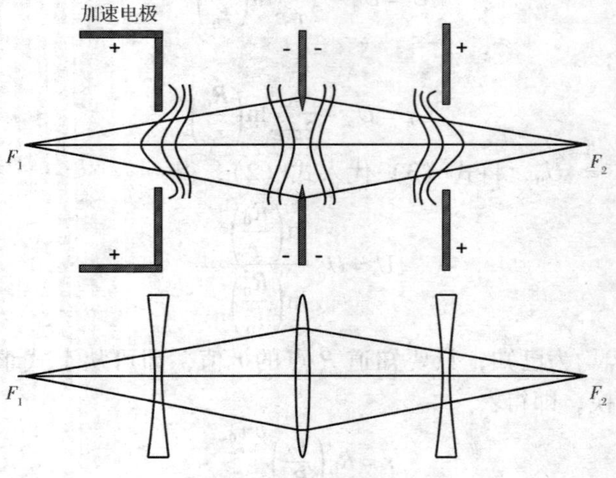

图 12－5　静电聚焦的原理

静电聚焦是通过改变电子枪中第一阳极的电势来实现的。从图 12－5 可以看出，电场主要分布在相邻两个电极之间；控制栅极与第一阳极之间；第一阳极与第二阳极之间。等势面向两极内部凸出。随着极间电压的改变，等势面的间距和弯曲亦随之改变。当等势面的曲率合适时（也就是极间电势差合适时），穿过极间的电子将在荧光屏上汇聚成细小圆点。

静电场的聚焦功能可以采有上述模拟描述法加以观测，即通过改变模拟电极间的电势差可以描述一系列不同曲率的等势线，如图 12－6 所示。模拟电极如图 12－7 所示。

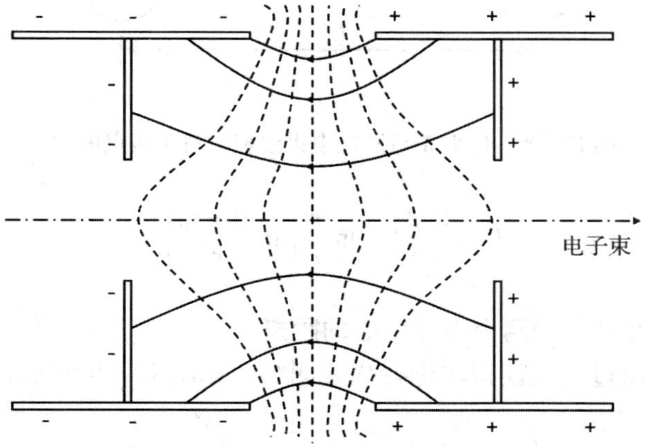

图 12－6　聚焦电场的等势面

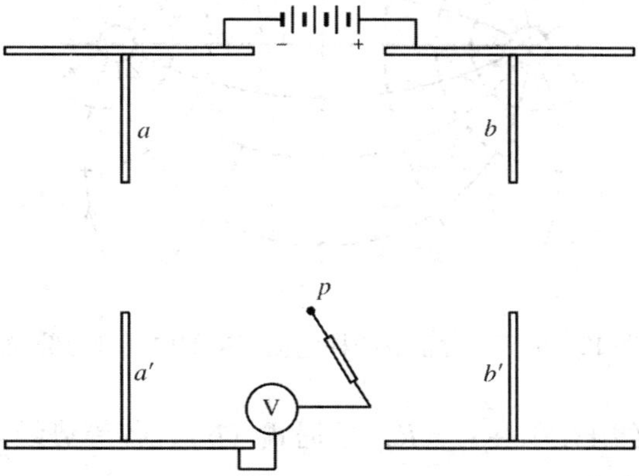

图 12－7　聚焦电场的模拟电极

3. 两个圆柱形电极模拟一对正负点电荷的电场

如图 12－8 是两个圆柱形电极置于导电纸上的情形。两电极间的电势差为 U_0，两电极的半径为 r_0，两电极中心间距为 R。则通过近似计算可以得到平面上 P 点的电势 U 为（以负电极处为电势零点）。

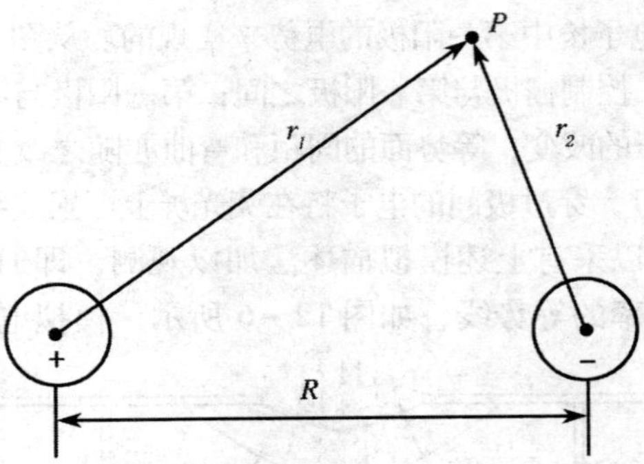

图 12－8　一对正负圆柱电极模拟一对正负点电荷

$$U=\frac{U_0}{2}\left(1+\ln\frac{r_2}{r_1}/\ln\frac{R-r_0}{r_0}\right) \tag{6}$$

式中 r_1，r_2分别为 P 点到两个电极中心的距离。

这一电场也可以通过上述模拟法进行描述测绘，如图 12－9 所示，

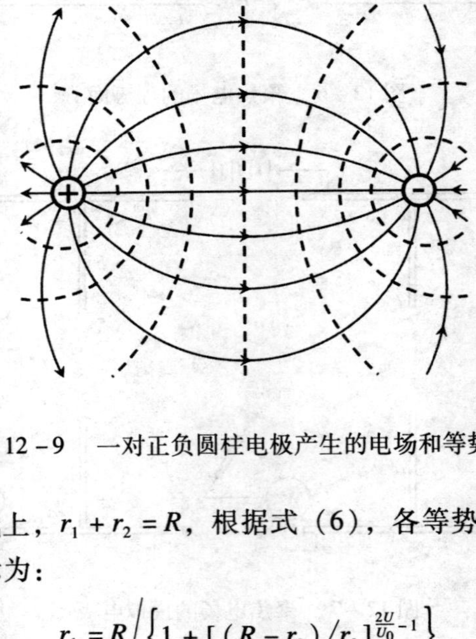

图 12－9　一对正负圆柱电极产生的电场和等势面

在两电极的连线上，$r_1+r_2=R$，根据式（6），各等势线与电极连线的交点到正电极中心的距离 r_1可表示为：

$$r_1=R\Big/\left\{1+[(R-r_0)/r_0]^{\frac{2U}{U_0}-1}\right\} \tag{7}$$

若取 $R=11.00\text{cm}$，$r_0=0.50\text{cm}$，r_1的值随 U/U_0的变化如表 12－2。

表 12－2　不同的 U/U_0值对应的 r_1值

U/U_0	0	0.1	0.2	0.3	0.4	0.5	0.6	0.7	0.8	0.9	1.0
r_1（cm）	10.50	10.11	9.48	8.49	7.12	5.50	3.88	2.51	1.52	0.89	0.50

【实验内容】

1．同轴圆柱形电极间电场的模拟

（1）按图 12－2 接线

（2）极间电压为 10V，分别找出电势为 1.0、2.0、3.0、4.0、5.0、6.0 伏等势线记录点，画出相应的等势线和电力线。

2．示波管的聚焦静电场的模拟

（1）更换极板。

（2）极间电压为 10V，分别找出 2.0V、3.0V、4.0V、5.0V、6.0V、7.0V、8.0V 的等势点，画出相应的等势线和电力线。

3．一对正负点电荷的电场的模拟

（1）更换极板。

（2）极间电压为 10V，分别找出 2.0V、3.0V、4.0V、5.0V、6.0V、7.0V、8.0V 的等势点，画出相应的等势线和电力线。

（3）通过式（6）计算上述各点在两电极连线上的理论值与实验值进行比较。本实验中 $R=11.00\text{cm}$，$r_0=0.50\text{cm}$。

【注意事项】

1. 下探针应始终和导电纸保持良好的接触，且下探针不宜太尖，移动要轻，防止划破导电纸。
2. 用上探针打点时，应注意观察伏特计电势的变化，及时调整到一定的电压。

【思考题】

1. 极间电压的正负交换一下，所作等势线会有变化吗？
2. 此模拟实验能否能交流电源供电，并用交流伏特计来测量？
3. 试找出同轴电极模拟实验产生误差的原因。

实　验　指　导

【预习要求】

1. 熟悉实验原理，明确 D、E、ε 与 δ、E、σ 之间的一一对应关系。
2. 熟悉仪器各部分的作用和使用方法。
3. 简要列出实验操作步骤。

【操作要点】

1. 调节电源输出电压即极间电压时，“电表指示”开关应指向“内侧”。
2. 测量导电纸上各点电势时，“电表指示”开关应指向“外侧”。
3. 实验中，同一等势线上的测量点应根据等势线的具体情况适当增加或减少，且分布均匀。

4. 模拟示波管聚焦电场时，应先描绘 5.0V 等势线，然后描绘 4.0、3.0、2.0、6.0、7.0、8.0V 等势线。

【数据处理】

1. 同轴圆柱形电极电场模拟的实验结果用作图法处理数据

（1）找出所有等势线的共同圆心，再用圆规画出一系列相应于各电势的同心圆，注意各等势线的测量点在相应同心圆的内外分布应均匀。然后对同轴电极进行理论值和测量值的比较，算出误差。

（2）作图 $\ln r - U$

$$\ln r = \frac{U}{U_0}\ln\frac{r_0}{R_0} + \ln R_0 = AU + B$$

观察实验数据与拟合直线的吻合关系。

2. 示波管的聚焦静电场的模拟时，在纸上画出相应的等势线和电力线

3. 一对正负点电荷的电场的模拟时

（1）对一定的 U 值，计算电极连线上的点的位置时，有关系 $r_1 + r_2 = R$，因此 U 的表达式为：

$$U(r_1) = \frac{U_0}{2}\left(1 + \ln\frac{R - r_1}{r_1} / \ln\frac{R - r_0}{r_0}\right)$$

对上式化简后可方便地求出各 U 值对应的 r_1 值，并仿照柱形电极与实验值比较。

表 12－3　不同的 U 值对应的 r 值

$U_{实验值}$（V）
U/U_0
$r_{实验}$（cm）
$r_{理论}$（cm）
$\Delta r = r_{理论} - r_{实验}$（cm）
$\frac{\|\Delta r\|}{r_{理}} \times 100\%$

（2）作图 $\ln\left(\frac{R}{r_1} - 1\right) - U$，

$$\ln\left(\frac{R}{r_1} - 1\right) = \frac{2U}{U_0}\ln\frac{R - r_0}{r_0} - \ln\frac{R - r_0}{r_0} = AU + B$$

观察实验数据与拟合直线的吻合关系。

【实验后的思考】

1. 为什么在实验中一定要保证电极的电导率远大于导电纸的电导率？
2. 在导电纸上能否模拟点电荷激发的电场？

12 Analogously Surveying and Mapping Electrostatic Field

Analogous method is commonly used in the area of physics, engineering, and other science. It is a systematic way of studying those quantities which are unable or difficult to measure directly. There are some ways of analogous, such as physical analogous, computer analogous, computer – physics analogous. It is hard to measure the electrostatic field by actual experiment methods. In this experiment we simulate the electrostatic field by steady electricity field. We draw a series of equipotent curves with different curvature in steady electricity field to simulate equipotent curves in the electrostatic field first. Then we can get the electric field line in the electrostatic field.

Objectives

Learning the method of analogously surveying and mapping electrostatic field.

Reinforcing the comprehension of the concepts of electric field and electric potential and their relationship.

Equipments

Electric field mapping instrument, type HLD – DZ – IV; conducting wires; compasses; triangle.

The electrostatic field mapping device is consisted of a frame, an in – step needles, plug – in electric poles, special electric source power (including meters).

1. The frame

There are two levels on the frame. The upper level fixing with a plot paper can be dotted by the needle. The lower level can place the board of different electric pole and on the board the in – step needles can be moved around. There are two bonding holes in the side of the frame to put in voltage.

2. In – step needles

The in – step needles have two spring flakes which are fixed on a steel rod with a seat. There are two chrome – platting needles at the ends of the two flakes. The lower needle presses on the electric conduction paper and it can detect the electric voltage by the meter connecting to it. Keeping in a straight line, the upper needle can dot different points corresponding to its voltage on the plot paper. There is a post head on the in – step needles.

3. Electric source

The electric source can provide an output voltage about 13V. When measuring, turn the "Meter Indication" switches to the "inner" position, the meter will show the voltage of its output, the "outer" position will show the voltage of the detecting needle.

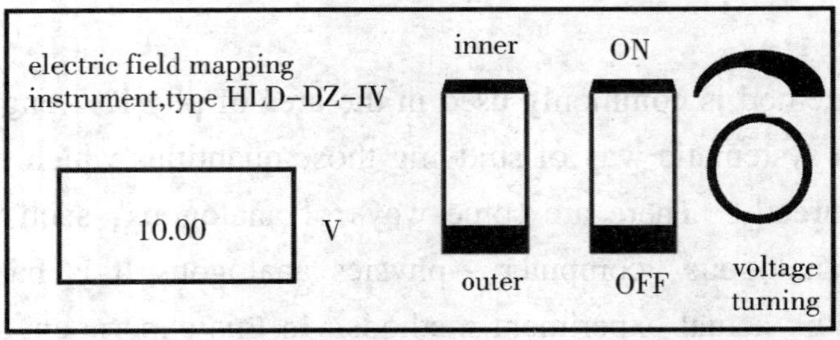

Figure 12 – 1 Electric field mapping instrument, type HLD – DZ – IV

4. The pole board

There are three kinds of pole board:

Concentric cable poles;

Oscilloscope focusing poles;

Poles of a pair of point charges.

Principle

Physical analogous is a method studying a physical system which is easy to be measured and modified, and getting the information of a similar system, which is the goal system. We analogously survey and map electrostatic field in this experiment.

Electrostatic field is a difficult quantity to be measured. But the rule of steady electricity field in the electrolyte is similar with the rule of electrostatic field, and it is much easier to be measured. Thus, we simulate the electrostatic field by steady electricity field.

According to the electromagnetic theory, when there is no free charge, in the isotropy dielectric there is a relationship between the electric displacement D and the electric – field intensity E:

$$D = \varepsilon E = \varepsilon_r \varepsilon_0 E$$

Where the ε_r is the relative permittivity, $\varepsilon_0 = 8.85 \times 10^{-12} C^2 \cdot N^{-1} \cdot m^2$. In the same way, the relationship between electric current density δ in the isotropy electrolyte and the electric – field intensity E is:

$$\delta = \sigma E$$

Here, the D, E, ε is corresponding with δ, E, σ. So we can simulate the electrostatic field by steady electricity field. Because the conductor in electrostatic field is a equipotent body and there is no electric field in the conductor, the conductivity of the pole must be much more great than the conductivity of the electrolyte, so as to keep the pole a equipotent. In this experiment, the pole is made of metal and the select conduction paper as electrolyte.

1. Simulating the electric – field of concentric cable poles

The cross section of a concentric cable is shown in figure 12 – 2, where *a* is the central pole and *b* is the concentric outer pole, *a* and *b* are all placed on the surface of conduction paper. There is a voltage difference *U* between the poles *a* and *b* (*a* is positive and *b* is negative). Electric current will flow along radial direction from the inner pole to the outer pole for the poles are Centro symmetry. The concentricity equipotent circles formed by the electricity field between the poles are similar to the equipotent circles formed by unlimited long concentric cable carrying charge, see figure 12 – 3.

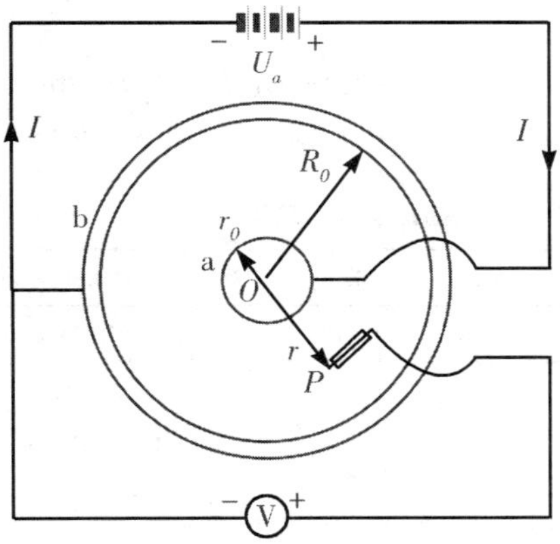

Figure 12 – 2 The circuit of simulating the electric – field of concentric cable poles

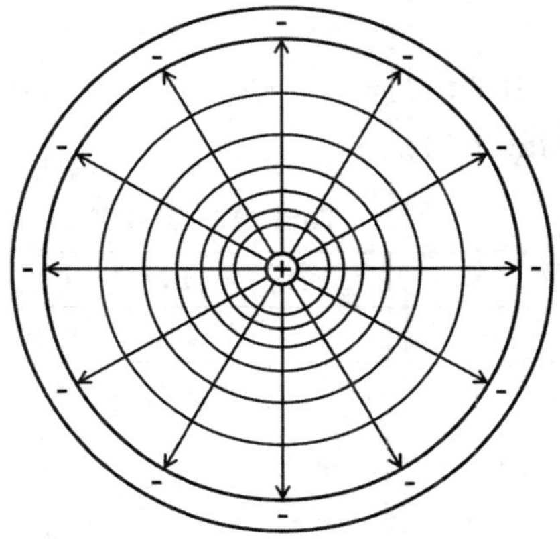

Figure 12 – 3 The concentricity equipotent circles of concentric cable poles

The electric – field intensity between the infinity long conductors of concentric cylinders carrying charge is:

$$E = \frac{dU}{dr} = -\frac{\lambda}{2\pi\varepsilon r} \qquad (1)$$

Where λ is the charge density in the line of cylinder; ε is the permittivity of dielectric; *r* is radius from a point to the center between the two conductors.

Say, the radius of inner pole is r_0 and its electric potential is U_a, the radius of outer pole is R_0 and its electric potential is U_b, then the potential of point P where from the center is *r* is:

$$U = \int_r^{r_0} E\mathrm{d}r + U_a = U_a - \int_{r_0}^{r} E\mathrm{d}r$$

$$U = U_a - \int_{r_0}^{r} \frac{\lambda}{2\pi\varepsilon r}\mathrm{d}r$$

$$U = U_a - \frac{\lambda}{2\pi\varepsilon}\ln\left(\frac{r}{r_0}\right) \tag{2}$$

When $r = R_0$:

$$U_b = U_a - \frac{\lambda}{2\pi\varepsilon}\ln\left(\frac{R_0}{r_0}\right) \tag{3}$$

Let $U_b = 0$, $U_a = U_0$, substitute (2) with (3), we get:

$$U = U_a \frac{\ln\left(\frac{R_0}{r}\right)}{\ln\left(\frac{R_0}{r_0}\right)} \tag{4}$$

Because R_0, r_0 and U_0 are all known quantities, we can get the potential U of point P for a certain value of r. Now transfer (4) into:

$$r = R_0\left(\frac{r_0}{R_0}\right)^{U/U_0} \tag{5}$$

If let $R_0 = 7.50\text{cm}$, $r_0 = 0.50\text{cm}$, r Changes with the value of U/U_0 as the following table.

Table 12 - 1 Different values r for different values of U/U_0

U/U_0	0	0.1	0.2	0.3	0.4	0.5	0.6	0.7	0.8	0.9	1.0
r (cm)	7.50	5.72	4.36	3.33	2.54	1.94	1.48	1.13	0.86	0.66	0.50

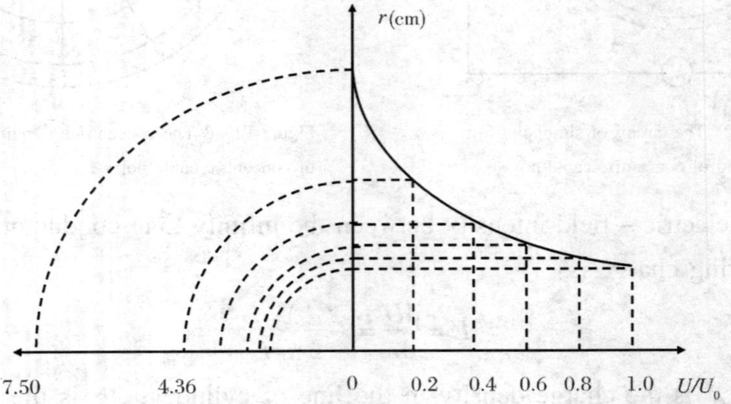

Figure 12 - 4 $r - U/U_0$

According to (5), we can easily get the radius r between the two poles with a certain electric potential U (figure 12 - 4).

2. Simulating electric – field of oscilloscope focusing poles

In the oscilloscope tube the electrons emits from cathode will become a beam when passing through grid and central hole of first anode, and go straightly against the fluorescence screen. The beam will make a macula on the screen because the electrons coming out from the hole having different directions. To get a clear macula on the screen we must focus the electrons into a fine beam. There are two focusing methods: electrostatic and magnetic.

The principle of electrostatic focusing is shown in figure 12 –5. Place two concentric cylinders on the path of electron beam and let the two cylinders have different electric voltage, forming a nonuniform electric field. When the electrons pass through the nonuniform field, they will change their moving direction because of the electric force acting on them, and finally get into a fine, clear point on the screen. This effect is similar with the lens on the light. So that we call this kind of focusing nonuniform electric field focusing electric – field or electrostatic lens.

The electrostatic focusing can work because the suitable electric potential of the first anode. Figure 12 – 5 shows that most part of electric field distributes within the space between two poles: grid and first anode; first anode and second anode. The equipotent surface bulges to the inner side of every pair of the poles; in different place the equipotent surfaces have different separation and curvature. Under the condition of suitable potential difference, electrons passed through the poles will focus into a fine point on the screen.

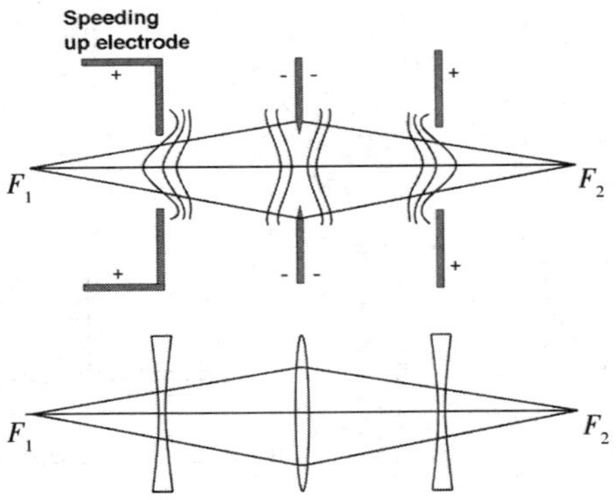

Figure 12 –5 The principle of electrostatic focusing

The electrostatic focusing field can be surveyed and mapped analogously and get a series of equipotent curves with different curvature, see figure 12 – 6. The shape of poles is shown in figure 12 –7.

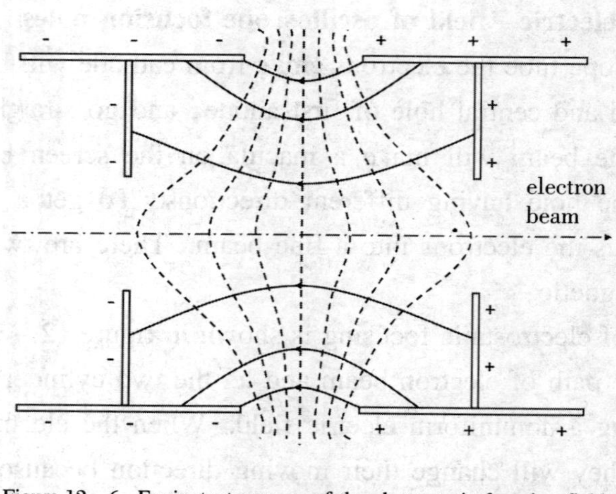

Figure 12 – 6 Equipotent curves of the electrostatic focusing field

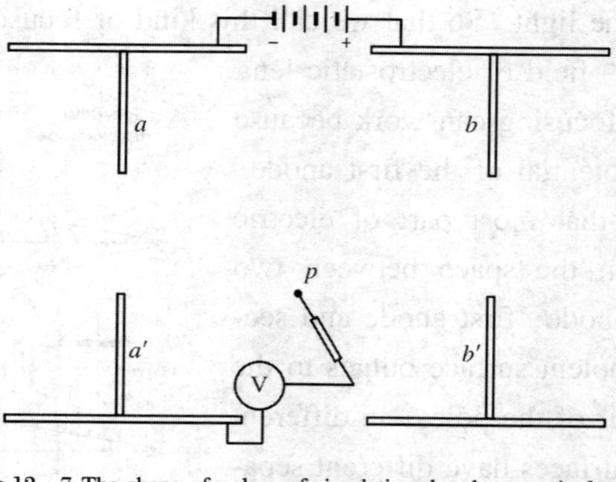

Figure 12 – 7 The shape of poles of simulating the electrostatic focusing

3. Simulating a pair of charge by two cylinder poles

The two cylinders are placed on the conduction paper, shown in Figure 12 – 8. The potential difference between the two cylinders is U_0, the two poles have a same radius r_0, and separation between the two poles is R_0. P is a point on the plane of the conduction paper, where has an electric potential U (the negative pole is the place of zero electric potential). We get:

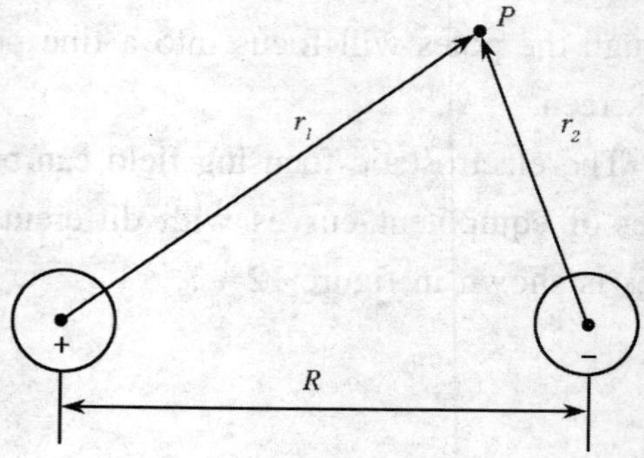

Figure 12 – 8 The two cylinders poles

$$U = \frac{U_0}{2}\left(1 + \ln\frac{r_2}{r_1}\Big/\ln\frac{R - r_0}{r_0}\right) \qquad (6)$$

Where, r_1, r_2 is the distance from the point P to the poles correspondingly.

This electric field also can be surveyed and mapped analogously, see Figure 12 -9.

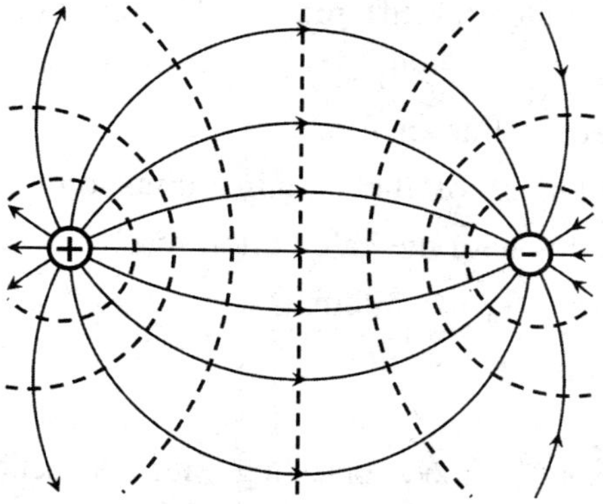

Figure 12 -9 The Equipotent curves of the two cylinders poles

In the line connecting the two poles, $r_1 + r_2 = R$, see equation (6), the distance r_1 from the positive pole center to the very chosen voltage point can be written as:

$$r_1 = R \Big/ \left\{1 + [(R - r_0)/r_0]^{\frac{2U}{U_0}} - 1\right\} \tag{7}$$

If $R = 11.00\text{cm}$, $r_0 = 0.50\text{cm}$, we can get the relation between $U/U_0 - r_1$ as following table.

Table 12 -2 Different valnes r_1 for different values of U/U_0

U/U_0	0	0.1	0.2	0.3	0.4	0.5	0.6	0.7	0.8	0.9	1.0
r_1 (cm)	10.50	10.11	9.48	8.49	7.12	5.50	3.88	2.51	1.52	0.89	0.50

Experiment

1. Simulating the electric field between concentric cylinders

(1) According to figure 12 -2 arrange the devices and connect the wires.

(2) Keep the electric potential difference 10V, find out the points of equipotent curve with voltage 1, 2, 3, 4, 5, 6 separately and dot with carefully. Then draw the equipotent curve and electric line of force.

2. Simulating the focusing electrostatic field

(1) Switch the pole board, connect the wires correctly.

(2) Keep the electric potential difference 10V, find out the points of equipotent curve with voltage 2, 3, 4, 5, 6, 7, 8 separately and dot with carefully. Then draw the equipotent curve and electric line of force.

3. Electric field of a pair of charge

(1) Switch the pole board, connect the wires correctly.

(2) Keep the electric potential difference 10V, find out the points of equipotential curve with voltage 2, 3, 4, 5, 6, 7, 8 separately and dot with carefully. Then draw the equipotential curve and electric line of force.

(3) Measure the distances of the certain points on the linking line of the two poles, and compare the values with the calculation results of equation (6). In this Experiment, the $R = 11.00\text{cm}$, $r_0 = 0.50\text{cm}$.

Notice

Always keep lower needle in good contacting with conduction paper. The lower needle should not be too sharp to tear conduction paper, and move it carefully and smoothly.

When dot, pay attention to any change of the voltage indicating by the meter and adjust it in time.

Problems

1. Will there be any change happen to the equipotential curves when switch the polarities of the poles?

2. Can this simulating experiment be done with a alternate current power source while measuring by a alternate meter?

3. Try to find the cause of errors in the simulating of concentric cable poles.

Additional instruction

Preparation Requirements

1. Understand the principle and the relationship between D, E, ε and δ, E, σ.
2. Be familiar with every part of the experiment and their usage.
3. Write down the experiment steps.

Operation Key points

1. When adjust the output voltage or the electric potential difference between the poles, "Meter indicates" switch should be on the side of "inner".

2. When the electric potential difference between needle and the negative pole is being detected, "Meter Indicates" switch should be on the side of "outer".

3. The dotted points on different curve should be different so that base on these can make a suitable curve.

4. When dot on the oscilloscope focusing field, it is better to dot out the curve of

5V, then the curves of 4, 3, 2, 6, 7, 8V.

Data processing

1. Data from the graph for the concentric cylinder poles' field

(1) We should find the center of the equipotent circles by compasses. The dots of a circle should distribute uniformly around the circle. Then, measure the radiuses and compare them with the theoretical values.

(2) Draw a graph with the relation $\ln r - U$.

$$\ln r = \frac{U}{U_0}\ln\frac{r_0}{R_0} + \ln R_0 = AU + B$$

We can check the corresponding relation between the experiment value with the line we draw.

2. Simulate the focusing electrostatic field. Draw the equipotential curve and electric line of force

3. For the pair of cylinders

(1) When calculate the position of a point on the two poles linking line for certain value of U, there is a relation: $r_1 + r_2 = R$. And then (6) can be written as:

$$U(r_1) = \frac{U_0}{2}\left(1 + \ln\frac{R - r_1}{r_1} \Big/ \ln\frac{R - r_0}{r_0}\right)$$

Table 12-3 Different values *r* for different values of *U*

U_e (V)	
U/U_0	
r_e (cm)	
r_t (cm)	
$\Delta r = r_t - r_e$ (cm)	
$\frac{\lvert\Delta r\rvert}{r_t} \times 100\%$	

According to the above equation we can get the value of U corresponding to r_1 one by one, and compare them with theoretical values.

(2) Draw a graph with the relation $\ln\left(\frac{R}{r_1} - 1\right) - U$,

$$\ln\left(\frac{R}{r_1} - 1\right) = \frac{2U}{U_0}\ln\frac{R - r_0}{r_0} - \ln\frac{R - r_0}{r_0} = AU + B$$

We can check the corresponding relation between the experiment value with the line we draw.

Problems after experiment

1. Why the conductivity of the pole must be much greater than the conductivity of the electrolyte in this experiment?

2. Can you simulate the electric field of a single charge in the select conduction paper?

实验十三　示波器的应用

示波器是一种能够形象地显示信号幅度随时间变化的仪器，它能把电信号的电压变化转换成直观的图形，用它既能直接观察电信号的波形，也能测定电信号的幅度、周期和频率等参数，还可以测量两个信号之间的相位差。借助于传感器，凡是电学量（如电流、电功率）和能转化为电信号的非电学量（如温度、压力、光强等）都可用示波器来观测。示波器是一种基本的电子测量仪器，有模拟式、数字式等种类，它在科学研究、工程实验、电子电工、仪器仪表等领域有着广泛的应用。本实验主要是学习使用示波器观察各种信号波形和测量正弦信号的电压、频率及相位差。

【实验目的】

1. 了解示波器的工作原理；
2. 掌握示波器面板上各控制开关与旋钮的作用，学会使用示波器；
3. 学会利用示波器观察电信号的波形，测量电压、频率；
4. 学会利用示波器测量交流电的相位差。

【实验仪器】

示波器（CS－4125 或 CS－4025 型）、DF1636A 信号发生器、220V/6V 变压器、电阻箱和电容器。

【实验仪器介绍】

1. 示波器的基本原理（以单踪示波器为例）

（1）基本结构　示波器主要由主机部分、X 轴系统、Y 轴系统三大部分组成。主机部分由示波管、高压电源、低电压源和校准信号发生器组成；X 轴系统由扫描信号发生器、X 轴放大器（水平偏向放大器）和增辉电路组成；Y 轴系统由输入耦合开关、衰减器和 Y 轴放大器（垂直偏向放大器）组成。各个组成部分之间相互联系，如图 13－1 所示。

（2）示波管　示波管是示波器的核心，如图 13－2 所示。它包括电子枪、偏转系统及荧光屏三部分。示波管的工作原理是：从热阴极发出的电子经过一连串的加速和聚焦电极（也称电子枪），形成一个窄而细的电子束。该电子束在高真空玻璃管中作高速运动，并打在荧光屏上，屏幕内表面覆盖有受到电子束打击能够发出荧光的物质，这样在荧光屏的中央就可形成一发光的亮点。该光点的亮度取决于电子束中电子数量的多少和速度的大小，光点的大小则取决于电子束的粗细。光点的亮度由示波器面板上的“辉度（INTENSITY）”旋钮来调节，光点的大小则由面板上的“聚焦（FOCUS）”旋钮来调节。

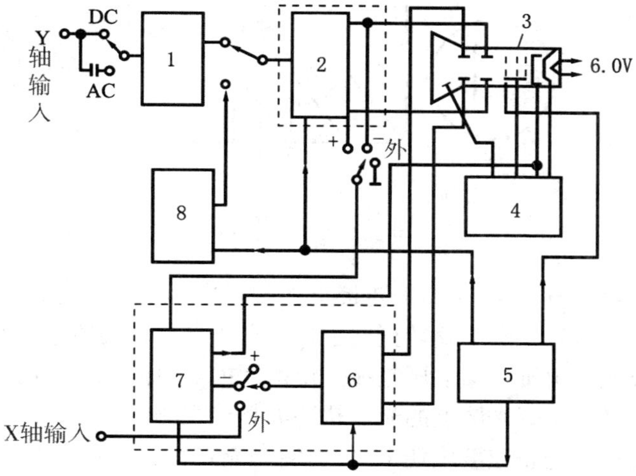

图 13－1　示波器的内部基本结构

1. 衰减器；2. Y 轴放大器；3. 示波管；4. 高压电源；5. 低压电源；6. X 轴放大器；7. 扫描信号发生器；8. 校准信号发生器

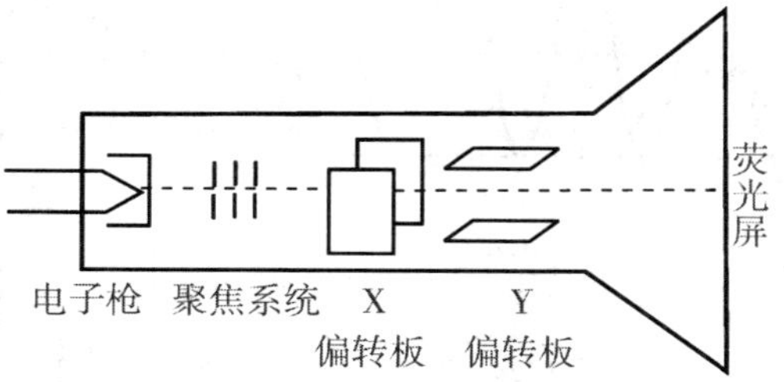

图 13－2　示波管的结构示意图

示波管内放置两对互相垂直的偏转板，当在两对偏转板上加上电压时，穿过其间的电子束在电场作用下就会同时在水平方向和垂直方向上发生偏转。电子束的偏转程度正比于加在偏转板上的电压。因此，当水平和垂直偏转板加不同的电压时，荧光屏上的光点可以到达屏幕上任何位置，屏上的光点位置就成了一个电压指示器。

（3）扫描原理　所谓扫描原理，即示波器显示波形的原理。示波器通常主要用来观测周期性电压信号的波形。如果仅把一个周期性的交变信号如正弦电压 $U_y = U_0 \sin\omega t$ 加到 Y 偏转板上而 X 偏转板不加任何电压信号，那么荧光屏上的光点只是作上下方向的正弦振动，当振动的频率较快时，在荧光屏上仅能看到一条竖直的亮线，如图 13－3（a）所示。

如果要在荧光屏上显示出上述正弦电压信号的波形，那么需要将光点沿 X 轴方向展开，所以必须同时在 X 轴偏转板上加上随时间作线性变化的电压，称扫描电压。扫描电压随时间变化的关系如同锯齿，故也称锯齿波电压，如图 13－3（b）所示。该电压的作用是使得光点在屏幕上做水平匀速运动。扫描电压是由示波器内扫描信号发生器（锯齿波发生器）产生的。如果单独把扫描电压加在 X 偏转板上而 Y 偏转板不加电压信号，若加快扫描电压的频率，那么也只能看到一条水平的亮线，如图 13－3（c）所示。

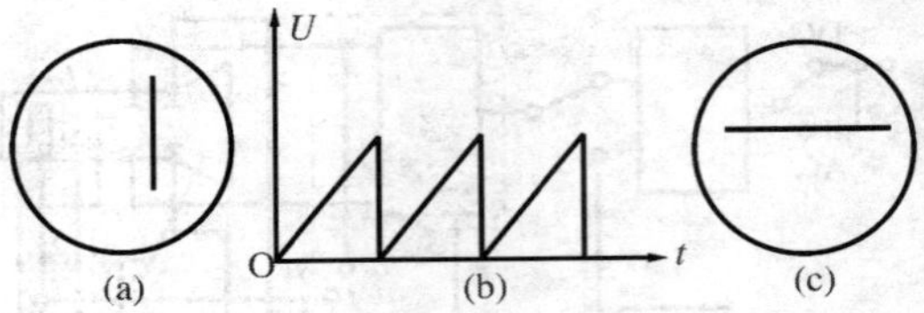

图 13－3　扫描原理

在 Y 偏转板上的电压信号与 X 偏转板上的扫描电压的同时作用下，电子束的垂直偏转正比于待观测电压信号，水平偏转正比于扫描电压，其运动轨迹是水平方向和垂直方向所产生的运动的合成。由于扫描电压是在一定范围内与时间成正比的线性电压，因此水平偏转也正比于扫描时间，荧光屏上的光点轨迹即是待观测电压信号随时间变化的图形。图 13－4 表示扫描电压 U_x 与正弦波电压 U_y 分别使光点在水平方向和垂直方向产生的位移的合成情况。

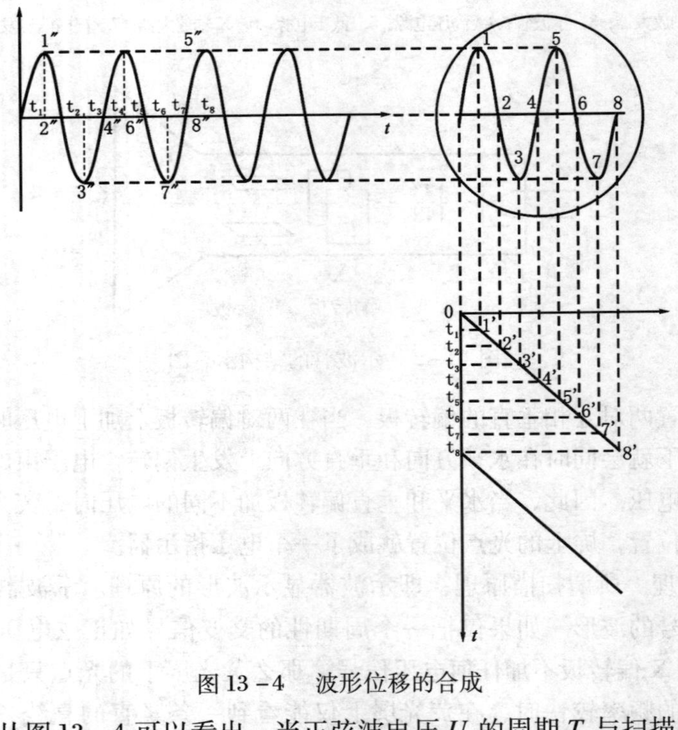

图 13－4　波形位移的合成

（4）同步　从图 13－4 可以看出，当正弦波电压 U_y 的周期 T_y 与扫描电压 U_x 的周期 T_x 完全相等时，正弦波电压变化一个周期，光点正好扫描一次，故电子束在每一个扫描周期里能在屏幕上描绘出一个周期的正弦电压波形。只有当每次扫描的图形完全重合时，屏幕上的波形才清晰、稳定。换言之，只有当扫描电压周期为待测正弦波电压周期的整数倍时，即：

$$T_x : T_y = n \text{ 或 } f_y : f_x = n \qquad n = 1, 2, 3\cdots$$

屏幕上才能显示出 n 个完整的待测正弦电压波形。如果 T_x 与 T_y 之间不存在上述整数倍关系，那么每次扫描显示的波形不能完全重合，便会出现波形的移动和重叠现象，因此无法在屏上观察到清晰、稳定的信号波形，也就无法进行观测。因而示波器中的扫描电压的周期必须可调，以满足周期成整数的条件。事实上，由于扫描电压与被测信号来自不同的信号源，周期间的整数倍关系不可能长时间内相对稳定，为了能清晰、稳定地观察被测信号，必须设法强行使扫描电压的周期与被测信号的周期成整数倍的关系，这种调整过程称为同步（也称“整步”）。因此，为在实验中得到清晰、稳定的波形，必须调节“扫描速度选择开关（SWEEP TIME/DIV）”和“同步（LEVEL）”旋钮，以使扫描电压的周期与被测信号的周期满足整数倍的要求。

（5）偏向放大器　示波器内部置有水平偏向放大器和垂直偏向放大器。由于示波管本身的 X 轴偏转板和 Y 轴偏转板的灵敏度不高（约 0.1 ~ 1mm/V），当加于偏转板的信号电压较小时，电子束不能发生足够的偏转，以致屏幕上光点位移过小，不利于观测。为了能观测较弱的电压信号，先使该电压信号经 Y 方向放大器放大后再加到 Y 轴偏转板上。X 方向放大器用来放大扫描信号或 X 轴的输入信号，以保证水平偏转板也有足够的偏转电压。

在 Y 轴放大器前还有衰减装置，其作用是将被观测的高电压信号衰减，其衰减程度由“垂直幅度衰减开关（VOLTS/DIV）”来控制。

2. CS－4125 型示波器

图 13－5 是 KENWOOD CS－4125 型双通道示波器前方面板图（CS－4025 示波器面板图及控制旋钮见附录）。

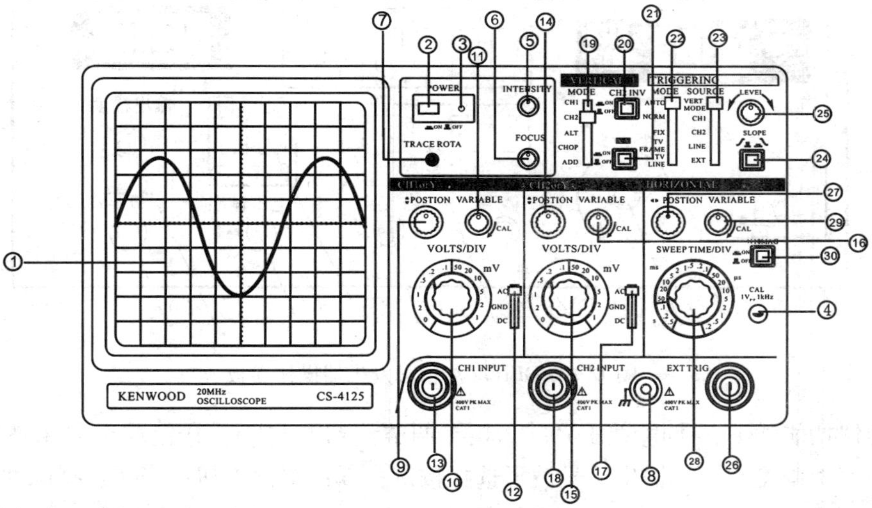

图 13－5 KENWOOD　CS－4125 型双通道示波器的面板

各旋钮功能如下：①CRT 荧光屏；②POWER 电源开关；③电源指示灯；④CAL 端：可得到峰－峰值 V_{p-p}为 1V 频率约 1KHz 的方波校准信号；⑤INTENSITY 光点亮度调节旋钮；⑥FOCUS 聚焦调节旋钮；⑦TRACE ROTA 调整水平亮线倾角旋钮；⑧GND 接地端；⑨ ♦ POSITION　CH1　垂直移位；⑩ VOLTS/DIV　CH1　垂直轴衰减旋钮；⑪ VARIABLE　CH1 垂直轴衰减微调旋钮；⑫AC－GND－DC　选择 CH1 X 轴输入信号的耦合方式；⑬CH1 INPUT　CH1 垂直轴输入端子；⑭CH2 垂直移位；⑮CH2 垂直轴衰减

旋钮；⑯CH2 垂直轴衰减微调旋钮；⑰选择 CH2 Y 轴输入信号耦合方式；⑱CH2 垂直轴输入端子；⑲VERTICAL MODE 垂直轴工作方式选择键 CH1 表示只显示 CH1 的输入信号、CH2 表示只显示 CH2 的输入信号、ALT 表示交替显示 CH1 和 CH2 的输入信号、CHOP 表示断续显示 CH1 和 CH2 的输入信号、ADD 表示显示 CH1 和 CH2 的输入信号的合成波形；⑳CH2 INV　CH2 输入信号极性反相按键；㉑X－Y 该按键被按下则 VERTICAL MODE 的设定将无效，而将 CH1 变为 Y 轴、CH2 变为 X 轴的 X－Y 轴示波器；㉒TRIGGERING MODE 触发方式选择键，一般情况下将此旋钮拔至 AUTO；㉓SOURCE 触发源选择开关，VERT MODE 表示触发信号源由⑲号旋钮选择，CH1 表示触发信号源为 CH1 之输入信号、LINE 表示触发电源的电压波形、EXT 表示由㉖旋钮之外输入信号作触发源；㉔SLOPE 选择触发极性旋钮；㉕LEVEL 调整触发电平旋钮；㉖EXT. TRIG 外触发信号输入端；㉗◀▶POSITION 波形水平移位；㉘SWEEP TIME/DIV，扫描时间切换旋钮；㉙VARIABLE 扫描时间微调旋钮，右旋至 CAL 位置时㉘旋钮指示值即为校正值；㉚×10MAG 显示波形水平扩大 10 倍旋钮。

3. DF1636A 信号发生器

DF1636A 信号发生器是一种可产生正弦波、三角波、方波和对称可调脉冲波的便携台式函数信号发器，其面板图如图 13－6 所示。

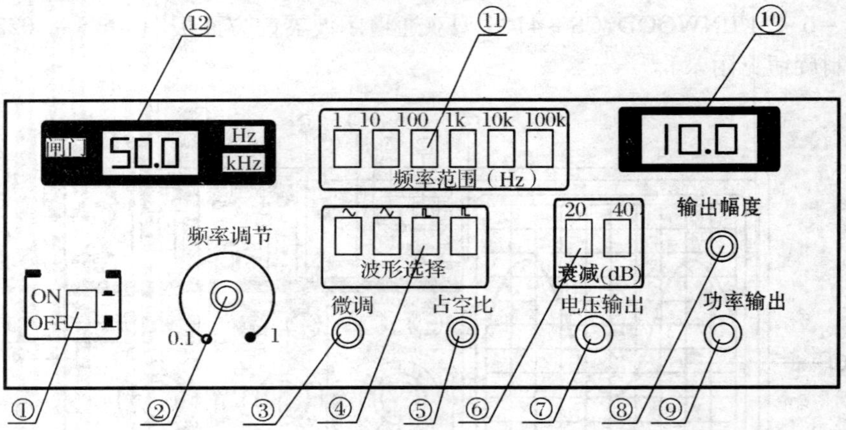

图 13－6 DF1636A 信号发生器操作面板

各旋钮功能如下：①电源开关；②频率调节旋钮，与③、⑪钮配合使用选择信号频率；③频率微调旋钮；④输出信号波形选择按键；⑤占空比旋钮：当④选择脉冲波时，该旋钮可改变输出信号占空比；⑥输出信号衰减开关：二只按钮同时按下时可以产生 60dB 的衰减；⑦电压信号输出端；⑧信号输出幅度调节旋钮；⑨信号功率输出端；⑩输出信号电压显示器：显示输出电压的峰－峰 V_{p-p} 值；⑪频率范围选择按键：与②③配合选择信号工作频率；⑫输出信号频率显示器：Hz 或 kHz 灯亮有效。

【实验原理】

1. 频率的测量

将待测信号从 *Y* 轴输入示波器，适当地调节示波器，使屏幕上显示出清晰、稳定的波形。

根据屏幕上的标尺刻度可知待测信号上相位相同（一个周期）的两相邻点之间的距离d_x，如图13－7所示，d_x以刻度“格（div）”为单位，则这两点之间的时间间隔即为待测信号的周期T_x：

$$T_x = d_x\ (div) \cdot t/div$$

式中t/div为示波器扫描每“格（div）”所使用的时间。

由周期与频率的关系：

$$f_x = 1/T_x$$

即可得到被测信号的频率f_x。

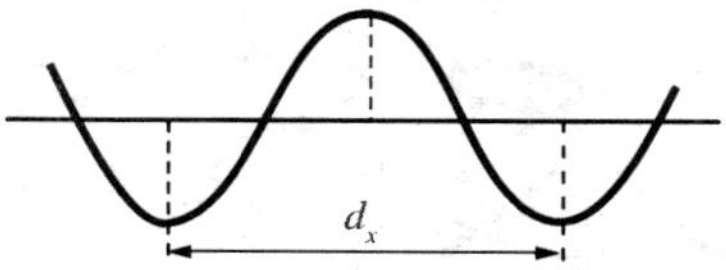

图13－7　正弦信号波形

2. 李萨如图形法测量频率

设有两个正弦波电压信号，其中一个的频率已知。将此两电压信号分别加在示波器的垂直偏转板和水平偏转板上，那么电子束将同时参与两个相互垂直方向的运动，它在荧光屏上的轨迹将由相互垂直的振动合成来决定。根据振动合成原理可知，当两互相垂直振动的频率成一定比例时，合成振动的轨迹一般为李萨如图形，如图13－8。在特殊情况下，当两电压信号的频率相同且相差为零或相差一定时，轨迹为直线或椭圆（若振幅相等，相差为$\pi/2$或$3\pi/2$时，则成为圆）。

将所得到的李萨如图形用方框围起来，并设：

n_x——方框水平方向的最大割点数目；

n_y——方框垂直方向的最大割点数目；

f_y——加在垂直偏转板上电压的频率；

f_x——加在水平偏转板上电压的频率。

则存在关系：

$$n_x : n_y = f_y : f_x \text{（此关系仅当李萨如图形闭合时才适用）}$$

例如对图13－9，因方框水平方向与李萨如图形的最大割点数为6（上下各3，即$n_x=6$），垂直方向与图形的最大割点数为4（左右各个2，即$n_y=4$），故

$$f_y : f_x = n_x : n_y = 6:4 = 3:2$$

因此，如f_x、f_y中有一个已知，即可得出另一频率。

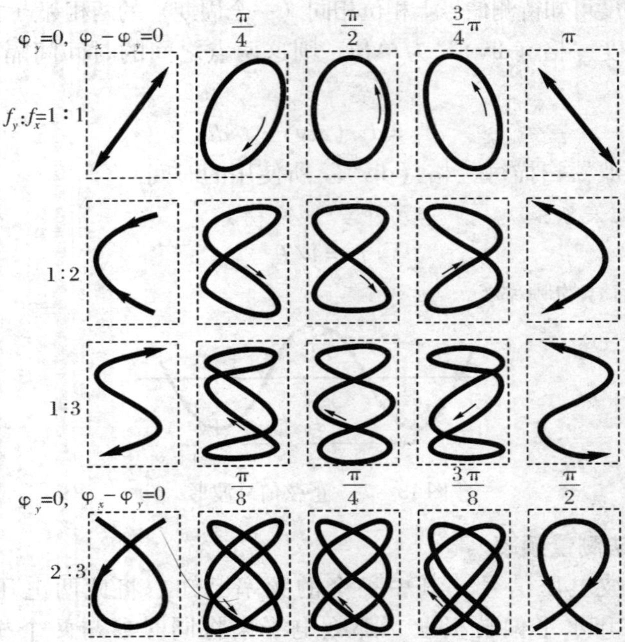

图 13－8　频率和相位满足不同条件时的李萨如图

3. 相位差的测量

用李萨如图形法还可以测量两个频率相同的简谐信号 U_1 和 U_2 之间的相位差 φ，如图 13－10。

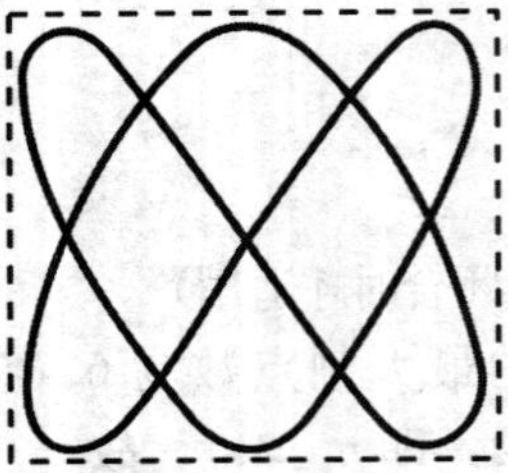

图 13－9　李萨如图的割点

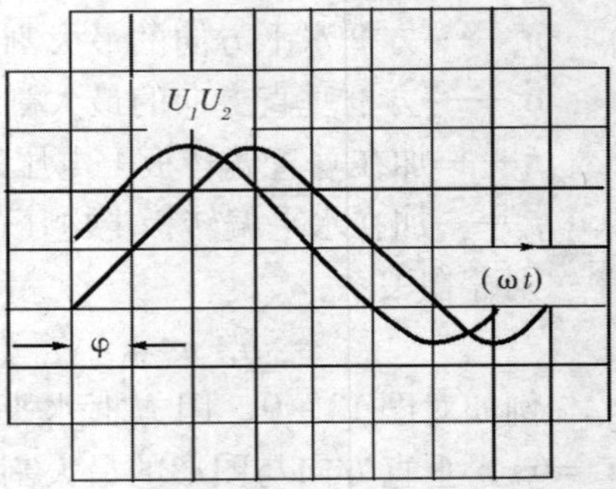

图 13－10　相位差 φ

设 U_1 加于示波器的垂直偏转板，U_2 加于示波器的水平偏转板，则：

$$U_y = U_1 = U_{m_1}\sin\omega t$$

$$U_x = U_2 = U_{m_2}\sin\ (\omega t + \varphi)$$

图 13－11 为此时在屏幕上显示的椭圆李萨如图形。如椭圆对称于坐标原点，当 $\omega t=0$ 时：

$$U_y=0,\ U_x=U_{m_2}\sin\varphi$$

U_x为椭圆与 X 轴相交两点间的距离 B 的 1/2。

当 $\omega t+\varphi=\pi/2$ 时：

$$U'_x=U_{m_2}$$

U'_x为椭圆的水平最大偏移 A 的 1/2。

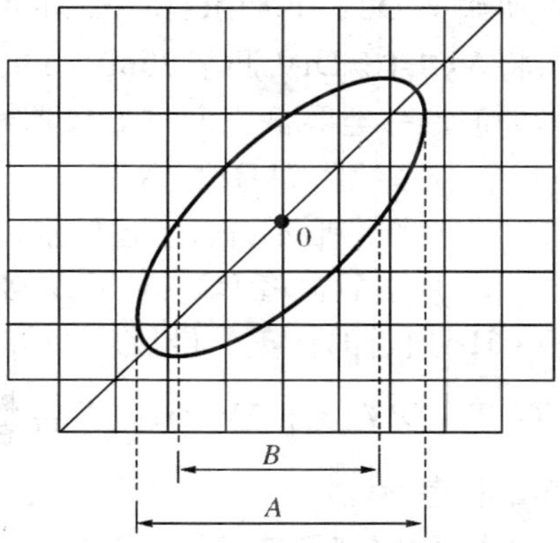

图 13－11　两个同频率垂直运动合成的李萨如图

将上述两式作比值，即 $\dfrac{U_x}{U'_x}=\dfrac{U_{m_2}\sin\varphi}{U_{m_2}}=\dfrac{B/2}{A/2}$，可得 $\sin\varphi=B/A$。

最后可得相位差：$\varphi=\sin^{-1}\dfrac{B}{A}$。

【实验内容】

1. CS－4125（或 CS－4025）型示波器使用前的检查

（1）预设面板上各控制旋钮（按键），设置如下。

MODE －－－－－－－－－－－－－－－－－－－－AUTO

SOURCE －－－－－－－－－－－－－－－－－－－－VERT MODE

VERTICAL MODE－－－－－－－－－－－－－－－－－－－－CH1（INV 为 OFF）

SLOPE －－－－－－－－－－－－－－－－－－－－＋

LEVEL －－－－－－－－－－－－－－－－－－－－中央

CH1 or Y 及 CH2 or X 部分：

▲▼POSITION －－－－－－－－－－－－－－－－－－－－中央

VARIABLE －－－－－－－－－－－－－－－－－－－－CAL

VOLTS/DIV －－－－－－－－－－－－－－－－－－－－5V/DIV

AC－GND－DC －－－－－－－－－－－－－－－－－－－－GND

HORIZONTAL 部分：

◀▶POSITION －－－－－－－－－－－－－－－－－－－－中央

VARIABLE －－－－－－－－－－－－－－－－－－－－CAL

SWEEPTIME/DIV －－－－－－－－－－－－－－－－－－－－0. 2mS/DIV

×10MAG －－－－－－－－－－－－－－－－－－－－OFF

（2）接通示波器电源，指示灯亮，稍等片刻，仪器进入工作状态。

（3）调整⑤INTENSITY 旋钮以获得清晰可见的亮度，接着调整⑥FOCUS 和⑦TRACE ROTA 旋钮以使亮线与水平刻度保持平行。

（4）将 PC－54 探头（信号接入线端头）接至 CH1 通道上，并将其 AC－GND－DC

按钮调至 DC，⑲VERTICAL MODE 调至 CH1，然后将 CH1 上探针接至④CAL 端子上，再将 VOLTS/DIV 调至 20mV/div，此时调节㉕LEVEL 旋钮使方波同步，然后将稳定、清晰的波形移至屏幕中间。如仪器性能基本正常，则显示的方波垂值幅度值约为 $1V_{p-p}$即 5div，频率约为 1kHz。

若校准波形不符合上述要求，不能做定量测量。

（5）然后再将 PC－54 探头接至 CH2 通道上，将⑲旋钮调至 CH2 后，如前述方法进行 CH2 通道的校正。最后将⑲旋钮调至 CH1，各通道的 AC－GND－DC 设于 AC，其 VOLTS/DIV 设于 5V/div，并将▲▼POSITION 和◀▶POSITION 调回中央。此种状态称为起始状态。

2. 观察波形并测量频率

（1）将 DF1636A 信号发生器接通电源，弹出⑥输出衰减按钮，④按键选择正弦波形，调节输出幅度使输出电压为 5V（峰－峰 V_{p-p}值），将其电压输出端与示波器的“CH1”输入端连接。

（2）旋转信号发生器的③频率调节旋钮，选定某一频率，适当调节示波器的㉘SWEEPTIME/DIV 旋钮及㉕LEVEL 旋钮（若待测信号的周期与扫描电压的周期不满足整数倍条件时调节该旋钮），使屏幕上得到清晰、稳定的 1 个、2 个或 3 个完整的正弦波波形。

（3）测量出正弦波形上两相邻且相位相同点的距离 d_x（该距离可从屏幕上的标尺上直接读出），然后从 SWEEP TIME/DIV 旋钮上读出此时每格的扫描时间 *t/div*（此时 VARIABLE 扫描时间微调旋钮必须右旋至 CAL 位置），根据由周期求频率的方法，测出该正弦波的频率，并将计算结果与信号发生器上所指示的频率值进行比较，填入数据表 13－1中。

表 13－1 频率的测量实验数据表

实验次数	1	2
被测长度 d_x（div）		
示波器“*t/div*”位置		
T_x（s）		
f_x（Hz）		
低频信号发生器的指示频率 f_0（Hz）		

3. 观察李萨如图形并测量频率

（1）接通 220V/6V 变压器，将其输出的 6V、50Hz 的电压接至示波器的“CH1”输入端，DF1636A 信号发生器的电压输出端与示波器的“CH2”输入端相接。

（2）调节 DF1636A 信号发生器的输出信号幅度旋钮使其输出电压为 5V（峰－峰 V_{p-p}值），示波器的 X－Y 旋钮必须按下。

（3）适当调节“CH1”的▲▼POSITION 和“HORIZONTAL”的◀▶POSIOTON 旋钮，再适当调节 CH1 的 *V/div* 旋钮和 CH2 的 *V/div* 旋钮，使屏幕上出现大小适中的李萨如图形。

（4）调节 DF1636A 信号发生器的频率，使屏幕上依次出现 f_y: f_x =1∶1、1∶2、1∶3 和 2∶3 时的李萨如图形，记下每种李萨如图形水平和垂直方向的最大割点数 n_x、n_y，根据公式 n_x: n_y = f_y: f_x 计算出未知频率 f_x（f_y =50Hz），并把 f_x 的计算结果与 DF1636A 信号发生器所显示频率值进行比较，填入数据表 13－2 中。

表 13－2　李萨如图形法测量频率实验数据表　　　**交流电频率 f_y =50Hz**

最大割点数之比 n_x: n_y	1∶1	1∶2	1∶3	2∶3
f_x（Hz）				
低频信号发生器的指示频率 f_0（Hz）				

4. 相位差的测量

（1）按图 13－12 连接电路。将 DF1636A 信号发生器的输出电压调为 5V（峰－峰 V_{p-p} 值），按下示波器的⑳X－Y 键。

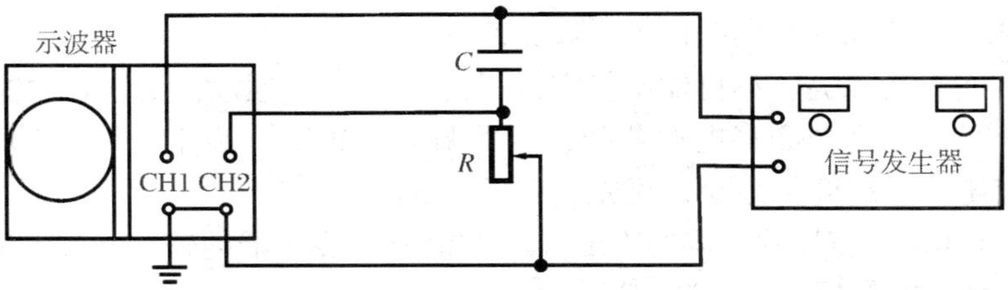

图 13－12　相位差测量电路图

（2）按表 13－3 所列数值分别调节电阻箱电阻 R 和 DF1636A 信号发生器的输出电压频率 f：

表 13－3　电阻 R 和频率 f 的取值　　　**电容 C =0.22μF**

R、f	取 值 大 小			
R（kΩ）	3	1.5	2	4
f（Hz）	200	400	400	200

（3）调节示波器的“VOLTS/DIV”开关和“VARIABLE 微调”旋钮，使椭圆大小合适，再调节“CH1”的▲▼ POSITION 和“HORIZONTAL”的◀▶POSIOTON 旋钮，使椭圆在屏幕中央。

（4）在 X 轴上读出 A、B 的数值（单位：div），并根据公式计算出相位差 φ。

（5）RC 串联电路中，电流与电压之间的相位差的理论值大小为：

$$\Phi = \text{tg}^{-1}\frac{1}{R\omega C}$$

将由上式计算的结果与式（4）中计算的结果相比较，并以 Φ 为标准值计算出百分误差 E，并将实验数据填入表 13－4 中。

表 13－4　测量交流电信号的相位差实验数据表　　$C=0.22\mu F$

R（kΩ）	3	1.5	2	4
f（Hz）	200	400	400	200
$\Phi=\text{tg}^{-1}\frac{1}{R\omega C}$				
B（div）				
A（div）				
$\varphi=\sin^{-1}\frac{B}{A}$				
$E=\frac{\lvert\varphi-\Phi\rvert}{\Phi}\times100\%$				

【注意事项】

1. 电子仪器的旋钮切忌用力硬拧，以免使其损坏。

2. 光点不可开得太亮，尽可能不使它久留在一点，以免损坏示波管上的荧光物质。

3. 暂时不使用时，务必将电源断开，然后调节“辉度”旋钮，使光点消失即可。在关机前，也应先使“辉度”最小，无光点后再关电源。

4. 注意 DF1636A 信号发生器的输出端及 220V/6V 变压器的输出端都不能短路。电路接好后，须经教师检查后，才能开始进行实验。

【思考题】

1. 使用示波器观察信号时，若屏幕上出现如下图形（图 13－13），哪些旋钮的位置不对？应如何调节？

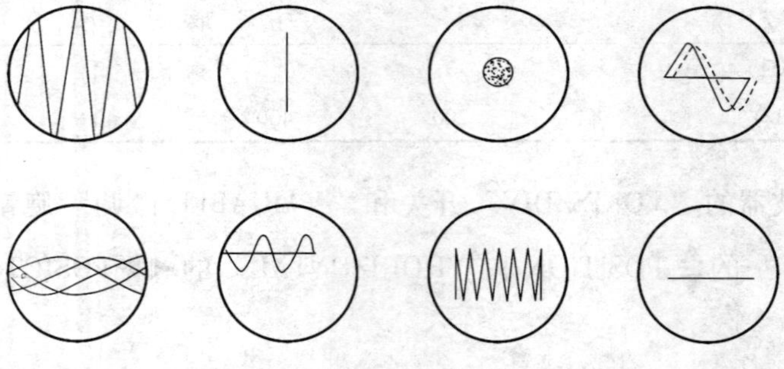

图 13－13　示波器屏幕波形图

2. 示波器正常，但屏幕上看不到亮线，这可能是由于哪些旋钮的位置不合适而造成的？怎样操作才能找到亮线？

3. 示波器正常，但 Y 轴输入信号后，屏幕上只有亮点而无图形，这可能是由于哪些原因造成的？你能自行解决吗？

4. 某信号频率为 500Hz，调节示波器可出现 2 个完整波形，若不再调节示波器面板上的旋钮，换另一待测信号，在屏幕上刚好出现 5 个完整波形，待测信号频率为多大？若再想让屏幕上的波形恢复到 2 个且不改变该信号，应怎样调节？

5. 观察李萨如图形时，能否用示波器的“同步”把图形稳定下来？为什么？

6. 给你一个标准电压，你可否利用示波器来测量信号的峰值？应如何进行测量？

7. 如果在实验中发现椭圆不是斜的而是正的，是什么原因？请给出解释。应如何解决？

实　验　指　导

【预习要求】

1. 掌握示波器的工作原理。

2. 熟悉示波器的内部方框图及各部分元件的作用。

3. 根据示波器的面板控制旋钮图 13－5，结合示波器内部结构方框图 13－1，了解其各旋钮的作用。

4. 了解 DF1636A 信号发生器各部分旋钮的作用。

5. 掌握用示波器测相位差的原理。

【操作要点】

1. 在掌握示波器使用时，可将示波器的面板控制旋钮分成 4 个部分：以屏幕方框右边缘线为界先将面板分成左右 2 个部分，再将右半部分分成 4 个功能块，如图 13－14 所示。

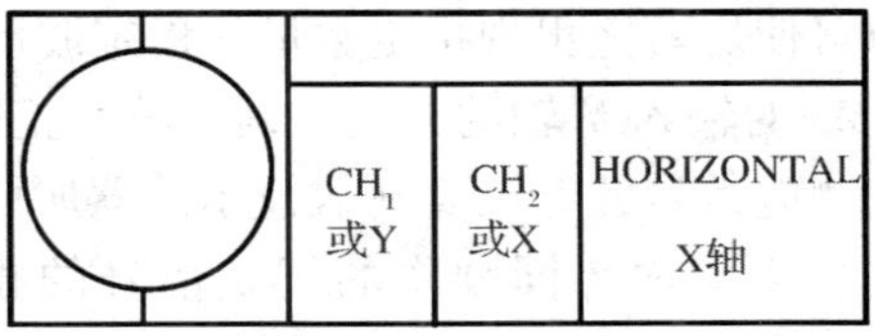

图 13－14　示波器面板分区

2. 右半部分上部主要是开机调试和整机功能控制旋钮。

3. 右半部分下部被分为 3 个主要部分。左侧两个部分分别为 CH1、CH2 两通道的 Y 轴系统，即垂直方向的控制旋钮；右侧为 X 轴系统，即水平方向的控制旋钮。垂直方向控制旋钮的调节将改变 y 轴方向信号的性质或变化情况；水平方向控制旋钮的调节将改变 X 轴方向信号的性质或变化情况，故应根据实验内容对 X、Y 两个方向的信号的具体要求来调节。

4. 垂直幅度衰减旋钮“VOLTS/DIV”和扫描速度选择旋钮“SWEEPTIME/DIV”各有一个微调旋钮即⑪和⑯，㉙旋钮在作定量测量时必须置于“校准”位置（即右旋至底），否则将影响测量的准确度。仅当用于定性观测实验时，这三个旋钮可用于连续调节。

5. 按图 13－12 连接线路时，要注意示波器的信号输入线是同轴电缆线，侧线接地，切不可将中心线接地，否则将无任何信号输入示波器。

6. 在测量椭圆与 X 轴的两交点间距离 B 及在水平方向最大偏移 A 时，为方便读数，尽可能减小读数误差，可采如下方法。

（1）测量 A 时可将“CH1orY”的耦合开关⑫拨至 GND 接地，此时屏幕上出一条水平光亮线段，再将此光亮线段上下移动，使其和水平标尺相重合，则此时在水平标尺上所读出的光亮线段的长度即为 A 值。

（2）测量 B 时将"CH1orY"的耦合开关⑫拨回至 AC，此时屏幕上出现的椭圆的中心必在水平标尺上。再将得到的椭圆向左或右稍加移动，使椭圆与水平标尺上的一个交点恰好经过水平标尺的某一刻度点，以此点为基准读出另一个交点与此点的坐标间隔数即为 B 的值。（注意在此过程中，切不可将椭圆在竖直方向作位置改变。

【报告要求】

1. 实验数据表格可参考以下内容。

（1）频率的测量　实验数据记录表格如表 13－1 所示。

（2）李萨如图形法测量频率　实验数据记录表格如表 13－2 所示。

（3）测量交流电信号的相位差　实验数据记录表格如表 13－4 所示。

2. 记录实验数据及数据处理过程中，必须注意有效数字的位数，请按实验一的要求完成。

【实验后的思考】

1. 若将方波、锯齿波分别输入示波器的 Y 轴来测量其周期，应如何调节示波器的旋钮方可使屏幕上出现指定数量的稳定波形？又如何测量其周期？

2. 对输入 Y 轴的方波、锯齿波、正弦波等常见电压信号，应如何调节及测量方可求出被测电信号的电压值（假设示波器所用的信号输入电缆的衰减系数已知）？

3. 对于两个同频率的正弦信号的相位差的测量，除实验介绍的李萨如图形法外，还有一种比较直观的测量方法——位移法。利用李萨如图形法可测量两电压信号的相位差，但不易判断相位差的正负符号，而位移法具有此判别功能。

将两个同频率而相位不同的正弦电压信号分别送入示波器的 CH1 和 CH2 通道（X－Y 键不要按下，调节 VERT MODE 的 ALT 或 CHOP 键），在屏幕上调出位置均关于 X 轴对称的两电压波形，如图 13－15 所示。

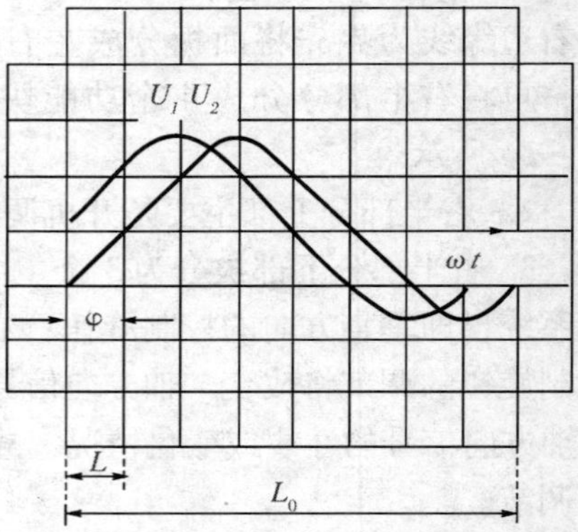

图 13－15　位移法测量相位差

由于 X 方向线段长度与相位成正比关系，故有：

$$\varphi = \frac{L}{L_0} \times 2\pi \ (360^\circ)$$

式中 L_0 为波形一个周期的长度，L 为两波形的水平位移差，其数值可从水平标尺上读出，并且由图中两波形的前后位置可判别出的相位超前或落后关系。

附　录

图 13－16 是 KENWOOD CS－4025 型双通道示波器前方面板图，其各旋钮功能可参见 CS－4125 型示波器。

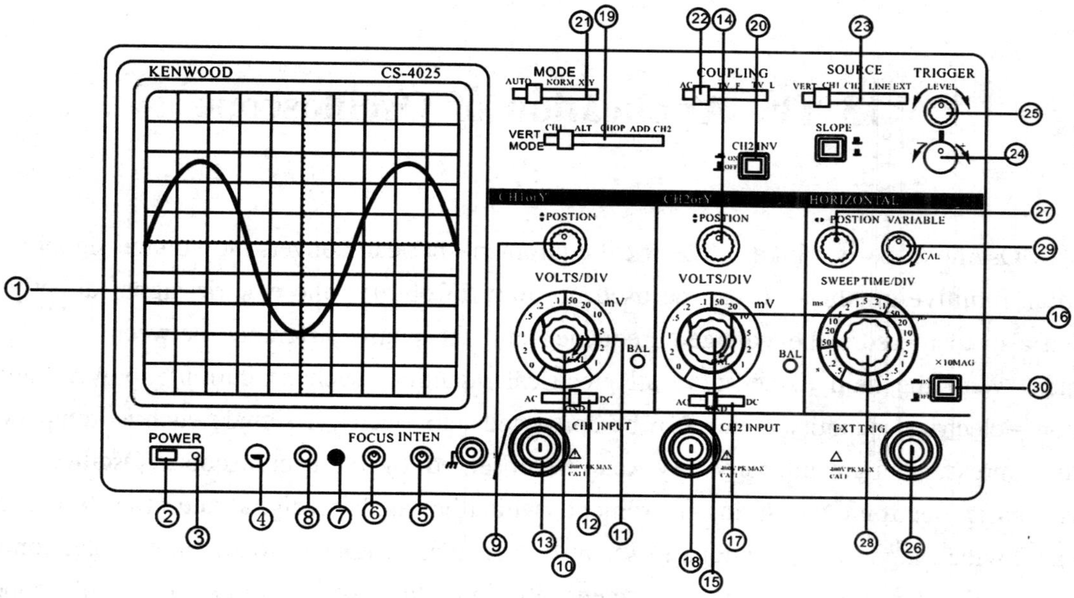

图 13－16 KENWOOD CS－4025 型双通道示波器的面板图

13 The Application of Oscilloscope

Oscilloscope is a kind of electrical equipment that can convert the voltage signal into an intuitive graphics. It can be used not only to observe the electric signal directly, but also to measure the voltage, frequency and the phase difference between two signals. With the help of sensors, all electrical quantity, such as current, power, and non – electrical quantity which can be converted into electrical signal, such as temperature, pressure, light intensity, etc, can be observed by the oscilloscope. Oscilloscope is a basic electronic measuring instrument, which has analog, digital and other types. It has a wide range of applications in scientific research, engineer, experiment, electronic fields and other areas. In this experiment, the important work is to use the oscilloscope to observe a variety of signals and measure their voltage, frequency and phase difference.

Objectives

1. Understand the working principle of an oscilloscope.
2. Master the function of buttons on the oscilloscope panel and learn to use the oscilloscope.
3. Observe the electric signal and measure voltage, frequency.
4. Learn to measure the phase difference between two alternate current.

Equipments

Oscilloscope (type CS – 4125 or CS – 4025), signal generator (type DF 1636A), a 220V/6V voltage changer, resistance box and capacitor.

Introduction of the equipments

1. The Basic working principle of the oscilloscope

(1) Basic structure

The primary components of an oscilloscope include: ① main frame: oscilloscope tube, high voltage source, low voltage source and standard signal generator; ②X – axes system: scanning signal generator, X – axis amplifier and lightening circuit. ③Y – axis system: input coupling switch, attenuator and Y – axis amplifier. The basic structure is showed in the figure 13 – 1.

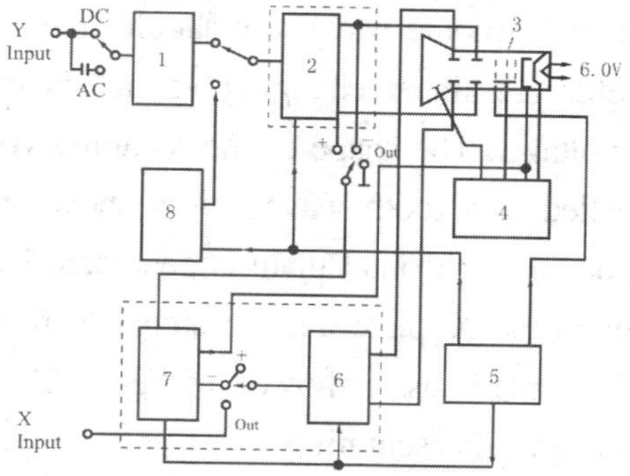

Figure 13 – 1 The basic structure of an oscilloscope

1. Attenuator 2. Y – axis amplifier 3. Oscilloscope tube 4. High voltage source
5. Low voltage source 6. X – axis amplifier 7. Scanning signal generator 8. Standard signal generator

(2) Oscilloscope tube

Oscilloscope tube is the most important part of the oscilloscope, as is shown is Figure 13 – 2. It includes electron gun, deflection system and luminescent screen. Its main principle is that the electrons coming from the cathode will be accelerated and focused by a series of electric poles and they form a fine electron beam after that. The electron beam moving with a high speed will strike the luminescent screen and make a light spot on the screen. Brightness of the spot is determined by the amount of electrons of the beam and their speed; and the brightness can be controlled by the button "Brightness" on the panel. The size of the spot is determined by the fineness of the beam, and also can be controlled by the button "Focus" on the panel.

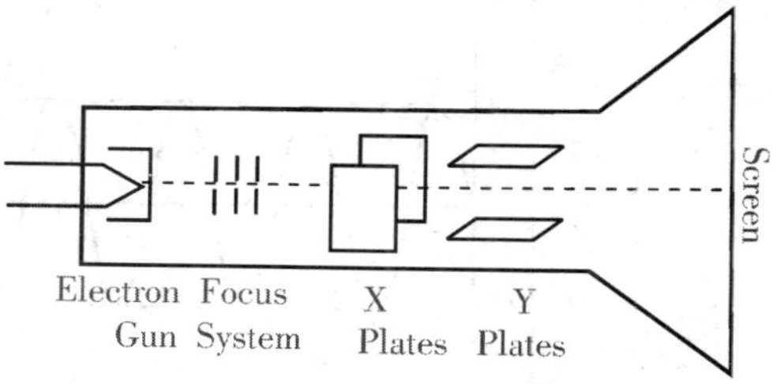

Figure 13 – 2 The basic structure of the oscilloscope tube

There are two pairs of perpendicular deflection plates in the oscilloscope tube. When the electron beam passes through the plates which have been connected to the voltage source, it will be deflected in two directions. Therefore, the deflection of the beam can be controlled by the voltage of the plate, and so as the spot. The spot can reach everywhere of the screen. So the place of the spot on the screen can be used to indicate the voltage.

(3) Scanning principle

The oscilloscope is usually used to observe a periodical signal. If a periodical signal, for instance a harmonic signal $U_y = U_0 \sin\omega t$, to the Y plate, and no signal to the X plate, the spot on the screen will only move vertically. When the frequency is big enough, a vertical bright line can be seen on the screen, as is shown in figure 13 – 3(a).

In order to get a normal harmonic wave, we should make the spot move in the X axis direction uniformly. Therefore, a linear changing voltage is put on the X plate, and this voltage is called scanning voltage. The shape of the scanning voltage with time is sometimes likes saw tooth, so it is called saw tooth wave, as is shown in figure 13 – 3 (b). The scanning signal being put on the horizontal plate is generated by the equipment itself. If the saw tooth wave is only put to the X plate and no signal to the Y plate, we can also see a horizontal bright line on the screen, as is showed in Figure 13 – 3 (c).

If the harmonic signal and the scanning signal are put to the Y and X plates respectively at the same time, the trace of the spot on the screen will be the signal to be observed. Figure 13 – 4 shows the trace of the spot making up by the harmonic signal and the scanning signal.

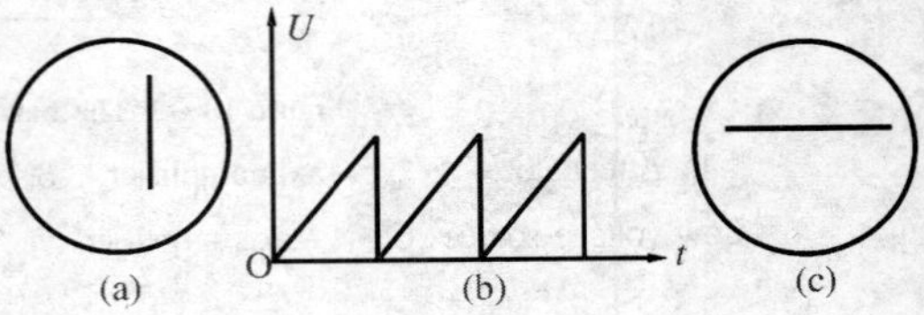

Figure 13 – 3 Scanning principle

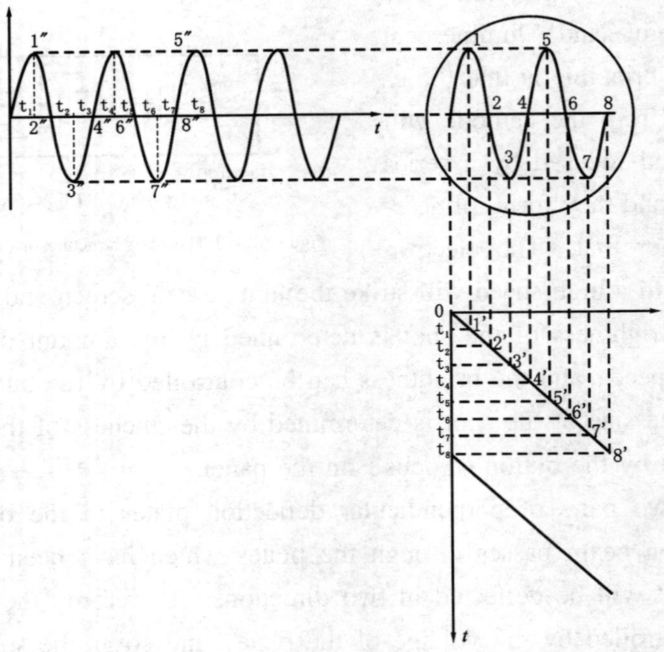

Figure 13 – 4 The trace of the spot making up by the harmonic signal and the scanning signal

(4) Synchronizing

The figure 13 – 4 shows that if the period T_y of the harmonic wave U_y equals to the period T_x of the scanning wave U_x, the oscilloscope will scan the harmonic wave in a period fully and the electron beam can draw a whole figure of the harmonic wave. Therefore, only if all the figures match with each other can we get a clear and stable wave picture on the screen. It can be expressed as: the period T_x of the scanning wave is n times bigger than that of the wave to be observed.

$$T_x : T_y = n \text{ or } f_y : f_x = n \qquad n = 1, 2, 3\cdots$$

There will be n periods of the wave shown on the screen. But if the condition is not satisfied, the wave figures will not overlap with each other, we can't get a clear and stable wave picture on the screen. Therefore, the period T_x or frequency f_x of the scanning wave can be changed to meet the condition. Actually, the condition between the waves can't be kept for a long time because they come from different sources. So we have to keep this condition at any time and make the figure clear and stable, this procedure is called synchronizing (or in – step). Adjust the knob "Scanning speed" or "Synchronizing" on the panel can make the scanning wave frequency be suitable to get a clear wave picture.

(5) Amplifier

There are two kinds of amplifier in the oscilloscope, X axis amplifier and Y axis amplifier. Because the sensitivity of the deflection of X axis and Y axis sometimes is not very large (about 0.1 ~ 1mm/V), if the signal that have been put on the deflection plate is too small, the displacement of the spot on the screen will be so small that it will be difficult to be observed. Therefore, for a weak signal, firstly it will be magnified by the Y axis amplifier and then is put to the Y axis deflection plate. For the same reason, X axis amplifier will magnify the scanning wave or the X axis signal to make sure that there are large enough deflection voltages.

There is also an attenuator before the Y axis amplifier to make the high voltage signal be small to a suitable level. It is controlled by the "Attenuator switch (VOLTS/DIV)".

2. CS – 4125 Oscilloscope

The front panel of KENWOOD CS – 4125 two – channel oscilloscope is shown in figure 13 – 5. (type CS – 4105 is shown in the appendix part). Following is the functions of its knobs.

①CRT Screen; ②Power; ③Power indicator; ④CAL end (can get a 1 volt $p-p$ rectangular wave with frequency of 1 kHz); ⑤INTENSITY Adjusting brightness; ⑥FOCUS; ⑦TRACE ROTA Adjusting the slope of horizontal line; ⑧GND; ⑨ POSITION (Vertical displace of CH1); ⑩VOLTS/DIV Vertical Attenuation of CH1; ⑪VARIBLE Vertical fine Attenuation of CH1; ⑫AC – GND – DC Coupling form of signal in CH1; ⑬CH1 INPUT; ⑭POSITION (Vertical displace of CH2); ⑮VOLTS/DIV Vertical Attenuation of CH2; ⑯VARIBLE Vertical fine Attenuation of CH2; ⑰AC – GND – DC Coupling form of signal in CH2; ⑱CH2 input end; ⑲ VERTICAL MODE (CH1 for only displaying signal of CH1; CH2 for only displaying signal of CH2; ALT for displaying signals of CH1 and CH2 alternate; CHOP for

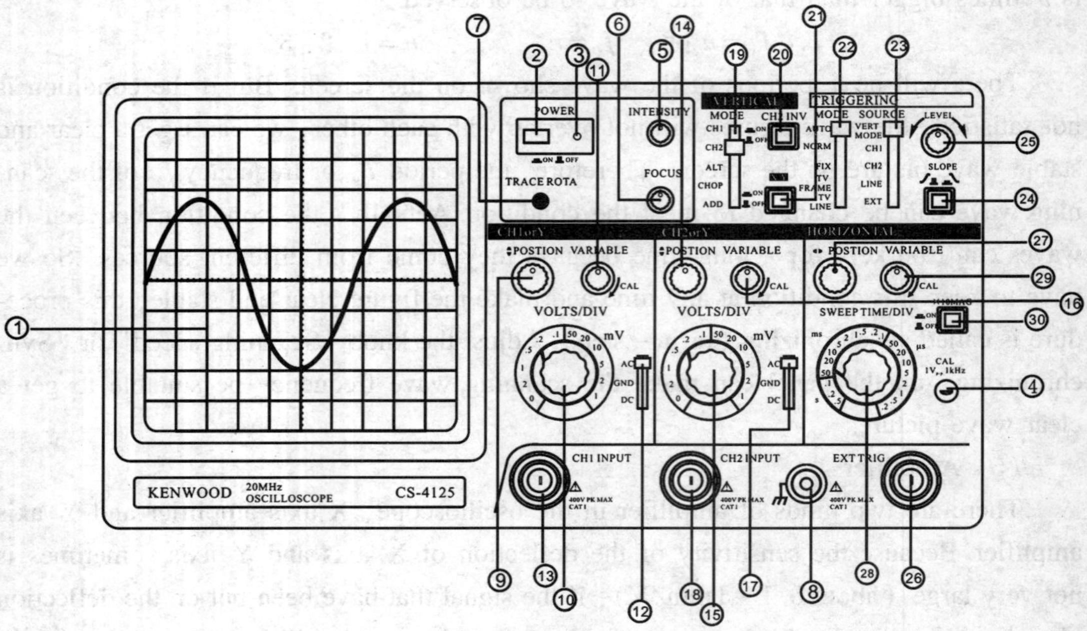

Figure 13 –5 The panel of KENWOOD CS –4125 Oscilloscope

displaying signal of CH1 or CH2 one in a time; ADD for the addition of signals of CH1 and CH2) ; ⑳CH2 INV for inverse polar of signal in CH2 ; ㉑X – Y Signal on CH2 will be put to X axis ; ㉒TRIGGERING MODE Usually in the state of AUTO ; ㉓ SOURCE for selection of triggering source ; ㉔ SLOPE ; ㉕ LEVEL ; ㉖ EXT. TRIG; ㉗ POSITION (Horizontal displace); ㉘SWEEP TIME/DIV for speed of scanning ; ㉙ VARIBLE For fine turning of scanning speed, the CAL position is standard position; ㉚ ×10MAG Enlarge the pattern on the screen for 10 times.

3. DF1636A Signal Generator

DF1636A is a portable function signal generator, which can produce harmonic wave, triangle wave, rectangular wave and symmetry pulse wave. Its panel is shown in figure 13 –6.

①Power switch; ②Frequency adjusting knob; ③Frequency fine adjusting knob; ④Selection for type of signal; ⑤Duty ratio; ⑥ Attenuation of output; ⑦Voltage output end; ⑧Amplitude of output; ⑨Power output; ⑩Displace of output voltage; ⑪ Frequency range selector; ⑫ Displace for output frequency (in unit of Hz or kHz).

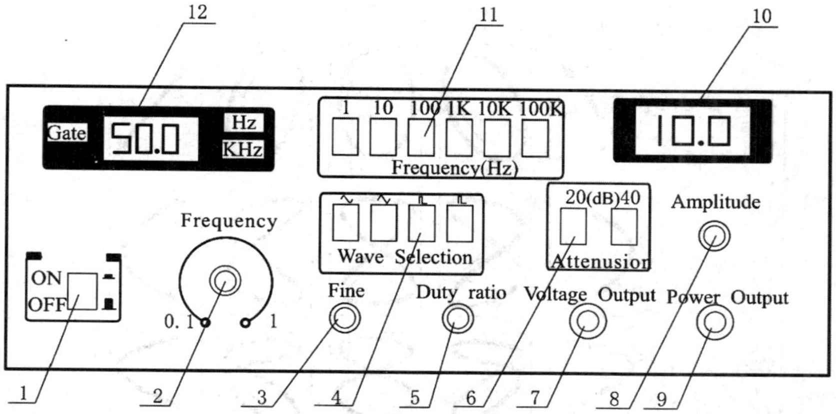

Figure 13 – 6 The panel of DF1636A Signal Generator

Principle

1. Measurement of frequency

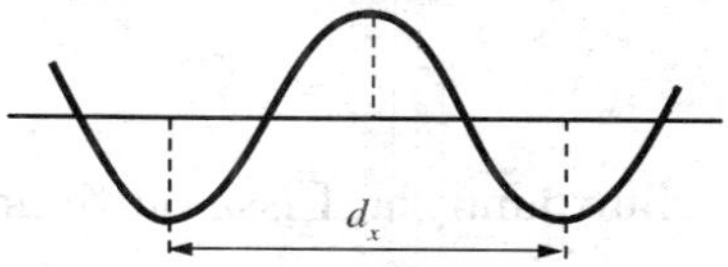

Figure 13 – 7 The wave of a harmonic signal

Firstly input the signal into the Y axis channel, and adjust the scanning frequency knob on the panel to get a clear and stable wave picture. The distance d_x between the two nearby points with the same phase can be get, as is shown in Figure 13 – 7. d_x is in the unit of screen scale (div); then The interval or the period of the signal can be determined by:

$$T_x = d_x \ (div) \ t/div$$

t/div is the time of scanning one division by the oscilloscope.

The frequency of the signal is:

$$f_x = 1/T_x$$

2. Measure the frequency by the Lissajou figure

There are two harmonic electric signals, and one's frequency is given. Now input the two signals to the Y and X channel of the oscilloscope separately, and the figure on the screen will be the result of the combination of the moving of the electron beam in the two perpendicular directions. It's called Lissajou figure, as is shown in Figure 13 – 8. When the two signals have the same phase or a certain phase difference, the figure will be a line or an ellipse. (For example, when the phase difference is $\pi/2$ or $3\pi/2$, it is a circle.)

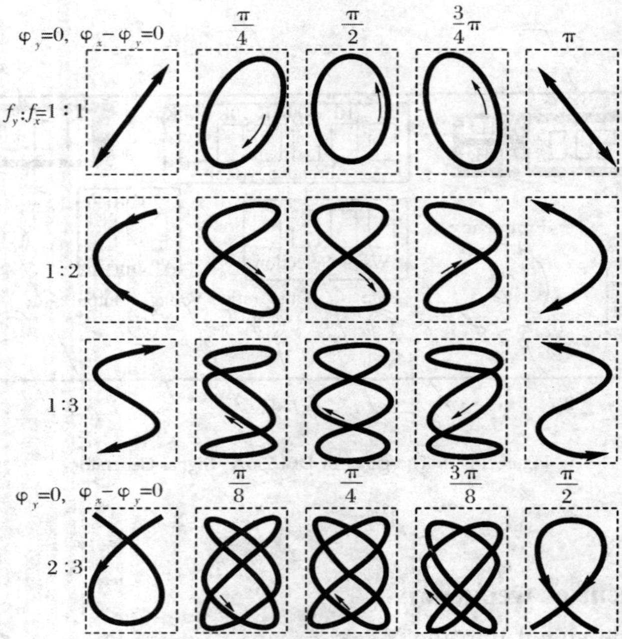

Figure 13 – 8 Lissajou figure

Surround the Lissajou figure with a rectangle, and let:

n_x——the biggest cut points of the horizontal line;

n_y——the biggest cut points of the vertical line;

f_y——the frequency in Y axis;

f_x——the frequency in X axis.

Then there will be a relation:

$$n_x : n_y = f_y : f_x \text{ (only for an enclosed figure)}$$

Take Figure13 – 9 for example, the biggest cut points of the horizontal line is 6, and that of the vertical is 4, thus

$$f_y : f_x = n_x : n_y = 6 : 4 = 3 : 2$$

If one of the frequency f_x or f_y is given, the other can be calculated.

Figure 13 –9 The cut point of the Lissajou figure

3. Measurement of phase difference

Figure 13 – 10 shows two harmonic signals U_1, U_2 with the same frequency but have a phase difference φ. The phase difference can be measured by the Lissajuo figure.

Suppose U_1 is put to the Y axis and U_2 to the X axis, then we have:

$$U_y = U_1 = U_{m_1} \sin\omega t$$

$$U_x = U_2 = U_{m_2} \sin\ (\omega t + \varphi)$$

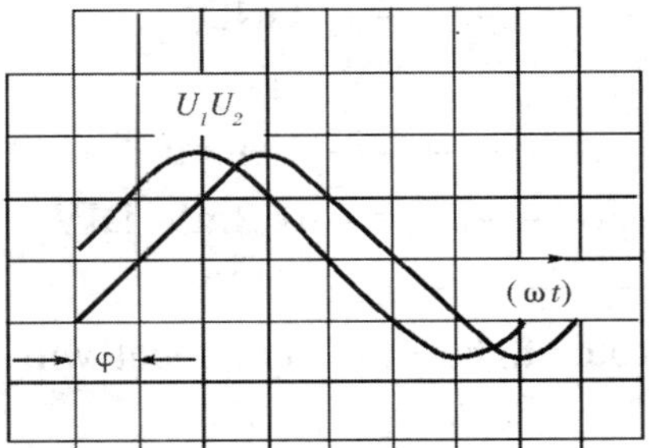

Figure 13 – 10 The phase difference φ

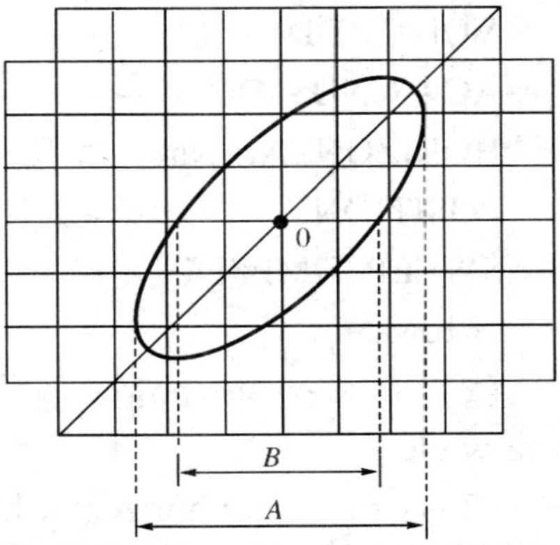

Figure 13 – 11 The ellipse of two harmonic signals

There will show an ellipse on the screen as in the figure 13 – 11. If the center of the ellipse is at zero of the coordinates, when $\omega t = 0$, we can get:

$$U_y = 0,\ U_x = U_{m_2} \sin\varphi$$

U_x is half of the distance B between the two crossing points of the ellipse with X axis.

For the same reason, when $\omega t + \varphi = \pi/2$, we also can get:

$$U'_x = U_{m_2}$$

U'_x is half of the cast shadow A on the X axis of the ellipse. Thus:

$$\frac{U_x}{U'_x} = \frac{U_{m_2} sin\varphi}{U_{m_2}} = \frac{B/2}{A/2}$$

And finally we can get $\sin\varphi = B/A$.

So the phase difference of the two harmonic signal is

$$\varphi = \sin^{-1}\frac{B}{A}$$

Experiment

1. Check the CS – 4125 (or CS – 4025) oscilloscope before the experiment

(1) Set the knobs on the proper position as follows.

MODE – AUTO

SOURCE – VERT MODE

VERTICAL MODE – CH1 (INV be OFF)

SLOPE – +

LEVEL – Central

CH1 or Y and CH2 or X part:

POSITION – Central

VARIABLE – CAL

VOLTS/DIV – 5V/DIV

AC – GND – DC – GND

HORIZONTAL part:

POSITION – Central

SWEEP TIME/DIV – 0. 2mV/DIV

×10MAG – OFF

(2) Turn on the power, and the power indicator light. The equipment will work for a while.

(3) Get a clear horizontal line with suitable brightness.

(4) Input the standard rectangle signal into CH1by using the PC – 54 probe, AC – GND – DC button to DC, knob ⑲ VERTICAL MODE to CH1, ⑩VOLTS / DIV to 20mV/div and connect the PC – 54 probe to the knob ④ CAL; then adjust the knob ㉕ LEVEL to make the rectangle wave clear and stable on the screen. If the oscilloscope is in its good condition, the voltage and frequency of the rectangle wave are $1V_{p-p}$ (about 5div) and 1 kHz.

(5) Input the standard rectangle signal into CH2, and calibrate it as in step (4).

2. Observing wave figure and measuring frequency

(1) Turn on the signal generator, and release the "output attenuation" button, then adjust the output voltage of harmonic wave to 5V (peak – peak value), then input the voltage to the CH1 of the oscilloscope.

(2) Adjust the frequency knob on the signal generator panel to get a certain signal, and then adjust the scanning speed of the oscilloscope to show 1, 2 or 3 whole harmonic wave on the screen.

(3) Measure the period of the signal and calculate the frequency, and fill the experiment data into the table 13 – 1.

Table 13 – 1 The experiment data of measuring the frequency

Measurement number	1	2
d_x (div)		
t/div		
T_x (s)		
f_x (Hz)		
f_0 (Hz) displaying on generator		

3. Observing Lissajou figure and calculate the frequency

(1) Link the 220 V /6V voltage changer output to the CH1, and connect the sig-

nal generator to CH2.

(2) Adjust the output voltage of generator to about 5V ($P-P$ value), and then press down the X – Y button of the oscilloscope.

(3) Adjust the oscilloscope properly to show the Lissajou figure on the screen.

(4) Adjust the frequency of the signal of the generator to show the Lissajou figure of f_y: f_x = 1 : 1, 1 : 2, 1 : 3, and 2 : 3, and calculate the unknown frequency. Compare the result with the frequency displayed on the generator, and fill the experiment data into the table 13 – 2.

Table 13 – 2 Measure the frequency by the Lissajou figure **f_y = 50 Hz**

n_x: n_y	1 : 1	1 : 2	1 : 3	2 : 3
f_x (Hz)				
f_0 (Hz)				

4. Measure the phase difference between two alternate signal

(1) Link the circuit according to the figure 13 – 12. Adjust the output of the signal generator to get a 5V ($p-p$ value) voltage. The button "X – Y" of oscilloscope should be pressed down.

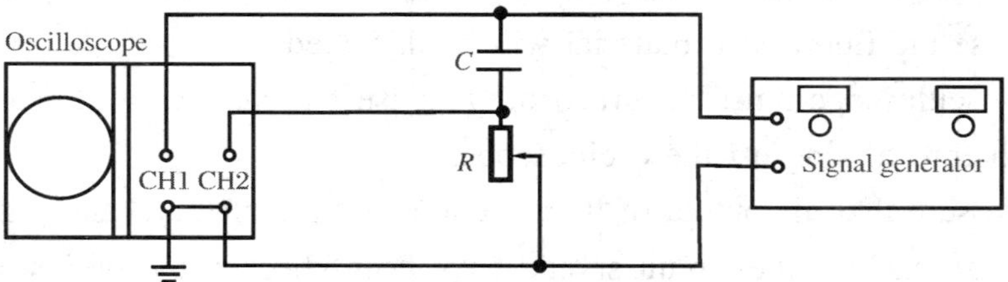

Figure 13 – 12 The circuit of measuring the phase difference

(2) The following table 13 – 3 contains four groups of the values of the resistance R and the frequency f. Adjust the resistor box and the signal generator to obtain them.

Table 13 – 3 The values of the resistance R and the frequency f **C = 0. 22 μF**

R、f	Value			
R (kΩ)	3	1.5	2	4
f (Hz)	200	400	400	200

(3) Adjust the "VOLTS/DIV" and "VARIABLE" knobs on the oscilloscope panel to make the ellipse in a suitable size. The ellipse should be in the center of screen by adjusting the "POSITION" knob.

(4) Measure the A, B values according to the X axis scale, and calculate the phase difference φ.

(5) The phase difference between the voltage and current in the RC circuit is determined by

$$\Phi = tg^{-1}\frac{1}{R\omega C}$$

Compare the results with that obtained by the step 4, and calculated the relative errors E, and fill the experiment data into the table 13 – 4.

Table 13 – 4 Measure the phase difference between the two alternate signals. $C = 0.22\mu F$

R (kΩ)	3	1.5	2	4
f (Hz)	200	400	400	200
$\Phi = tg^{-1}\frac{1}{R\omega C}$				
B (div)				
A (div)				
$\varphi = \sin^{-1}\frac{B}{A}$				
$E = \frac{\lvert\varphi - \Phi\rvert}{\Phi} \times 100\%$				

Notice

1. Adjust the knobs of the electric equipments properly.

2. The spot on the screen can't be too bright and stay the same place for a long time; otherwise the fluorescent material will be damaged.

3. The oscilloscope must be turned off if it isn't used. Screw down the "Brightness" knob before turning off the oscilloscope.

4. Make sure that the output of the generator or the voltage changer isn't short circuit. Before turning on, the circuit should have been checked by the instructor.

Problems

1. If the following figure 13 – 13 is shown in the screen when observing signals, which knob should be adjusted?

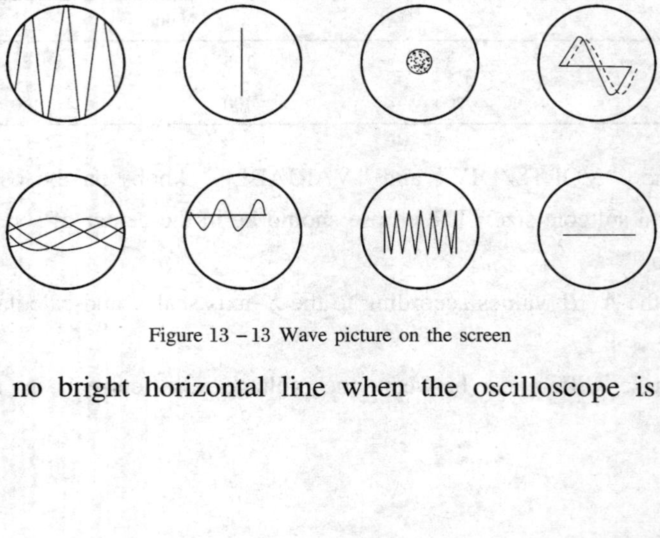

Figure 13 – 13 Wave picture on the screen

2. There is no bright horizontal line when the oscilloscope is working, which

knobs should be adjusted to get the line?

3. The oscilloscope is working, but there is no figure on the screen except a bright spot after input a signal to the Y axis, what is the reason?

4. It can be shown 2 whole waves on the screen for a signal of 500Hz, and switch this signal with another one of an unknown frequency, and 5 whole waves can be shown on the screen, what is the unknown frequency? How to adjust the oscilloscope to show 2 whole waves on the screen again?

5. Can we adjust the "Synchronizing" knob to make the figure stable when composing Lissajou figure?

6. How to measure the voltage by the oscilloscope if a standard signal is given? How to?

7. If you find the ellipse on the screen is not inclined to but straightly standing, what is the reason? How to make it inclined?

Additional Instruction

Preparation requirements

1. Understand the working principle of an oscilloscope.
2. Learn the function of every part of the oscilloscope panel.
3. Understand the function of all the knobs according to figure 13 – 1 and figure 13 – 5.
4. Understand the function of all knobs of the DF1636A signal generator.
5. Understand the principle of measuring phase difference of two alternate signals.

Operating key points

1. The oscilloscope panel can be divided into four functional parts, as is shown in Figure 13 – 14.

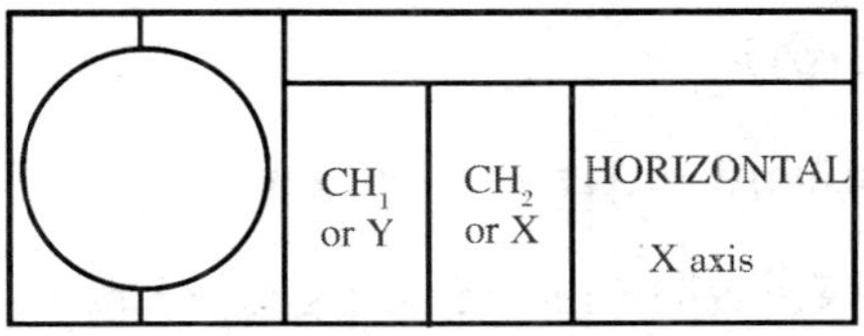

Figure 13 – 14 The four functional parts of the oscilloscope panel

2. The upper part on the right hand contains knobs of whole function controllers of the equipment.

3. The lower part of the right is divided into three main sections. The two left sections are for the CH1, CH2 channels separately. They contain the knobs of controlling the vertical Y – axis; the right section is for the X – axis, and it contains the knobs of controlling the horizontal X – axis. The adjustment of the vertical knobs will change the parameters of the signal that

has been put into Y – axis, and the horizontal knobs for the signal of X – axis. It should be based on the requirements of the experiment to regulate the knobs of X, Y section.

4. The vertical amplitude attenuation knob "VOLTS / DIV" and the scanning speed knob "SWEEP TIME / DIV" all have "fine turning" in their centers, that is knob ⑪, ㉙ and ⑯. They must be placed in the "calibration" position when doing the quantitative measurements; otherwise, they will affect the accuracy of the measurement. Only when used for the qualitative experiments can they be adjusted continuously.

5. When connect the circuit of figure 13 – 12, the side line of the coaxial cable should be grounded and not connected to the central line; otherwise there will be no signal put into the oscilloscope.

6. The following two methods can be used when measuring the value *A* and *B*.

(1) Ground the "CH1" or "Y" channel, the ellipse will become a horizontal straight line, and make the line to coincide with the horizontal scale. Its length is the value of *A*.

(2) Show the ellipse on the screen again, and now the X axis goes through the center of the ellipse and intersects the ellipse in two places. The distance between the two cross – points is the length of *B*.

Report requirements

1. Data table

(1) The experiment data table of measuring the frequency can be referred to table 13 – 1.

(2) The experiment data table of measuring the frequency by the Lissajou figure can be referred to table 13 – 2.

(3) The experiment data table of measuring the phase difference between the two alternate signals frequency can be referred to table 13 – 4.

2. Pay more attention to the requirement of the data digit, as is discussed in Exp. 1.

Problems after experiment

1. When rectangle signal or saw – tooth signal are linked to the Y axis, How to make the figure clear and stable? How to measure their periods?

2. When rectangle signal, saw – tooth and harmonic signal are linked to the Y axis, How to measure their voltages (if the coefficient of the attenuation is given)?

3. For the two sinusoidal signals with the same frequency and the different phase, that is, they have a phase difference. Lissajous graphics is a proper method to measure

the phase difference, but there is another more intuitive measurement method – displacement method. The Lissajous graphics method is not easy to determine the phase difference "plus or minus", while the displacement method with this discriminated function.

Input the two sinusoidal signals, which have the same frequency, but the different phase, into the CH1 and CH2 channel of the oscilloscope respectively (X – Y key can't be pressed, and adjust the "ALT" or "CHOP" of the VERT MODE.), and the two X – axis symmetrical voltage waveforms can be seen on the screen, as is shown in Figure 13 – 15.

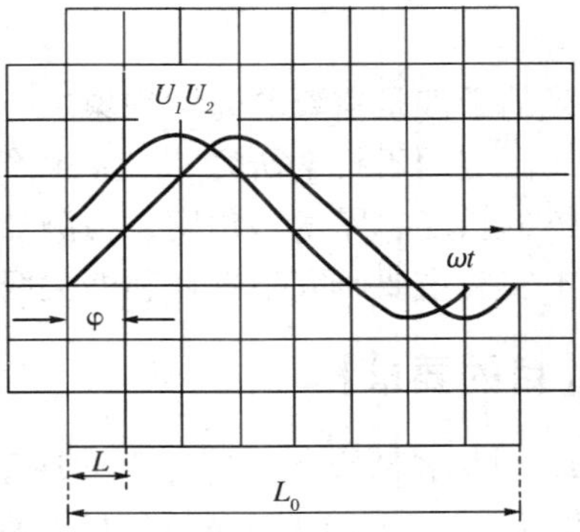

Figure 13 – 15 Measure the phase difference by the displacement method

The length of X direction is proportional to the phase, and we have:

$$\varphi = \frac{L}{L_0} \times 2\pi \ (360°)$$

Where L_0 is the length of a whole waveform, and L for the horizontal displacement of the two waveforms, whose value can be read out by the level scale. The phase difference "plus or minus" can be determined by the map.

Appendix

The panel of *Kenwood* CS – 4025 two traces oscilloscope is shown in figure 13 – 16; please refer to the CS – 4125 for knobs and their functions.

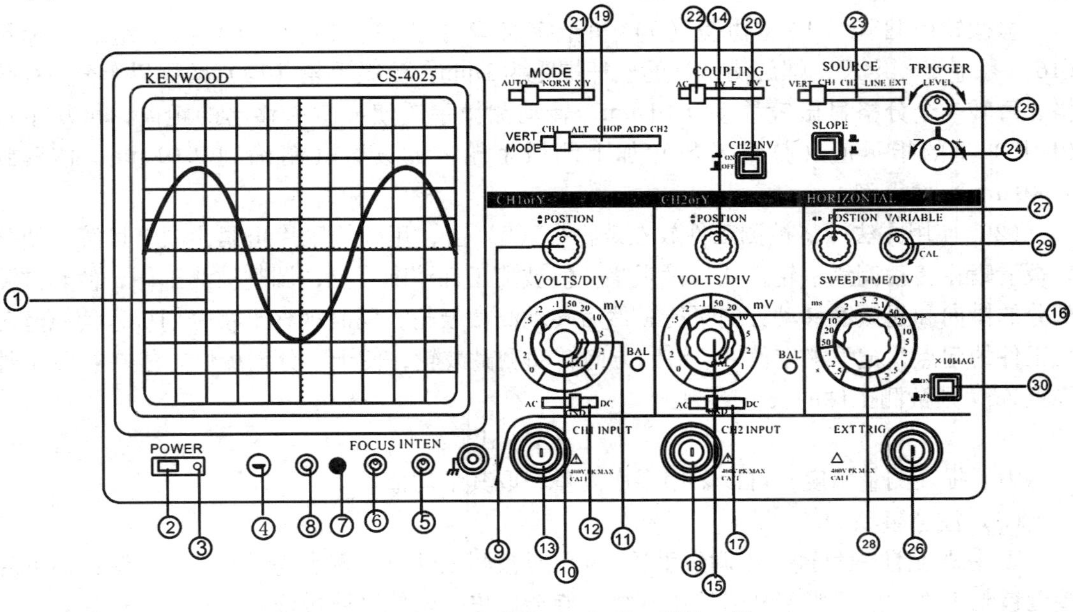

Figure 13 – 16 The panel of Kenwood CS – 4025 oscilloscope

实验十四　光 的 干 涉

光的干涉是重要的光学现象之一，它为光的波动性提供了重要的实验证据。光的干涉现象广泛应用于科学研究、工业生产和检测技术中，如用作测量光波波长，精确地测量微小物体的长度、厚度和角度，检测加工工件表面的光洁度等。

本实验主要研究牛顿环和劈尖两个典型的光的等厚干涉现象。

【目的要求】

1. 观察和研究光的等厚干涉现象及特点。
2. 练习用干涉法测量透镜的曲率半径、物体的微小直径（或厚度）。
3. 学习正确使用读数显微镜及钠光灯。

【仪器器材】

JCD－Ⅱ型读数显微镜、GP20Na－B 型钠光灯、牛顿环、劈尖。

【仪器概述】

1. JCD－Ⅱ型读数显微镜

（1）结构简介　图 14－1 是 JCD－Ⅱ型读数显微镜简图。它由两个主要部件构成：一是用来观看被测物体放大像的带十字叉丝的显微镜；另一个是用来读数的螺旋测微计装置。

显微镜由目镜（1）、物镜（6）和十字叉丝［装在目镜筒（1）内］组成。标尺（16）是毫米刻度尺，螺旋测微计的丝杆螺距为 1mm，测微手轮（15）的周界等分 100 分格，每转一个分格显微镜移动 0.01mm。转动测微手轮使显微镜移动的距离，可从主尺（标尺）上的指示值（整毫米数）加上测微手轮上的读数（精确到 0.01mm，估读到 0.001mm）而得到。

（2）使用方法　将被测工件放在载物台（8）上，用弹簧压片压紧，调节目镜，使分划板清晰；转动调焦手轮，从目镜观察，使被测工件清晰可见；调整被测工件，使其被测部分的横向和显微镜移动方向平行；然后转动测微鼓轮，同时观察十字分划板，使纵丝对准工件的起点，记下测定值 A，沿同方向转动测微鼓轮，使十字分划板恰好停在被测工件终点 A'，则被测工件的长度 L 为：

$$L = A' - A$$

为了提高测量精度，可以采用多次测量，取其平均值。

（3）注意事项

① 在眼睛注视目镜，用调焦手轮（4）对被测工件进行调焦前，应先使物镜镜筒下降接近被测工件，然后眼睛从目镜中观察，旋转调焦手轮使镜筒慢慢向上移动，这就避免了

两者相碰挤坏被测工件的危险。

② 防止螺距误差。由于螺杆和螺母不可能完全紧密接触，当螺旋转动方向改变时，它们的接触状态也将改变，因此移动显微镜使其从相反方向对准同一目标的两次读数将不同，由此产生的测量误差称为螺距误差。为防止螺距误差，在测量时应向同一方向转动测微鼓轮使叉丝和各目标对准。

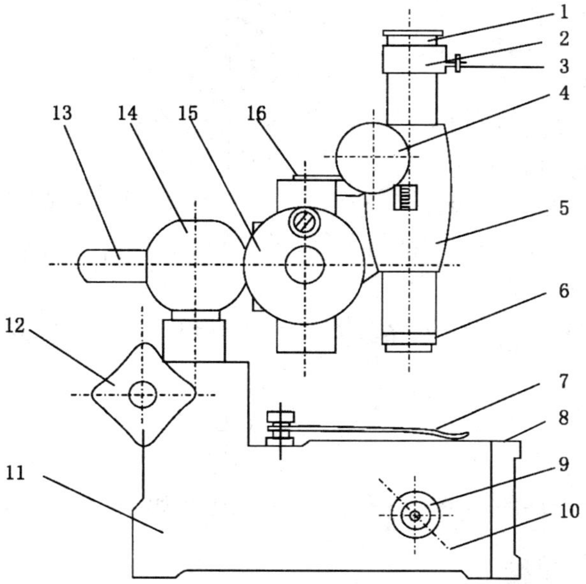

图 14－1　读数显微镜示意图

1. 目镜；2. 锁紧圈；3. 锁紧螺钉；4. 调焦手轮；5. 镜筒支架；6. 物镜；7. 弹簧压片；8. 载物台；9. 旋转手轮；10. 反光镜；11. 底座；12. 旋手；13. 方轴；14. 接头轴；15. 测微手轮；16. 标尺

2. 钠光灯简介

钠蒸汽放电时，发出的光在可见光范围内有两条强谱线：589.0nm 和 589.6nm，通常称为钠双线。因两条谱线很近，通常取平均值 589.3nm 作为该单色光源的波长。由于它的强度大，单色性好，是最常用的单色光源。使用钠光灯时应注意：

（1）钠光灯必须与扼流线圈串接使用，否则易被烧毁。

（2）灯点燃后，需等待一段时间才能正常使用（起燃时间约为 5～6 分钟），又因为忽燃忽熄容易损坏，故点燃后不要轻易熄灭它。另一方面，在正常使用下也有一定消耗，使用寿命只有 500 小时左右，因此在使用时必须注意节省，使用时间尽量集中。

（3）在点燃时不得撞击或震动，否则灼热的灯丝易损坏。

【实验原理】

利用透明薄膜上下两表面对入射光的依次反射，入射光的振幅将分解成有一定光程差的几个部分。两束反射光相遇时的光程差取决于薄膜的厚度，同一干涉条纹所对应的薄膜厚度相同，这就是所谓的等厚干涉。这是一种获得相干光的重要途径，它被多种干涉仪所采用。牛顿环和劈尖干涉都是典型的等厚干涉。

1. 牛顿环

将一块平凸透镜（其曲率半径 R 较大）的曲面置于光学平板玻璃上，在透镜凸面和平板玻璃面间就形成一层空气薄膜，其厚度从中心接触点到边缘逐渐增加。当以平行单色光垂直入射时，入射光将在此薄膜上下两表面反射，产生具有一定光程差的两束相干光。由于以接触点为圆心的同一圆周上各点的光程差相同，因此干涉图样是以接触点为中心的一系列明暗交替的同心圆环——牛顿环。其光路图见图14－2。

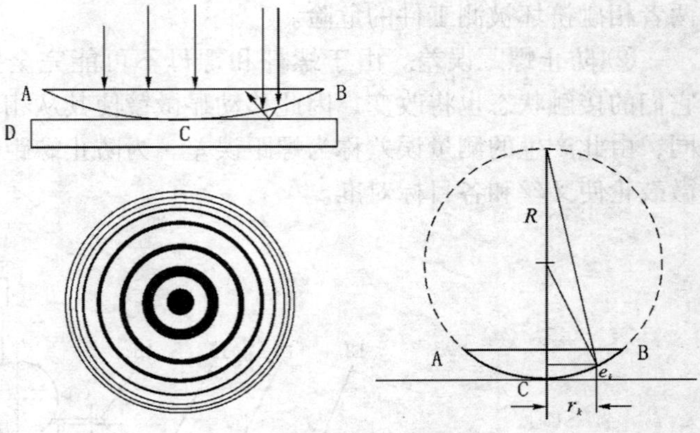

图14－2　牛顿环光路图

由光路图分析可知：与第 k 级条纹对应的两束相干光的光程差（空气折射率近似为1）为：

$$\delta_k = 2e_k + \frac{\lambda}{2} \qquad (1)$$

其中 e_k 是第 k 级暗环的半径，$\frac{\lambda}{2}$ 是由于光从光疏媒质射到光密媒质的交界面上反射时，发生半波损失引起的。

由图14－2可知

$$R^2 = {r_k}^2 + (R - e_k)^2$$

由于空气薄膜厚度 e 远小于透镜的曲率半径，则可略去二级小量 e^2，于是有

$$e_k = r_k^2 / 2R \qquad (2)$$

由干涉条件可知，当光程差为半波长的奇数倍时，干涉条纹为暗条纹，由式（1）有：

$$\delta_k = 2e_k + \frac{\lambda}{2} = (2k+1)\frac{\lambda}{2} \qquad k=0, 1, 2, 3\cdots \qquad (3)$$

将（2）式代入（3）式可得

$$r_k^2 = kR\lambda \qquad k=0, 1, 2\cdots \qquad (4)$$

如果已知入射光的波长为 λ，并测得第 k 级暗条纹的半径 r_k，则由式（4）可算出透镜的曲率半径 R。但实际观察牛顿环将会发现，牛顿环中心不是一点，而是一个不甚清晰的暗或亮的圆斑。其原因是透镜和玻璃平板接触时，由于接触压力引起形变，使接触处为一圆面。实验中，镜面上可能有微小灰尘、油膜等存在，从而引起附加的光程差，这都会给测量带来较大的系统误差。因此难以准确判定级数 k 和测出 r_k。

我们可以通过取两个暗条纹半径的平方值之差，来消除附加光程带来的误差。

由式（4）可得

$$R = \frac{r_m^2 - r_n^2}{(m-n)\lambda} \qquad (5)$$

又因暗环圆心不易确定，故将上式改写成

$$R=\frac{(r_m-r_n)(r_m+r_n)}{(m-n)\ \lambda}$$

可以用 x_n、x_m 表示圆环半径 r_n、r_m 在显微镜上的相应位置上的坐标，如图 14－3，注意到：

$$x_m-x_n=r_m-r_n$$

$$x_m-x_n{}'=r_m+r_n$$

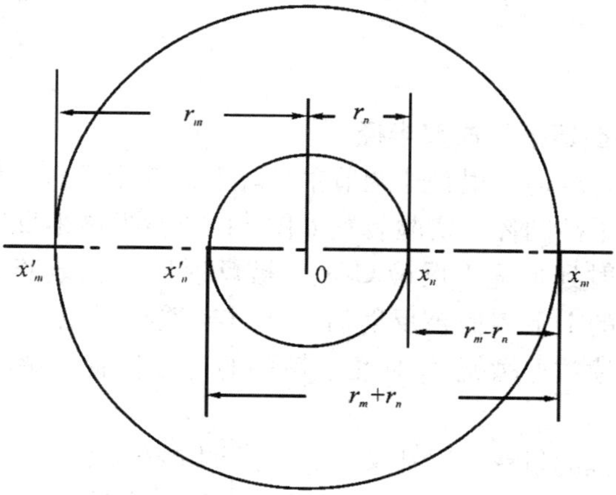

图 14－3　半径 r_n、r_m 示意图

所以

$$R=\frac{(x_m-x_n)\ (x_m-x_n{}')}{(m-n)\ \lambda} \tag{6}$$

可见，在实验中不必知道环的中心位置，只需读出 m、n、n'的位置坐标 x_m、x_n、$x_n{}'$即可。

2. 劈尖

将两块光学平板玻璃叠在一起，在一端插入一薄片（或细丝等），则在两玻璃板间形成一空气劈尖。

当用单色光垂直照射时，和牛顿环一样，在劈尖薄膜上下两表面反射的两束光发生干涉，其光程差由式（1）表示。产生的干涉条纹是一簇与两玻璃板交接线平行且间隔相等的平行条纹（图 14－4）。

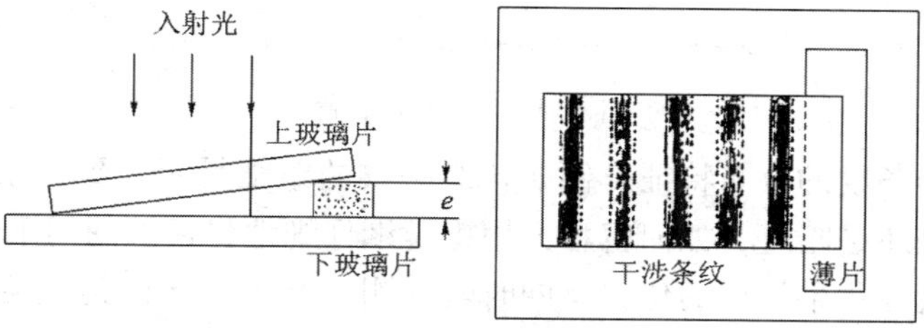

图 14－4　劈尖形成的干涉条纹

显然，当 $\delta_k=2e_k+\frac{\lambda}{2}=(2k+1)\frac{\lambda}{2}$时，为干涉暗条纹。

与第 k 级暗条纹对应的薄膜厚度为：

$$e_k = k\lambda/2 \tag{7}$$

可知两相邻明（或暗）条纹所对应的空气薄膜厚度为： $\Delta e = \lambda/2$。

若两玻璃板交线处到薄片处的劈尖面上共有 K 条干涉条纹，则待测薄片厚度为：

$$e = K\lambda/2 \tag{8}$$

在干涉条纹很多的情况下，不必直接数总条纹数，只要测出单位长度上的干涉条纹数 n 及劈尖的长度 L，则总条纹数 $K = nL$ 代入式（8）得：$e = nL\lambda/2$，即可求出薄片厚度或细丝直径等微小量。

【实验内容】

1. 用牛顿环测量平凸透镜的曲率半径

（1）调整实验装置。均匀且很轻地调节牛顿环上的三个螺丝，使牛顿环的中央部分出现一组同心的无畸变的干涉圆环。实验装置如图 14-5，半透半反镜 G 和水平方向成 45°角，从钠光灯发出的光射到半透半反镜 G 上，被反射后垂直地照射到牛顿环装置 N 上。在牛顿环装置空气薄膜的上下表面被反射后，透过半透半反镜 G 射入读数显微镜。调节牛顿环装置的位置，使读数显微镜 A 对准牛顿环的中心位置。调节钠光灯的高度，使显微镜视场中亮度最大。

（2）调节读数显微镜的目镜，看清叉丝，然后调节镜筒高度：先把镜筒往下调，使物镜慢慢靠近被观察物而不接触。做这一步时眼睛必须在旁监视，一定不能将镜头碰到物体，以免碰坏牛顿环和镜头。然后眼睛通过目镜观察视场，使镜筒缓慢上升，直至在显微镜整个视场都能观察到清晰的干涉条纹。

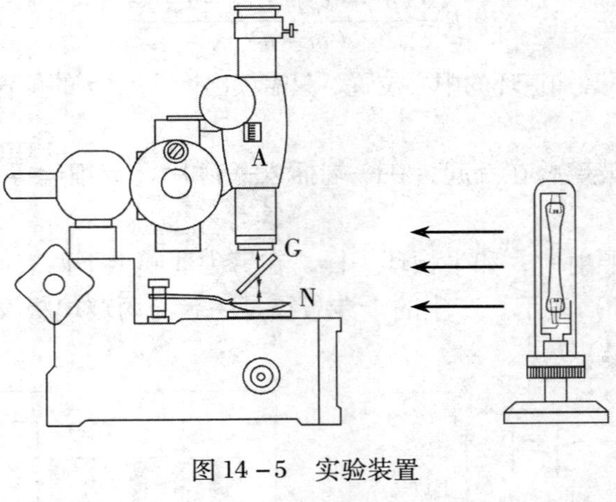

图 14-5　实验装置

（3）观察干涉条纹的分布特征。例如各级条纹的粗细是否一致？条纹间隔是否均匀？牛顿环中心是亮斑还是暗斑，如何解释？用擦镜纸仔细地将接触的两个表面擦干净，可使中心呈暗斑。轻微变化牛顿环的位置，同时转动测微鼓轮（显微镜也会一起动），使显微镜交叉点在牛顿环中心暗环上，同时观察整个调节范围是否能测出最大直径，如果条件不满足，还需对仪器或牛顿环位置进行调整。

（4）测牛顿环半径 R。根据式（6），λ 为入射光的波长，$\lambda = 589.3\text{nm}$，为了避免转

动部件的螺旋间隙产生的螺距误差，实验过程中鼓轮只能沿一个方向转动：旋转微调鼓轮（或微调螺丝），使显微镜中的十字叉丝从中心 O 开始向右数 27 条暗条纹，然后退两环到 25 条，使竖直的叉丝与第 25 条暗环相切，记下 r_{25} 的坐标 x_{25}，再到第 5 条暗环，记下 r_5 的位置 x_5，经过中心 O 到左边第 5 条暗环，记下 x_5'，继续到左边第 25 条暗环，记下 x_{25}'。最后再从第 27 条暗环，退回到第 25 条，由左至右重复记录下另一次测量数据。将数据填入表 14－1 中，并代入式（6）计算 R。

表 14－1 牛顿环半径测量数据（单位：mm）

	x_{25}'	x_5'	x_5	x_{25}	R	$\overline{R}$
从右向左						
从左向右						

2. 劈尖干涉法测微小厚度（或微小直径）

（1）将被测薄片（或细丝）夹在两平板玻璃的一端，另一端直接接触，形成劈尖，然后置于显微物镜载台上。用显微镜观察，描绘劈尖干涉的图样。

（2）调节叉丝劈尖放置方位，使镜筒移动方向与干涉条纹相垂直，以便准确测出条纹间距。

（3）用读数显微镜测出 20 条暗条纹间的垂直距离 l，再测出劈尖棱边到薄片所在处的总长度 L，将数据填入表 14－2 中，求出薄片厚度 d。

表 14－2　劈尖测量数据（单位：mm）

l_0	l_{20}	$n=\frac{20}{l_{20}-l_0}$	L	$e=nL\lambda/2$

【思考题】

1. 透射光的牛顿环是如何形成的？如何观察？它与反射光的牛顿环在明暗上有何关系？为什么？

2. 若改用白光照射，能否看到牛顿环和劈尖的干涉条纹？此时的干涉条纹有何特征？

3. 为什么读数显微镜测量的是牛顿环的直径，而不是牛顿环放大像的直径？如果改变显微镜筒的放大倍率，是否会影响测量结果？

4. 在本实验中若遇到下列情况对实验结果是否有影响？为什么？

（1）牛顿环中心是亮斑而非暗斑；

（2）读数显微镜的十字叉线的横线未通过圆环的中心，因而测量的是弦而非真正的直径。（图 14－6）

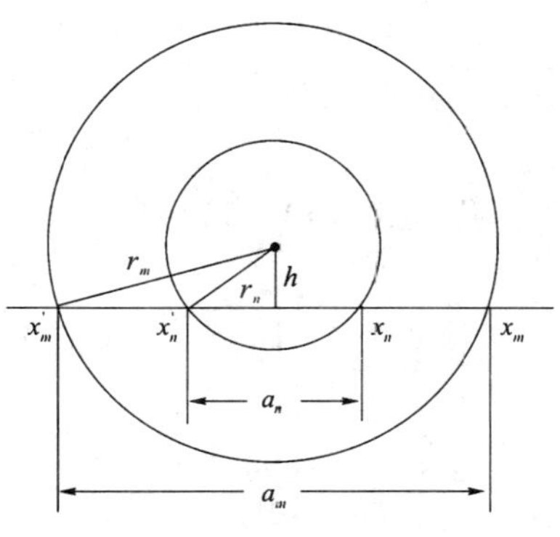

图 14－6　测量示意图

实 验 指 导

【预习要求】

1. 何谓等厚干涉？如何应用光的等厚干涉测量平凸透镜的曲率半径和薄片厚度？

2. 如何使用读数显微镜？使用读数显微镜应注意哪些问题？

3. 认真阅读实验讲义，根据实验原理和实验内容，写好预习报告。

【操作要点】

1. 使用读数显微镜测一组数据时，只能从一个方向开始单向移动，中途不能反转，以免引入螺距误差。

2. 由于读数显微镜具有一定的量程范围，实验中应使牛顿环的中央区域所对应的刻度大致在读数显微镜标尺的中心。

3. 桌面要平稳，不能振动，读数显微镜不能摇晃，更不可数错暗纹的圈数，否则要重测。

4. 实验完毕后应将牛顿环装置上的三个螺丝松开，以免牛顿环装置变形。

5. 钠光灯关闭后，必须稍停片刻才能重新启动。

【实验后的思考】

1. 牛顿环的中心斑在什么情况下是暗的？在什么情况下是明的？

2. 透射光能否形成牛顿环？它和反射光所形成的牛顿环有何区别？

3. 如何用劈尖干涉检测光学平面的表面质量？

4. 测量牛顿环半径过程中，要求显微镜中叉丝的纵线与干涉环相切，因为干涉条纹有一定的宽度，因此相切的“程度”要一致。实际上在实验中可以由十字叉丝的横线与干涉环相割所得弦长来代替半径测量。如图 14－6 所示，r_m、r_n为第 m 级、第 n 级干涉圆环的半径，a_m、a_n为相应的弦长，h 为弦到圆心的距离，则

$$r_m^2 = (a_m/2)^2 + h^2$$
$$r_n^2 = (a_n/2)^2 + h^2$$

两式相减得：

$$r_m^2 - r_n^2 = (a_m^2 - a_n^2)/4 + h^2$$

代入式（14－5）有

$$R = (a_m^2 - a_n^2)/4(m-n)\lambda$$

14 Interference of Light

Interference of light is one of the very important optical phenomena, which provided the strongly proof for the wave character of light. Interference of light has been widely applied in the field of science research, industry production, and detection technology. The principle of interference of light can be used for measuring the length, thickness and angles of micro – object, examining surface roughness of work piece and so on.

Newton′ s ring and optical wedge, as two typical equal – thickness interferences of light, have been studied in this experiment.

Objective

1. Study the equal thickness interference of light.
2. Measure radius of optical lens and minute thickness.
3. Learn to use reading microscope and sodium lamp.

Equipments

Reading microscope, type JCD – II; Sodium lamp, type GP20Na – B; Newton′s ring; wedge

Introduction of Equipments

1. Reading microscope, type JCD – II

(1) Introduction of structure

Figure 14 – 1 is the configuration of JCD – II reading microscope. Reading microscope is made up of two main parts: One is the microscope with cross filament which is used to watch the enlarge image clearly, the other is used to read the number of the screw micrometer equipment.

Microscope is made up of three parts: ocular glass (1), object lens (6) and cross filament (installed in ocular tube). The measuring scale (16) is a millimeter dividing ruler. The pitch of lead screw of the micrometer is 1mm. The circumference of the micrometric hand wheel is divided into 100 mesh division, so the microscope will move 0.01mm when the micrometric hand wheel turn one mesh division. The sum of the indicated value on the measuring scale (whole millimeter) and the numerical reading from the micrometric hand wheel is the moving distance of microscope.

(2) Operation

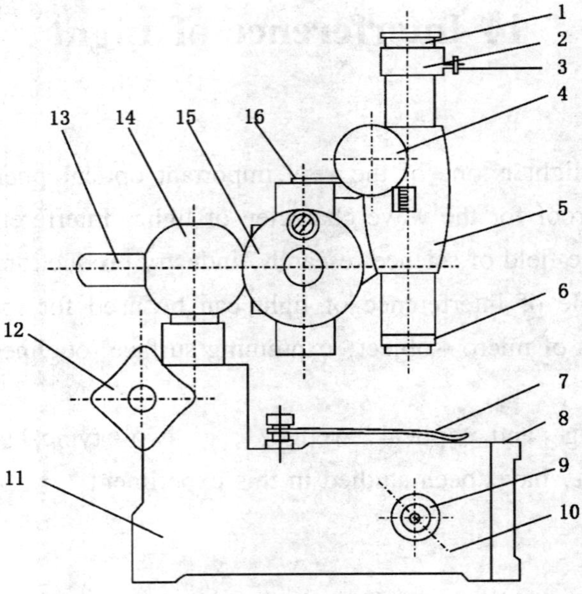

Figure 14 – 1 Configuration of the reading microscope

1. ocular glass; 2. clamp ring; 3. locking screw; 4. focus hand wheel; 5. cylinder holder; 6. objective lens; 7. spring presser bit; 8. object stage; 9. rotary hand wheel; 10. reflecting mirror; 11. base frame; 12. hand wheel; 13. square shaft; 14. junction shaft; 15. micrometric hand wheel; 16. measuring scale

Put the measured workpiece on the object stage (8) and compress it by the spring presser bit. Adjust the ocular glass of reading microscope in order to see cross filament clearly. And then turn the focus hand wheel to see the measured workpiece clearly. Adjust the measured workpiece in order to make the workpiece's landscape orientation parallel to the moving direction of the microscope. Turn the micrometric hand wheel and make the longitudinal filament aim at the starting point of the workpiece. Write down the measured value A. Turn the micrometric hand wheel along the same direction to make the longitudinal filament stop at the end point A' of the workpiece. Then the length of the workpiece is:

$$L = A' - A$$

In order to improve the measurement accuracy, we can measure the length many times and use the average value as final result.

(3) Matters need attention

① Fall the objective lens tube down and approach the measured workpiece before measurement. In order to avoid damaging the workpiece, you should turn the focus hand wheel to move the objective up slowly while you observe from the ocular glass.

② Prevention screw – pitch error. Since the screw rod and the screw nut can't contact tightly, when the movement direction of the screw change, the contact state of them will change at the same time. So the reading will different when you measure the same object from forward and backward direction. You should turn the micrometric hand wheel along same direction to make the longitudinal filament aim all measured target in order to avoid screw – pitch error.

2. Sodium lamp

There are two strong lines (589.0nm and 589.6nm) between visible regions when sodium vapors discharge. Since two spectrum lines are very close to each other, the Sodium lamp can be regarded as a good monochromatic source in experiment and the correspondence wavelength is 589.3nm, which is the average of the two spectrum lines. Pay attention when you use Sodium lamp:

(1) Sodium lamp should be concatenated with protective choke.

(2) After Sodium lamp is ignited, wait a minute before you use it regularly. Since the lamp is easily damaged if you turn on it several times in short time, so do not extinguish the lamp before you finish experiment.

(3) Don't strike or shake the lamp when it is igniting, otherwise the glowing filament will be easily damaged.

Principle

The incident light will be splitted into several beams when it is reflected by the two layers of a transparent film and the difference of the optical path of the beams are determined by the thickness of the film and the incident angle. The place of the film where produce the same interference fringe will have the same thickness, and this conclusion is called equal thickness interference.

1. Newton's ring

Put a plane convex lens on an optical sheet glass with curved side (of great radius) facing and contacting the sheet glass. Then, an air film will be formed between plane convex lens and optical sheet glass, and its thickness increases gradually from the middle to edge. When a parallel monocolour light normally incident on them, the two beams of the reflection light will produce interference fringes. The fringes are alternately dark and bright concentric rings with a black round area in the center, this fringes are called Newton's rings. See figure 14 – 2.

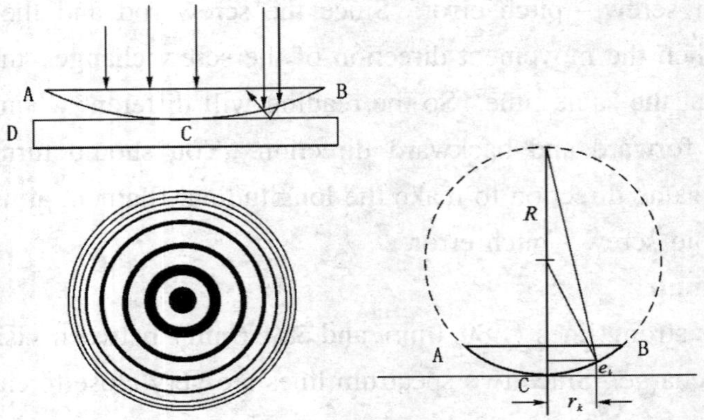

Figure 14 -2 Optical path of Newton ring

According to the light path analysis, the optical path difference between the two beams of the coherent light of order k fringe can be expressed as:

$$\delta_k = 2e_k + \frac{\lambda}{2} \qquad (1)$$

Where, the e_k is the radius of the kth dark ring, $\lambda/2$ is half wave loss when the light reflect from the optically denser medium.

There is a relation: $R^2 = r_k{}^2 + (R - e_k)^2$

Because the thickness of air film is greatly less than radius of lens, the e_k^2 can be omitted:

$$e_k = r_k^2/2R \qquad (2)$$

From interference conditions we know interference fringe is black if optical path difference equal to odd multiple of half wavelength, so we can get:

$$\delta_k = 2e_k + \frac{\lambda}{2} = (2k+1)\frac{\lambda}{2} \qquad k = 0,\ 1,\ 2,\ 3\cdots \qquad (3)$$

Substitute (2) to (3)

$$r_k^2 = kR\lambda \qquad k = 0,\ 1,\ 2\cdots \qquad (4)$$

We can calculate the radius of the lens R from the known wavelength of incident light λ and the radius of the order k of fringe r_k. Because of the deformation caused by contacted pressure and its make the contact place to be a circle surface, the center of Newton's ring will be a black round spot. If there are minute dust, oil film on the mirror surface, additional optical path difference maybe produced. All these will cause system error for measuring. So it's difficult to determine series k and measure r_k accurately.

We can use radius square value difference of two black fringes to eliminate additional error.

$$R = \frac{r_m^2 - r_n^2}{(m-n)\lambda} \qquad (5)$$

It is not easy to determine the center of the fringe, we rewrite the above equation:

$$R = \frac{(r_m - r_n)(r_m + r_n)}{(m-n)\lambda}.$$

Let x_m, x_n denote the coordinates of the fringes with radius r_m, r_n shown on the reading microscope (see Figure 14 −3), then:

$$x_m - x_n = r_m - r_n$$

$$x_m - x_n' = r_m + r_n.$$

The radius R:

$$R = \frac{(x_m - x_n)(x_m - x_n')}{(m-n)\lambda} \qquad (6)$$

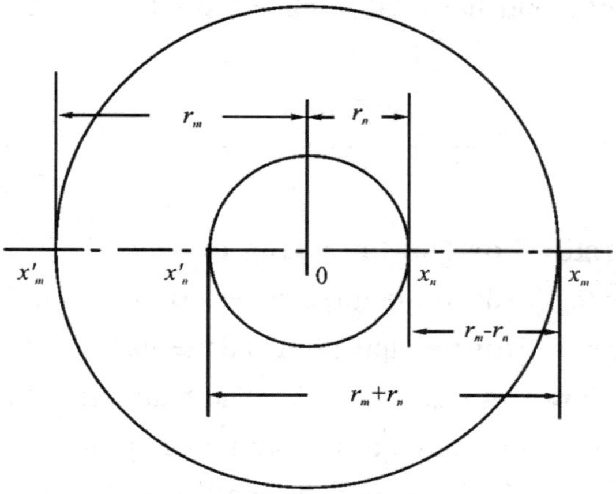

Figure 14 −3 Fringes with radius r_m, r_n

We need not know the center position of fringe as long as to read site coordinates x_m、x_n、x_n' in experiment.

2. Optical wedge

Pile two optical sheet glasses up, insert a flake film (or filament) at one end, the air between the two sheets will form a wedge.

When parellel monochromatic light normally incident on the wedge, just like it happening in Newton's ring, there will appear interference fringes. The figure of the fringes is made up by equidistant parallel black and bright stripes, which are parallel with edge of glasses, as shown in figure 14 −4.

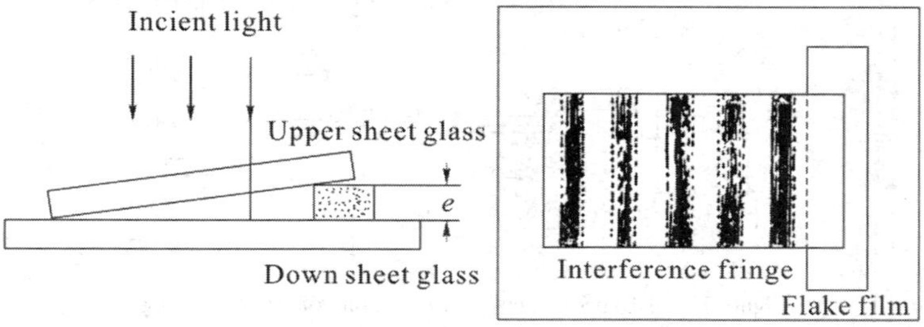

Figure. 14 −4 Interference stripe of wedge

Black stripe will appear at the place where the optical path difference is:

$$\delta_k = 2e_k + \lambda/2 = (2k+1)\ \lambda/2 \qquad (k=0,\ 1,\ 2,\ \cdots)$$

The film thickness e_k corresponding with the order k black stripe is:

$$e_k = k\lambda/2 \tag{7}$$

And the thickness difference between two adjacent dark stripe is: $\Delta e = \lambda/2$.

If there are K stripes totally, the maximal thickness of flake will be:

$$e = K\lambda/2 \tag{8}$$

If there are too many stripes, it is better to measure the number n of interference stripes per unit length first, and then the wedge's length L. The total stripe number can be determined by $K = nL$, and the total thickness of flake or the diameter of thread et al. will be:

$$e = nL\lambda/2$$

Experiment

1. Measure lens radius by Newton's ring method

(1) Adjust the device. Adjust the three screws on Newton's ring gently to make a set of orthoscopic concentric fringes appearing in the central of Newton' s ring. The experimental device is shown in figure 14 – 5. The half splitted mirror G is inclined about 45° with the horizon, the light from Sodium lamp will be reflected by the half splitted mirror G, then normally incidence on the lens of setup N (also called Newton' s ring setup). The beam reflecting on the two surface of air film will back into the lens and half of them will incident into the reading microscope. Move the position of the Newton's ring setup and make the reading microscope aim to the center of Newton's ring. Adjust Sodium lamp's height, half splitted mirror and position of Newton's ring setup, to make the field of view of the reading microscope bright.

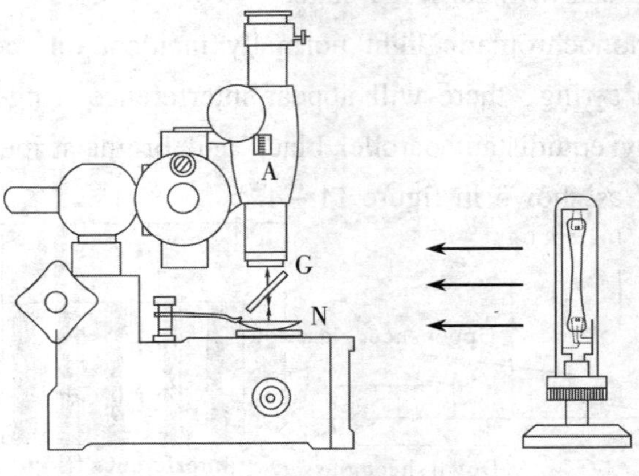

Figure 14 – 5 Experimental configuration of Newton's ring

(2) Adjust the ocular glass of reading microscope in order to see cross filament

clearly. And then adjust the cylinder's height slowly until the fringes can be seen clearly in the whole field of view of the reading microscope.

(3) Observe distribution properties of the interference fringes. Such as: Are the fringes' breadth in different order same? The fringe's interval is unique or not? How to explain a bright/black spot appeared at the center of Newton's ring? If two contact surfaces of the lens and the mirror are cleaned carefully, a black spot will be appeared at the center of the Newton's ring.

(4) Measure radius of the optical lens R. The radius R of Newton's ring can be gotten according to equation (6), where, λ is the wavelength of incident light, $\lambda = 589.3\text{nm}$, $m = 25$, $n = 5$. To avoid pitch error, the screw drum should be turned in same direction during the measuring process: Turn micrometric hand wheel to move cross filament from centre to 27th black stripe, then back to 25th black strip and we can get four coordinates data for one measurement, they are: x_{25}', x_5', x_5 *and* x_{25}, respectively. And we can measure the four data as another measurement in the opposite direction. Now we can fill all measuring data into table 14 – 1 and calculate R according to equation (6).

Table 14 – 1 Data of Newton's ring (unit: mm)

	x_{25}'	x_5'	x_5	x_{25}	R	$\overline{R}$
From right to left						
From left to right						

2. Measure the minute thickness with optical wedge

(1) Put a thin film at one side of two glass plates to form optical wedge, then place them on the objective stage of the microscope to observe the interference fringes. Draw a figure of them. Replace the thin film with another and observe the fringes' difference.

(2) In order to measure the fringe spacing accurately, the moving direction of cylinder should be normal to the interference fringe by adjusting cross filament and optical wedge.

(3) Measure vertical distance l of 20 black fringes and the total length L of wedge. Fill data into table 14 – 2 and calculate flake thickness d.

Table 14 – 2 Data of Optical wedge (unit: mm)

l_0	l_{20}	$n = \frac{20}{l_{20} - l_0}$	L	$e = nL\lambda/2$

Problems

1. How is the Newton's ring formed by transmitted light? How to observe it? What is the difference between the fringes formed by reflected light and transmitted light of Newton's ring? Why?

2. For a beam of white incidence light, what will happen to the interference fringes if the incidence light is a white light?

3. The measured results by microscope are actual diameter of the Newton's ring rather than image's diameter, why? Is there influence on the measurement results if the magnification of the microscope is changed?

4. In the following cases, what will happen to the measurement result? And why?

(1) The center of Newton's ring is bright spot rather than dark spot;

(2) The horizontal line (filament) of reading microscope doesn't cross the center of fringes circle, the measuring result of length is of a chord rather than of the diameter, see figure14 -6.

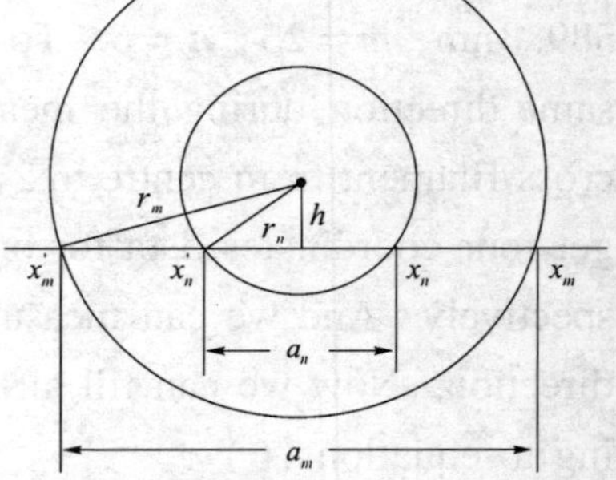

Figure 14 -6 Detection sketch

Additional Instruction

Preparation Requirement

1. What is equal thickness interference? How to measure radius of curvature and flake thickness by equal thickness interference?

2. How to use reading microscope? What should be pay attention to during experiment?

3. Read experiment book carefully and write preparation report according to experiment principle and experiment contents.

Operation Key Points

1. To avoid pitch error, turning the screw drum of the microscope in the same direction during you measure a series of data.

2. It's better to place the Newton's ring in the middle of view field approximately.

3. Keep the desk and microscope stable, and count the fringes correctly.

4. After finish experiment, loosen the three screws of Newton's ring setup to avoid

deformation.

5. After shut Sodium lamp down, it should wait for a moment for a new start.

Problems after experiment

1. In which condition will the centre spot of Newton's ring be black? In which condition will it be bright?

2. Can transmitting light form Newton's ring? What difference between two Newton's rings, which formed by transmitting light and reflected light?

3. How to measure surface quality of optical flat by wedge?

4. During the detection procedure, the vertical line of the cross hair of the microscope should be contacted with interference ring and behaved as tangent line of the interference ring. Since the interference fringes have definitive width, the tangency degree should be kept in same level. Actually, the transverse line of the cross symbol can be used to cut the interference ring to get the chord instead of the radius measurement, and this will help to avoid the determination of the tangency degree.

As shown in figure 14 -6, r_m、r_n is the radius of the mth and n^{th} of the interference ring, and the a_m、a_n are the corresponding chords, h is the distance between the chord and center of the circle, and then:

$$r_m^2 = (a_m/2)^2 + h^2$$
$$r_n^2 = (a_n/2)^2 + h^2$$

Subtracted with each other:

$$r_m^2 - r_n^2 = (a_m^2 - a_n^2)/4 + h^2$$

Substitute equation (14 -5) into last equation:

$$R = (a_m^2 - a_n^2)/4(m-n)\lambda$$

实验十五　分光计的调节和使用

光线在传播过程中，遇到不同媒质的分界面时，会发生反射和折射，入射光与反射光、入射光与折射光之间就有一定的夹角。通过对一些角度的测量，再利用反射定律、折射定律等光学定律可求出折射率、色散率、光栅常数、光波波长等光学量。因此精确测量角度在光学实验中显得非常重要。分光计就是一种用来精确测量角度的光学仪器；另外分光计和其他一些光学仪器如分光光度计、摄谱仪和单色仪等在结构上有很多相似之处，是这类仪器的一种雏形。

【目的要求】

1. 了解分光计的构造及各部件的作用。
2. 掌握分光计的调节原理及调节方法。
3. 用分光计测定三棱镜的顶角。

【仪器器材】

JJY 型分光计及其附件、汞灯、玻璃三棱镜、平面镜。

【仪器概述】

1. 分光计的构造

分光计由五个部分组成：平行光管、望远镜、载物平台、读数圆盘和底座。图 15－1 是其外形图。

（1）三角底座　底座（20）是整个分光计的支架。其中心有一垂直方向的转轴，即仪器的中心轴。望远镜、载物平台、刻度盘及游标盘均可绕该轴转动。

（2）望远镜　阿贝式自准直望远镜（10）用来确定光线传播的方向。它由阿贝目镜 b（消色差目镜 c、全反射棱镜 d、分划板 e）和消色差物镜 f 组成，见图 15－2。物镜 f 装在镜筒的一端，阿贝目镜 b 装在镜筒的另一端套筒 g 中。在目镜套筒的侧面开有一小孔 a，小孔旁装有一小灯泡，它发出的光经棱镜全反射后照亮分划板上小十字缝，并沿望远镜筒向外传播。调节手轮（11）可改变目镜 c 和分划板 e 的相对位置，即对目镜调焦，从而能看清分划板上的黑刻线。分划板的正视图见图 15－3。由于棱镜 d 在视场中挡掉一部分光线，故呈现出它的阴影。松开螺钉（9），沿望远镜光轴方向移动目镜筒可改变物镜与分划板之间的距离，使分划板能同时位于物镜和目镜的焦平面上。

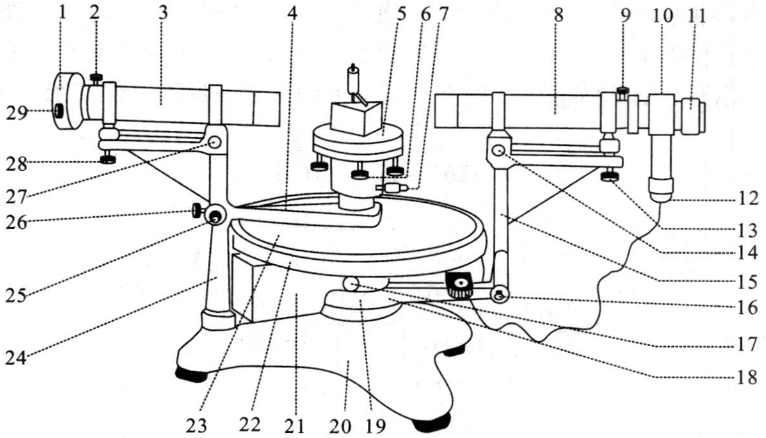

图 15-1　JJY 分光计外形图

1. 狭缝装置；2. 狭缝装置锁紧螺钉；3. 平行光管部件；4. 制动架（一）；5. 载物平台；
6. 载物平台调平螺钉（3 只）；7. 载物平台锁紧螺钉；8. 望远镜部件；9. 目镜锁紧螺钉；
10. 阿贝式自准直目镜；11. 目镜视度调节手轮；12. 照明小灯；13. 望远镜光轴高低调节螺钉；
14. 望远镜光轴水平调节螺钉；15. 支臂；16. 望远镜微调螺钉；17. 望远镜与读数圆盘锁紧螺钉；
18. 望远镜锁紧螺钉（在另一侧）；19. 制动架（二）；20. 底座；21. 转座；22. 刻度盘；23. 游标盘；
24. 立柱；25. 游标盘微调螺钉；26. 游标盘锁紧螺钉；27. 平行光管光轴水平调节螺钉；
28. 平行光管高低调节螺钉；29. 狭缝宽度调节手轮；

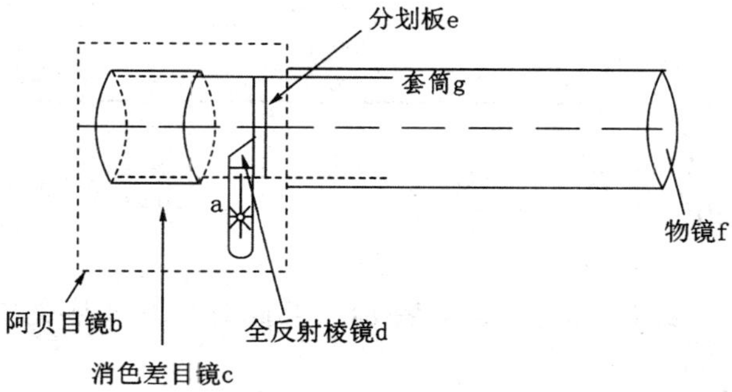

图 15-2　望远镜示意图

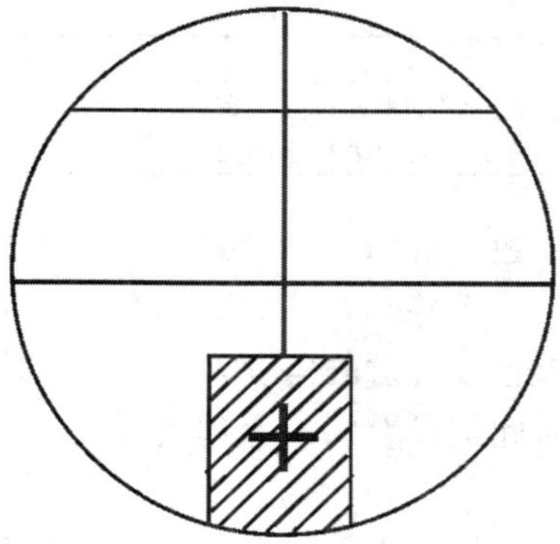

图 15-3　分划板正视图

调节螺钉（14）可改变望远镜光轴水平位置，调节螺钉（13）可改变望远镜光轴高低位置，使之垂直于仪器中心轴。螺钉（18）用来固定望远镜，松开它，望远镜可绕中心轴转动，当它锁紧后，可用螺钉（16）微调望远镜的水平位置。

（3）载物台　载物台（5），用以放置平面镜、三棱镜、光栅等光学元件。松开螺钉（7），平台（5）可绕中心轴转动，也可根据需要升高或降低。调节到所需位置后，再用螺钉（7）将之固定。用螺钉（6）（由三只呈正三角放置的螺钉组成）调节平台水平，使之与中心轴垂直。

（4）读数圆盘　读数圆盘用来指示望远镜或载物台旋转的角度。它由刻度盘（22）和游标盘（23）组成。锁紧螺钉（17），松开（18），望远镜可与刻度盘一起转动。刻度盘刻有 720 等分的刻线，每小格为半度；游标盘通过螺钉（7）与平台相连。松开螺钉

(26)，游标盘可带动平台绕中心轴转动。当（26）锁紧时，可用螺钉（25）微调游标盘位置。在游标盘相隔 180°位置有两个对称游标 A 和 B，各有 30 个分格，它和刻度盘上 29 个分格的角度相等，故此游标可读到 1′。读数方法按游标原理。例如图 15－4 所示的读数为

$$116° + 12' = 116°12'$$

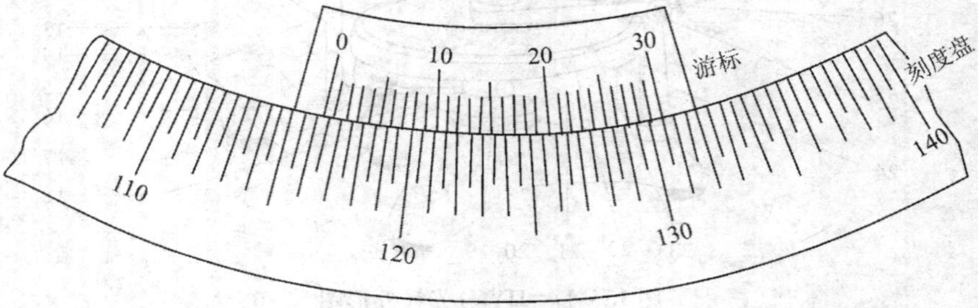

图 15－4　读数示意图

读数时可手持照明放大镜协助读数。为了克服由于刻度盘与游标盘可能不是严格同轴所造成的偏心误差，每次测量必须分别读出 A、B 游标的读数，然后取平均值作为望远镜或载物台转过的角度。例如：

	游标 A	游标 B
望远镜的初始位置读数	330°45′	150°42′
望远镜转过 θ 角后读数	90°47′	270°46

差数：游标 A：$(360° + 90°47') - 330°45' = 120°2'$

游标 B：$270°46' - 150°42' = 120°4'$

望远镜转过的角度 θ 为：120°3′。

从此例可看到，望远镜转动时，游标 A 经过 0°位置，要特别注意此时计算角度的方法。

(5) 平行光管　平行光管用来获得平行光束。由狭缝（1）和装在管的另一端的物镜组成。狭缝可由螺钉（29）来调节宽度，调节的范围是 0.02～2mm。螺钉（2）是狭缝固定螺钉，松开（2）可以调节狭缝和物镜的距离，并可转动狭缝。当狭缝正好位于物镜的焦平面时，从狭缝射进来的光经物镜后就成为平行光。平行光管与分光计的底座固定在一起。螺钉（27）、(28）分别用来调节平行光管的水平和高低位置。

2. 仪器调节

分光计的调节标准有三条。

(1) 望远镜能接受平行光。

(2) 平行光管能发出平行光。

(3) 望远镜光轴和平行光管光轴垂直于转轴；待测元件的光学面平行于转轴。

为保证这些条件，必须对分光计进行下述调节，其中望远镜的调节最为重要，其他调节均以望远镜为准。

(1) 目测粗调　调节螺钉（13)，使望远镜镜筒大致水平；调节螺钉（28)，使平行

光管大致水平；调节螺钉组件6，使载物平台与转轴大致垂直。

（2）望远镜目镜的调焦　点亮望远镜中的照明小灯，从望远镜中可看到视场下方有一绿色暗斑。旋转目镜调焦手轮（11），使分划板上的“╪”形黑刻线成像清晰，如图15－3所示。此时分划板已在目镜的焦平面上，不能再动目镜调焦手轮（11）。

（3）望远镜物镜的调焦　物镜调焦的目的是使分划板同时又在物镜的焦平面上，即望远镜能接收平行光（望远镜聚焦无穷远）。可采用自准直法对物镜调焦：在载物平台上，放一反射面对着望远镜，且与望远镜主光轴大致垂直的平面镜（平面镜的放置方法如图15－5所示）。松开螺钉（9），沿望远镜光轴方向移动目镜筒改变分划板与物镜距离，使分划板同时又在物镜的焦平面上，那么透过十字缝的光线，经物镜出射后必为平行光。平行光被平面镜反射回望远镜中去，经物镜再次聚焦，在分划板上就形成清晰的亮“十”字反射像。此时观察者若把头左右移动，发现亮“十”字像与黑刻线无相对位移，即不存在视差，这时分划板已处在物镜的焦平面上，望远镜聚焦于无穷远处，即望远镜发出平行光也可接收平行光。如果通过目镜只看到叉丝是清晰的，而反射的亮“十”字像却是模糊的，说明目镜和物镜的焦平面还没完全重合，此时应该反复移动目镜筒，直到亮“十”字像清晰并消除视差，拧紧螺钉（9），此后目镜筒不能再伸缩移动。

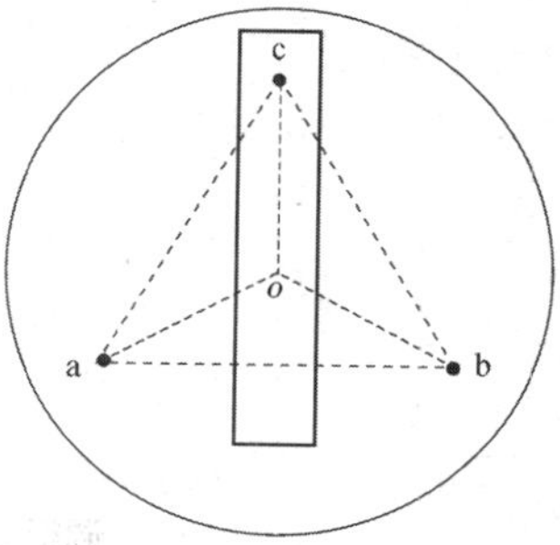

图15－5　平面镜放置示意图

（4）调节望远镜光轴垂直于仪器中心轴，平面镜平行于转轴　焦距调好后，利用二分之一渐近调节法，调节平面镜的倾斜度，使其反射面与分光计转轴平行；调节望远镜的高低位置［通过螺钉（6）及（13）实现］，使望远镜光轴垂直于仪器中心轴，此时十字像正好与十字缝处于对称的位置，如图15－6所示。

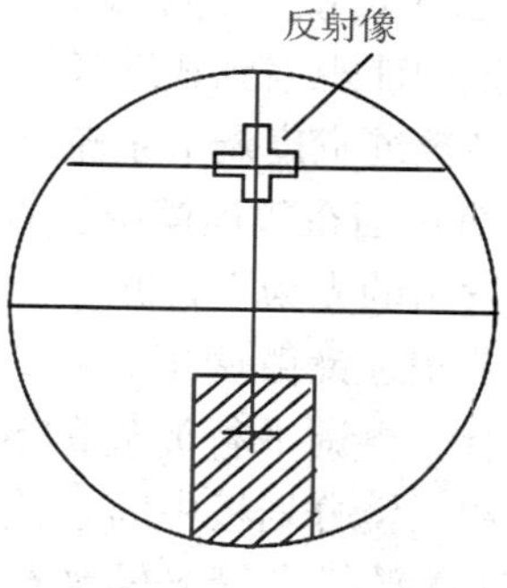

图15－6　望远镜光轴与镜面垂直时的视场图

①转动载物平台，分别使平面镜的两反射面反射回来的亮“十”字像均能进入望远镜。

② 将平面镜其中一反射面正对望远镜筒，此时亮“十”字像可能与分划板上方的十字线不重合，如图15－7（a）所示，调节平台调平螺钉a（b）（图15－5），使高度差减少一半，如图15－7（b），再调望远镜高低调节螺钉（13），改变望远镜的倾斜度，使亮“十”字像与上方的十字叉丝重合，见图15－7（c）。

③ 转动载物台，使平面镜另一反射面正对望远镜筒，方法同上调节平台调平螺钉b（a）及望远镜高低调节螺钉（13），使亮“十”字像与上方的十字叉丝重合。

④ 重复上述步骤（2）、（3）反复调节，直到从两个反射面反射的亮“十”字像均与分划板上方十字叉丝重合为止，此时望远镜光轴垂直于仪器中心轴，平面镜的反射面平行于仪器转轴。

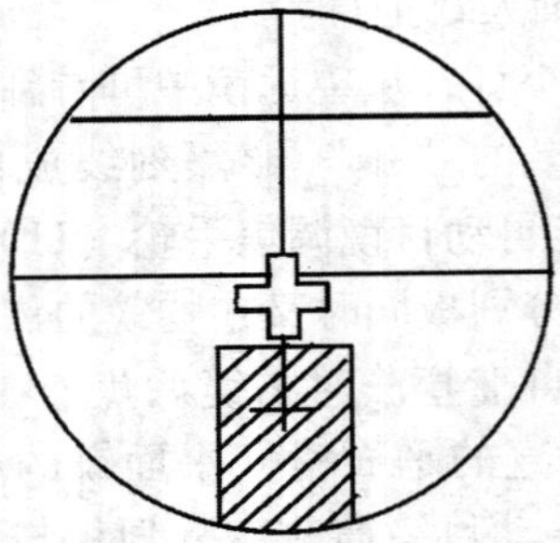

（a）平面镜反射的十字像的初始位置

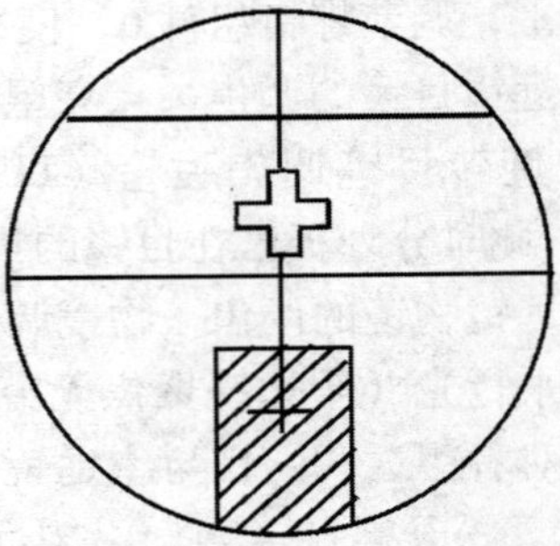

（b）调平台高低调节螺钉 a（b）使十字像上移一半

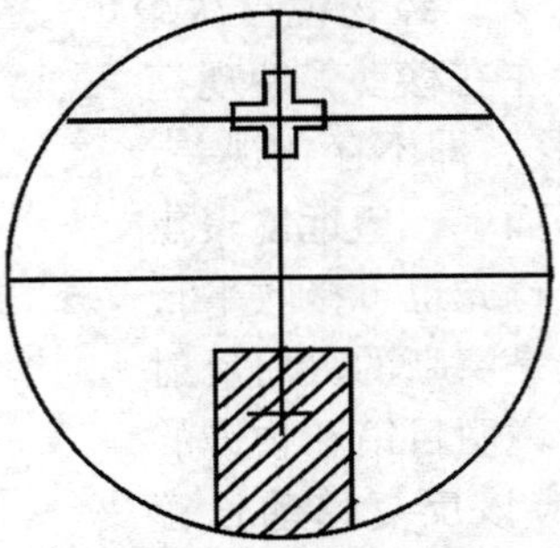

（c）调望远镜高低调节螺钉，使十字像位于分划板上横线处

图 15－7　二分之一渐近法原理图

（5）调节平行光管，使其光轴垂直于仪器中心轴并出射平行光

① 移去平面镜，点亮汞灯照亮平行光管的狭缝。转动已调好的望远镜对准平行光管，使从望远镜中能观察到狭缝像。

②调节狭缝宽度调节手轮（29），使狭缝宽约为 1mm。松开螺钉（2），前后移动狭缝装置（1），当在望远镜中能看到清晰的狭缝像，且与分划板上的叉丝无视差时，说明平行光管发出的光为平行光。

③调节望远镜微调螺钉（16），使狭缝像与分划板中央竖线重合，再调平行光管的高低调节螺钉（28），使狭缝像位于望远镜分划板中间且被中央横线平分，如图 15－8 所示。这样平行光管光轴平行于望远镜光轴，也垂直于仪器中心轴。此时分光计基本调节完毕，后续调节依实验要求而定。

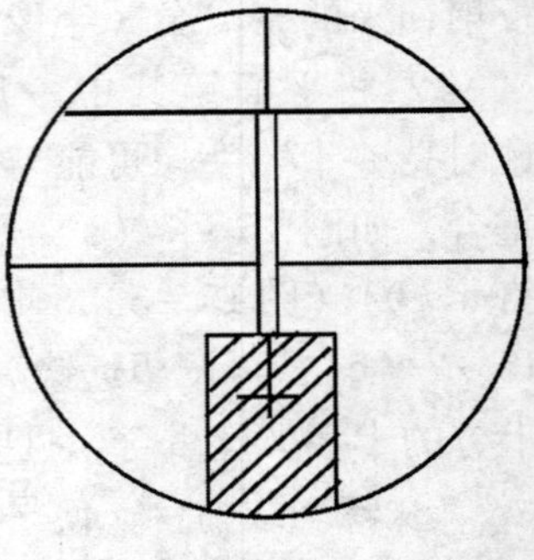

图 15－8　平行光管的调节

【实验原理】

本实验是利用自准直法测量三棱镜的顶角。其原理如图 15－9 所示。三棱镜待测顶角 A 的两个侧面 AC 及 AB 是光学面，BC 为毛玻璃面，称为三棱镜的底面。用自准直法调节望远镜光轴既能垂直于 AC 面又能垂直于 AB 面，测出望远镜从 AC 面转到 AB 面过程中所经角度 α，则可根据

$$\angle A = 180^\circ - \alpha$$

求得三棱镜顶角 A。

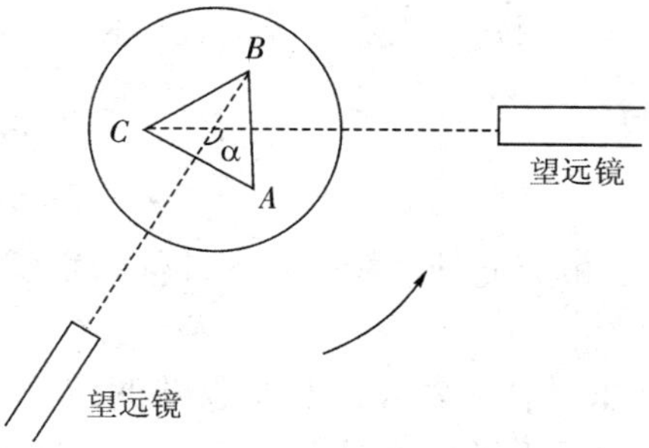

图 15-9　三棱镜顶角测量原理图

【实验内容】

1. 调节分光计

调节分光计的要求是使平行光管发出平行光；望远镜接受平行光，即使望远镜聚焦无穷远；平行光管和望远镜光轴垂直于仪器中心轴。调节前应对照仪器，熟悉分光计各部件及各螺钉作用，并先用眼睛估计各部件位置，使尽量合适，然后分别对各部分进行调节。

（1）调目镜焦距　调节手轮（11），使分划板位于目镜焦平面上。此时应能清楚地看到图 15-3 所示图像。

（2）用自准直法调节望远镜使之聚焦无穷远　将平面镜如图 15-5 放置于平台上，慢慢左右转动平台，在望远镜中找到平面镜反射回来的亮十字像。松开螺钉（9），前后移动镜筒以改变物镜焦距，使亮十字像清晰。

（3）使望远镜光轴垂直于仪器中心轴，用二分之一渐近法调节平台螺钉（6）中的 a、b 和望远镜高低调节螺钉（13），使平面镜无论哪个镜面和望远镜相对时，视场中都能看到图 15-7 所示图像。

（4）调节平行光管，取下平面镜，将望远镜对准平行光管，利用已调好的望远镜为标准，调节平行光管产生平行光，且使平行光管光轴垂直仪器中心轴

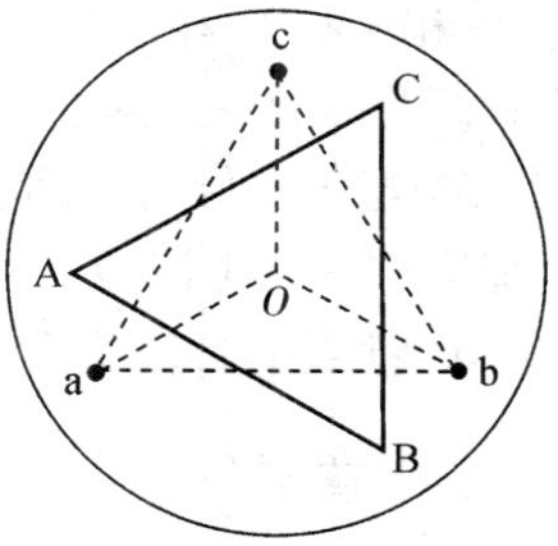

图 15-10　三棱镜放置示意图

2. 测三棱镜顶角

（1）如图 15-10 所示将三棱镜放置在平台上。调节螺钉（6）中的 c 和 a，使望远镜光轴既能垂直于 AC 面，又能垂直于 AB 面。

（2）测量 α 角三次，取其平均值，进而求出三棱镜顶角 A。

【思考题】

1. 什么是分光计的偏心误差？如何消除此误差？
2. 两次读数间，如果游标经过 0°位置，应该如何计算角度？

实 验 指 导

【预习要求】

1. 对照图 15－1，初步了解分光计的构造及各部件的主要作用；熟悉分光计各螺钉的作用。

2. 分光计调整的具体要求是什么？参阅教材仪器调节中的操作要点，写出具体实验步骤，要求简明扼要、前后顺序正确且条理清楚。

3. 在平台上如何放置平面镜？测量三棱镜顶角时，三棱镜又如何放置？为什么？

【操作要点】

1. 本实验的重点是掌握分光计的调节方法。调节前必须先弄清分光计各螺钉的作用，然后再按正确的操作步骤反复练习，直到能较熟练地调节好分光计后，再进行测量。

2. 当望远镜目镜的焦距调节好后，要用自准直法使望远镜聚焦无穷远，这是通过改变物镜与分划板之间的距离来达到此目的，而不能再改变目镜的焦距。当望远镜已聚焦无穷远，在此后整个实验过程中不得再改变望远镜的焦距。

3. 为了调节方便，平面镜按图 15－5 所示位置放在平台上。这样，每次对一个镜面调节时，只需调螺钉 a（或 b），对另一镜面调节时，就只要调螺钉 b（或 a）了。

4. 三棱镜应按图 15－10 放置在平台上。使三棱镜的底面 BC 与平台调平螺钉（6）中的 a、b 连线垂直，即与 c（调平面镜中没有调过的那只螺钉）在平台上与中心轴的连线 co 相平行。当 AC 面对着望远镜时调节螺钉 c，使之符合图 15－6；当 AB 面对着望远镜时调节螺钉 a，再使之符合图 15－6。如此反复，直至两个面均符合为止。

5. 当十字像与分划板上的叉丝位置相差很小时，应利用望远镜的微调螺钉（16），使之精确对准后再读数。

6. 读数前必须认准 A、B 游标。

【注意事项】

1. 分光计是精密仪器，使用时必须多加爱护。旋动螺钉时不要用力过猛，以免损坏螺纹，尤其当某部件被锁紧螺钉锁紧后，不能再用力转动此部件，以免损坏仪器。

2. 严禁用手擦拭或触摸各光学面。如三棱镜的光学面、平面镜面及望远镜与平行光管的镜头等。如果上面有灰尘，必须用擦镜纸擦拭。

3. 三棱镜、平面镜等用完后随即放入盒内，用时再取出，以免打碎。

【实验后的思考】

1. 通过实验，你能总结出分光计调节过程中的几个主要步骤是什么吗？其中关键步骤又是什么？

2. 测量三棱镜顶角前必须使望远镜既能垂直于 *AC* 面又能垂直 *AB* 面。实验中如何知道是否已达到该要求？

3. 调节望远镜光轴垂直于仪器中心轴，在调这一步时为什么要用二分之一渐近法调？

4. 如图 15 – 5 及图 15 – 10 那样放置平面镜及三棱镜有何好处？

5. 为什么取两游标读数的平均值可以消除分光计的偏心误差？

图 15 – 11 表示分光计存在偏心差的情形。图中外圆表示读数圆盘外盘，中心在 O，内圆表示平台，其中心在 O′（画中夸大了两者之差），两个游标与平台固联在其直径两端。它们与读数圆盘外盘圆弧相接触。通过 O′点的虚线表示两游标零线的连线。假定平台从 φ_1 转到 φ_2，实际转过角度为 θ。而外盘上的读数 φ_1、φ_1'；φ_2、φ_2'。计算得到的转角

$$\theta_1 = \varphi_2 - \varphi_1,\ \theta_2 = \varphi_1' - \varphi_2'$$

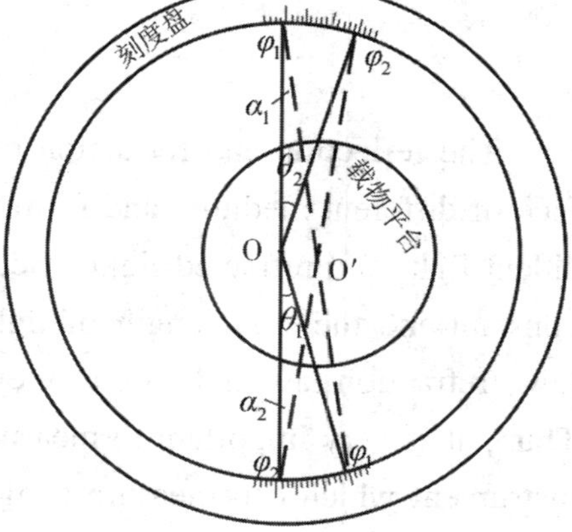

图 15 – 11　偏心误差

根据几何定理有

$$\alpha_1 = \frac{1}{2}\theta_1,\ \alpha_2 = \frac{1}{2}\theta_2$$

而

$$\theta = \alpha_1 + \alpha_2$$

故平台实际转过的角度

$$\theta = \frac{1}{2}(\theta_1 + \theta_2) = \frac{1}{2}[(\varphi_2 - \varphi_1) + (\varphi_2' - \varphi_1')]$$

由上式可见，两个游标读得的平均值即为平台实际转过的角度。因此使用两个游标的读数装置可以消除偏心误差。

15 Spectrometer

The reflection and refraction of light will happen when the light reaches the interface of different medium and form angles between incident light and reflected light, incident light and refracted light. Because the refractive index, dispersive power, grating constant and the wavelength of light are related to the optical law such as: reflection law, refraction law and so on, they can be deduced from the measured certain angles. Thus, it is very important to measure angles precisely. Spectrometer is a kind of optical instrument which can measure angles precisely. There are a lot of similarities of structures between spectrometer and other optical instruments, for instance: spectral photometer, spectrograph, monochromator. Spectrometer is an embryonic form of such instruments.

Objectives

1. Know the structure of spectrometer and the functions of every part.
2. Master the adjusting principle and adjusting method.
3. Measure the apex angle of a triple prism by spectrometer.

Equipment

Spectrometer (type JJY) and its accessory, mercury lamp, glass triple prism, plane mirror.

Introduction of equipment

1. Structure of spectrometer

Spectrometer is made up of five parts: collimator, telescope, object stage, reading dial and base frame. Figure 15 – 1 is the experimental configuration of spectrometer.

(1) Triangle base frame

Triangle base frame is the basement of spectrometer. There is a vertical rotation axis in the middle of the base frame, the telescope and the reading dial can be turned around it.

(2) Telescope

Abbe autocollimator can determine the propagating direction of light. The Abbe autocollimator is made up of Abbe ocular b (achromatic ocular c, total reflection prism

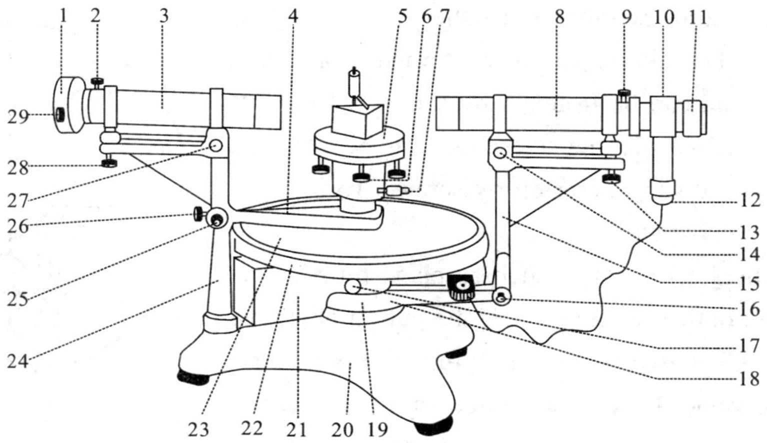

Fignre 15 – 1 Experimental configuration of spectrometer

1. Slit setting; 2. Fixing screw of slit setting; 3. collimator; 4. Brake application (1); 5. Object stage; 6. Adjusting screws of object stage; 7. Fixing screw of object stage; 8. Telescope; 9. Fixing screw of ocular; 10. Abbe autocollimator; 11. Turning knob of ocular; 12. Lighting lamp. 13. Vertical adjusting screw of telescope axis; 14. Horizontal adjusting screw of telescope axis; 15. Supporting arm; 16. Fine adjusting screw of telescope; 17. Fixing screw of reading dial; 18. Fixing screw of telescope (on the other side); 19. Brake application (2); 20. Base frame; 21. Turning frame; 22. Reading dial; 23. Vernier plate; 24. Supporting post; 25. Fine adjusting screw of vernier plate; 26. Fixing screw of vernier plate; 27. Horizontal adjusting screw of collimate; 28. Vertical adjusting screw of collimate; 29. Turning knob of slit width;

f, reticule e) and achromatic objective lens f. (figure 15 – 2). The achromatic objective f is set at the one end of the lens tube and the Abbe ocular b is set at the sleeve g located on another end of tube. There is a small hole on the side of the ocular sleeve and a small lamp is set beside the small hole. The light illuminates the opening sign " + " on the reticule after total reflection by prism, and propagates outside along the axis of telescope. The black sign " + " on the reticule can be seen clearly if the knob is turned properly to adjust the relative position between ocular c and reticule e. This step is called focusing of ocular. Figure 15 – 3 is the front view of reticule. Loosen screw (9), then move ocular tube along the axis of telescope to change distance between object lens and reticule, thus the reticule can be on the focal plane of ocular and object lens at the same time.

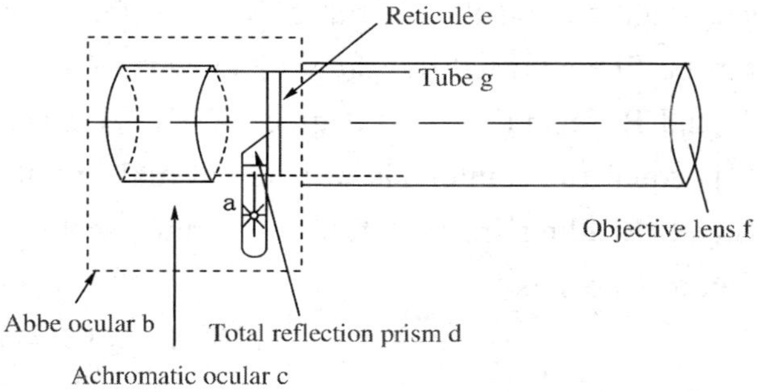

Figure 15 – 2 Sketch map of telescope

The position of the telescope can be changed horizontally by adjusting screw (14). The position of the telescope can be changed vertically by adjusting screw (13), and make the telescope be normal to the central axis of spectrometer. The telescope can be turned around the central axis, and can be fastened by screw (18). If screw (18) is locked, the telescope can be changed horizontal position in microstep by screw (16).

(3) Objective stage

The optical elements, such as plane mirror, triple prism and optical grating, can be put on objective stage (5). Objective stage can be turned around central axis freely and its position can be adjusted upper or lower if the screw (7) is unlocked. Its horizontal attitude can be adjusted with screws (6) (there are three screws in the top of a regular triangle).

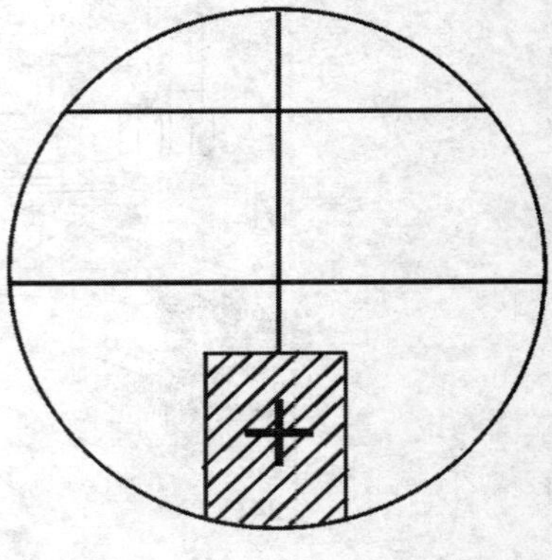

Figure 15 – 3 Front view of reticule

(4) Reading dial

Reading dial can be used to determine the relative position of telescope and objective stage. It is made up of dial (22) and vernier plate (23). The telescope can rotate along with the dial when the screw (17) is locked. There are 720 equal intervals marked on the external dial. Every interval is half degree. The vernier plate connects with the platform through screw (7). The vernier plate can be turned along with object stage around the central axis when the screw (26) is unlocked. The internal dial can be fine adjusted by screw (25) when screw (26) is locked. There are two verniers A and B separating 180 degree on vernier plate. Every vernier has 30 equal intervals. It is equal to 29 intervals of external dial. So the vernier plate can read 1 minute of an angle. Reading method follows vernier scale principle. For example, the reading of Figure 15 – 4 is:

$$116° + 1 \times 12' = 116°12'$$

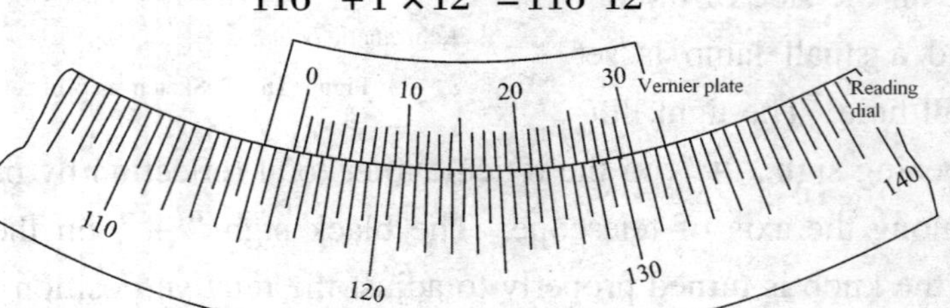

Figure 15 – 4 Reading method

The lightening amplifying – lens maybe helpful for getting vernier reading. The reading of vernier plate *A* and vernier plate *B* must be read respectively, and the average of them is adopted in order to avoid the decentration error because the internal dial and the external dial may not be concentric. For examples:

	Vernier A	Vernier B
The initial reading of telescope	330°45′	150°42′
The final reading after the telescope turning θ	90°47′	270°46

Difference: Vernier A: (360° +90°47′) −330°45′ =120°2′

Vernier B: 270°46′ −150°42′ =120°4′

So the angle θ the telescope turned is: 120°3′.

It should pay more attention for the readings in the case of the vernier plate have passed the 0 position.

(5) Collimator

Collimator is used to obtain collimated light. It is made up of slit (1) and an objective lens setting at the other end of the tube. The width of the slit can be adjusted by screw (29) . The adjusting range is from 0. 02mm to 2mm. Screw (2) is used to fix the slit setting. The distance between the slit and the objective lens can be adjusted when screw (2) is unlocked. The light passing through objective lens will be collimated light if the slit is on the focal plane of the objective lens. The collimator is fixed with the base frame. Its horizontal position and the vertical position can be adjusted by screw (27), screw (28) respectively.

2. Adjustment of spectrometer

There are three adjustment standards:

(1) The telescope can accept parallel light.

(2) The collimator can emit parallel light.

(3) The optical axis of telescope and collimator are vertical to the central axis of instrument. The optical face of measured component is parallel to the central axis of instrument.

In order to achieve these qualifications we should adjust spectrometer as follows. The adjustment of telescope is most important, other adjustments are according to telescope.

(1) Coarse adjustment by eyes

Adjust screw (13) to level telescope approximately; adjust screw (27) to level collimator approximately; adjust screws (6) to make objective stage be vertical to rotation axis approximately.

(2) Ocular focusing of telescope

Turning on the light located in telescope, a green speckle will be seen from the telescope. The symbol "+" sculpted on the reticule can be seen clearly while you turn the focusing knob of ocular (11) (as shown in figure 15 −3) . It is means the reticule is located on the focal plane of ocular, the focusing knob of ocular (11) can not be moved now.

(3) Object lens focusing

The aim of the object lens focusing is to make the reticule locate on the focal plane of object lens, the telescope can receive the parallel light (the focal point of the telescope is on the infinite). The autocollimator method can be used to focus the object lens. Putting a reflective mirror on the stage and making the reflecting surface face to the telescope, the reflecting surface is normal to the optic axis (as shown in figure 15 – 5). Loosening screw (9), changing the distance between the reticule and the object lens by moving ocular focusing tube along the optic axis, the reticule will be on the focal plane of object lens. The light transmitted from the cross slit is parallel light which is collimated by the object lens. The parallel light will be reflected to the telescope by reflecting surface and focused by object lens. A clear symbol "+" of reflective image can be seen on the reticule. There are no optical parallax and the focal point of telescope is on infinite, when the observer moved his head from one side to another and has not find relative movement between the bright "+" symbol and black sculpted lines. If you see a clear cross hair side and a ambiguous "+" symbol image, it means the focal plane of the object lens and ocular have not coincided each other, repeat the preceding steps as described before until a bright "+" symbol can be seen and optical parallax has been erased, then keeping the ocular tube stable by tightening the screw (9).

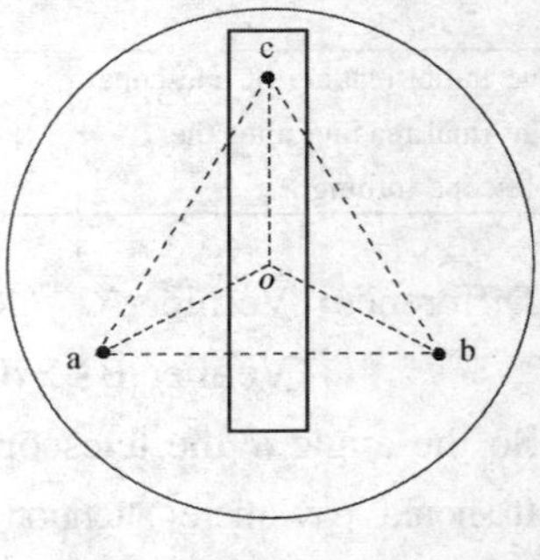

Figure 15 – 5 Schematic plan of mirror placement

(4) Adjusting the axis of telescope to be vertical to the central axis of spectrometer

After focusing of the objective lens, the method of half – approach can be applied for adjusting (screw 6 and 13) the normal line of the flat mirror and the vertical position of the telescope, getting the "+" image staying at the upper cross center symmetric with the "+" slit (figure 15 – 6).

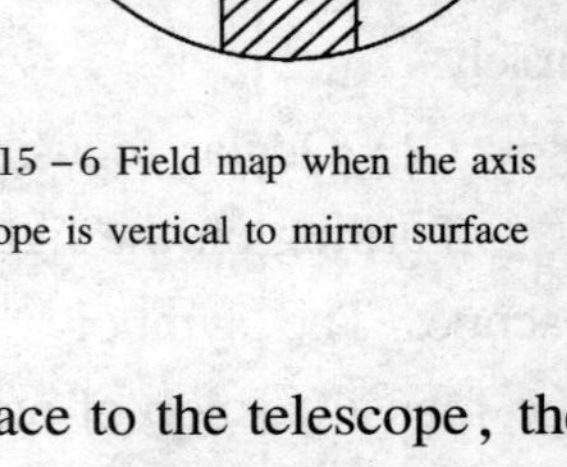

Figure 15 – 6 Field map when the axis of telescope is vertical to mirror surface

① Turning optical stage until the bright reflective sign "+" from two planes of the mirror can enter into telescope respectively.

② When one of the reflective surface of the plane mirror face to the telescope, the bright "+" image may not match together with the upper cross center (figure 15 – 7 (a)). Adjusting stage screw a (or b) (figure 15 – 5) to reduce half difference in height (figure 15 – 7 (b)). Adjusting the telescope screw (13) to change the vertical position of the telescope until the “ + ” image staying at the upper cross center symmetric with the “ + ” slit (figure 15 – 7 (c)) .

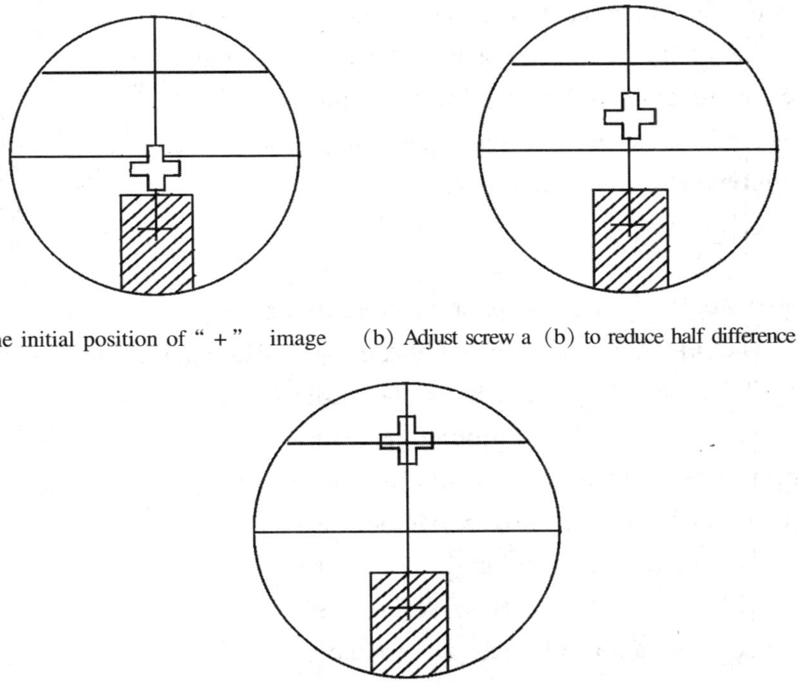

(a) The initial position of “ + ” image (b) Adjust screw a (b) to reduce half difference in height

(c) Adjust telescope screw 12 getting the “ (” image staying at the upper cross center

Figure 15 – 7 Schematic diagram of half – approach

③ Turning object stage, adjusting screw 6 and 12 to get the “ + ” image staying at the upper cross center when the other reflective surface of the plane mirror face to the telescope.

④ Repeat this adjusting for many times, the optical axis of telescope will be vertical to the central axis when the bright “ + ” images reflecting from two reflective surface of the plane mirror are all on the upper cross center.

(5) Adjusting collimator, the optical axis of collimator is normal to the central axis of the instruments, and the emergent light of the collimator is parallel.

① Moving the plane mirror out the stable, switching on the mercury lamp and illuminating the slat of the collimator. Moving the prepared telescope and facing to the collimator, the image of the slit can be seen from the telescope.

② Adjusting the width of the slit to 1mm by turning handle (29) . Loosing screw 2, and moving slit from back and forth until a clearly image of the slit can be seen from the telescope, and coincided with the cross hair of the reticule.

③ Adjusting micrometer screw (16) of the telescope, and making the image of the slit coincide with the central line of the reticule, Re – adjusting the low – high tuning screw (28), and setting image of the slit locate at the central of the reticule of the telescope, as shown in Figure 15 –8. Then, the optical axis of the collimator is parallel to the optical axis of the telescope, and vertical to the central axis of the instruments. After all, the adjusting process of the photometer is finished; the subsequent steps should be done as requirements of the experiment.

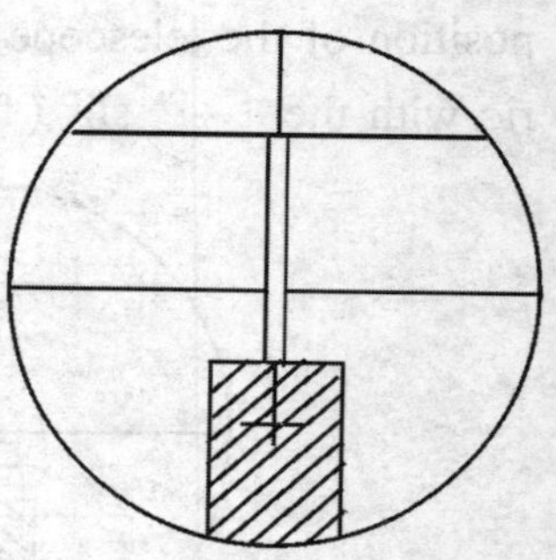

Figure 15 –8 Adjustment of collimator

Principle

The apex angle of the triple prism is measured by auto-collimator. The principle is shown in figure 15 –9. The apex angle A is the inner angle of the optical surface AC and AB of the triple prism. The base surface BC is a ground glass surface. The optical axis of telescope should be adjusted not only normally to the surface AC but also normally to the surface AB. The angle α that the telescope turned from surface AB to surface AC can be measured by auto collimating – method. Then the apex angle A can be obtained according to $A = 180 - \alpha$.

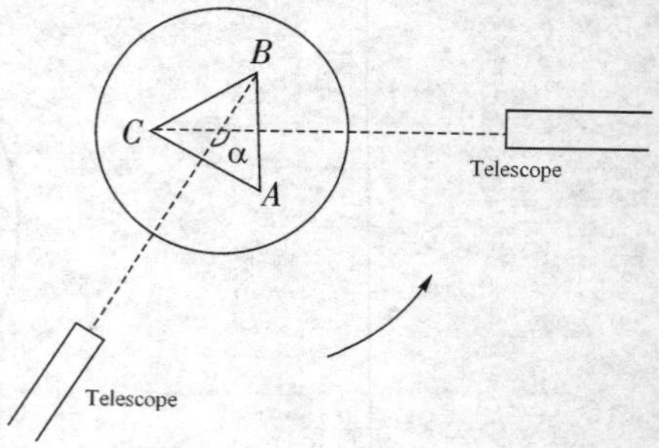

Figure 15 –9 Schematic diagram of measuring A

Experiment

1. Adjusting the spectrometer

There are several objectives for the adjusting of a spectrometer: collimate tube can produce collimated light; collimated light can make distinct image by the telescope; the optical axis of telescope is vertical to the central axis. Thus, it is necessary to be familiar with the screws and every part of the spectrometer.

(1) Adjust the focal length of the ocular. That is to turn the knob to make the picture of figure 15 –3 can be seen clearly.

(2) Put the flat mirror on the object stage as the position showing in Figure 15 –5. Turning the stage slowly to make the bright " + " can be seen through the ocular, then, loose the screw (9) and adjust the objective lens tube to make the " + " clearly.

(3) Adjust the object stage and the telescope to make the axis of telescope to be normal to the central axis. It's a process of application of the half – approach method. After adjusting the screw of object stage (just a and b) and screw (13) of telescope for several times, the picture of Figure 18 –7 can be seen for either side of the flat mirror.

(4) Adjust the collimating light tube. Take the flat mirror from the object stage and make the telescope aim at the collimator, Adjust the collimator to output the collimated light and make the optical axis of collimator normal to the central axis of the instruments.

2. Measure the apex angle of a triple prism.

(1) Place a glass triple prism on the position of the object stage as shown in figure 15 - 10. Adjust the screw (just a and c), to make the telescope being vertical to surface AC and surface AB when the two surface are facing the telescope.

(2) Measure the angle α for three times. Take the average for result and calculate the apex angle A at last.

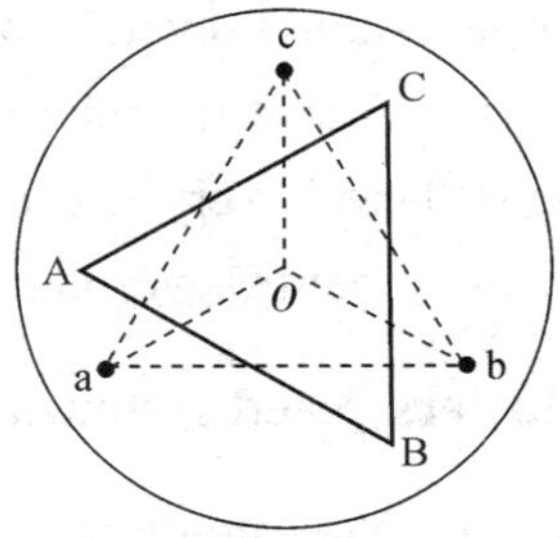

Figure 15 - 10 Putting schematic plan of triple prism

Problems

1. What is the decentration error of a spectrometer? How to remove it?

2. During a measurement for an angle, if one of the vernier plate passed the 0° position, then, how to calculate the angle?

Additional Instruction

Preparation Requirement

1. Understanding the structure of spectrometer and the main function of every part according to figure 15 - 1. Familiar with the functions of each screws of spectrometer.

2. What are the specific requirements when adjusting spectrometer? Write out specific steps of experiment.

3. How to put plane mirror on object stage? How to put triple prism when you measure the apex angle of triple prism?

Operation Key Points

1. The focal point in this experiment is how to adjust spectrometer? You should make clear the function of each screw before adjusting. Practice several times according to right experiment steps until you can adjust spectrometer expertly, and then you can do experiment.

2. If you want to adjust the telescope focusing infinite distance by self - collimation after ocular focal length setting has finished, you should adjust the focus of object lens instead of the focus of ocular. Don't adjust the focus of the telescope during exper-

iment when the telescope has been focused infinite distance.

3. In order to adjust easily, the plane mirror should put on object stage as shown in figure 15 – 5. then you only adjust one screw a (or b) for one face of the plane mirror and adjust b (or a) for another face.

4. Put the triple prism on the object stage as shown in figure 15 – 10. The base surface BC is not only vertical to line *ab* but also parallel to line *co*. Adjust screw c when surface AC is facing the telescope; Adjust screw a when surface *AB* is facing the telescope. Adjust several times until you can see as shown in figure 15 – 6 from the telescope

5. Adjust the micrometer adjusting screw 15 of telescope and vernier plate before reading if there is very small difference between " + " image and cross filament on the reticule.

6. Identify vernier A and B before reading.

Matters Need Attention

1. You should take good care of the spectrometer during experiment because it is a precise instrument. Don't overexert when turning screw to avoid vitiate thread. Especially not turn the part which is tightened by locking screw to avoid vitiate instrument.

2. Don't clean or touch the optical face of triple prism, plane mirror, telescope and collimator by hand directly. You should clean them by lens wiping paper if there is dust on them.

3. Put triple prism and plane mirror into box after finishing experiment.

Problems after experiment

1. Can you summarize several main steps of adjusting spectrometer after experiment? What is the key step?

2. The telescope should be vertical to face AB and face AC, how to get it.

3. Why the half – approach should be used when you adjust the telescope normal to the central axis of instrument?

4. What are the advantages of putting plane figure 15 – 5 and 15 – 10?

5. Why the average value of the reading from tration error of spectrometer?

Figure 15 – 11 shows that the spectrometer exists decentration error. The excircle is the outer dial of reading dial and its center is in O. The inner circle is object stage and its center is in O′ (The difference between O and O′ is overdrawn in figure). The two verniers and the stage are fixed on the two terminal points of diameter and

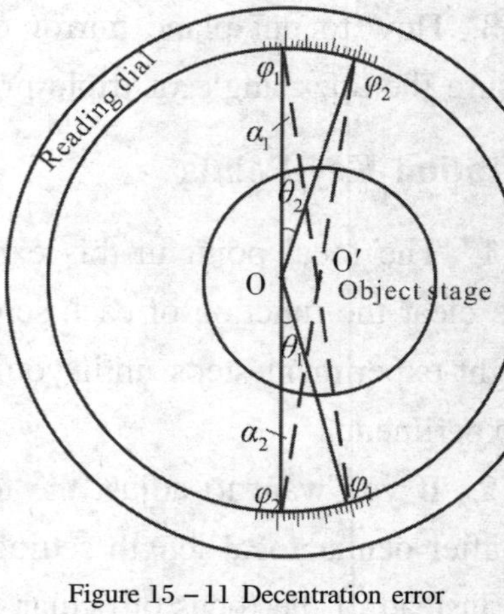

Figure 15 – 11 Decentration error

they contact with the circle arc of outer dial. The broken line via O' is the link of zero line of two verniers. The actual turning angle is θ if the stage turns from φ_1 to φ_2. But the reading from the outer dial are φ_1, φ_1'; φ_2, φ_2', so the turning angle from calculation is:

$$\theta_1 = \varphi_2 - \varphi_1, \quad \theta_2 = \varphi_1' - \varphi_2'$$

According to geometry theorem, we get:

$$\alpha_1 = \frac{1}{2}\theta_1, \quad \alpha_2 = \frac{1}{2}\theta_2$$

And

$$\theta = \alpha_1 + \alpha_2$$

So the actual turning angle is:

$$\theta = \frac{1}{2}(\theta_1 + \theta_2) = \frac{1}{2}[(\varphi_2 - \varphi_1) + (\varphi_2' - \varphi_1')]$$

From above we know the average value of the reading from two verniers is the actual turning angle of stage. So the reading device employing two verniers can remove the decentration error.

实验十六　衍射光栅

光栅是根据多缝衍射原理制成的一种分光元件。它能产生谱线间距较宽的匀排光谱，且分辨本领比棱镜大。光栅不仅适用于可见光，还能用于红外和紫外光波段，常用在光谱仪上。

光栅按结构可分为：平面光栅、阶梯光栅和凹面光栅等几种；按衍射方式，光栅又分为透射式和反射式两类。本实验采用透射式平面刻痕光栅。

对光栅的衍射现象进行研究，不仅有助于加深对光的波动性的理解，而且也有助于进一步学习近代光学实验技术，如光谱技术、晶体结构分析、全息照相、光学信息处理等。

【目的要求】

1. 观察光线通过光栅的衍射现象，加深对光的干涉、衍射及光栅分光作用基本原理的理解。

2. 进一步熟悉分光计的调节和使用。

3. 测定透射光栅的光栅常数及光的波长。

【仪器器材】

分光计、光栅、汞灯。

【实验原理】

透射式平面刻痕光栅是在光学玻璃上刻划大量相互平行，等宽等间距的刻痕而制成的。当光入射到光栅时，刻痕把光漫反射而不透光，只有刻痕间的平滑部分能透光。因此，光栅实际上是一排密集的，等宽等距的狭缝，实验教学上用的一般是复制光栅。

图 16－1 中，a 表示缝宽，b 为缝间距，$(a+b)$ 叫做光栅常数。如以平行光束投射在这样一系列等宽等距的狭缝上，光通过每一狭缝都要产生衍射，而通过各缝的衍射光彼此间又要产生干涉。因此，经透镜会聚在焦平面 E 上的条纹应是衍射和干涉的总效果。

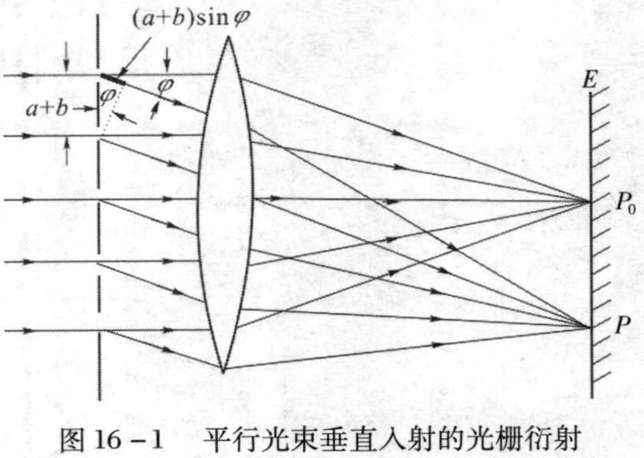

图 16－1　平行光束垂直入射的光栅衍射

1. 垂直入射时的光栅公式

若以单色平行光垂直投射在光栅上，相邻两缝在衍射角 φ 方向上的光程差满足下列条件时：

$$(a+b)\sin\varphi = k\lambda \qquad k=0,\ \pm1,\ \pm2\cdots \tag{1}$$

在透镜的焦平面上产生明条纹。式中 k 为条纹级数，λ 为光波波长。通常称式（1）为光栅公式。

如果入射光不是单色光，则由式（1）可以看出只有 $k=0$ 即 $\varphi=0$ 处，各色衍射光仍叠加在一起，组成中央明条纹。其余各级由于光的波长不同，其衍射角也各不相同，于是复色光被分解为单色光，并按波长大小顺序依次排列为彩色光谱。这就是光栅的分光原理。图 16－2 所示的是汞灯复色光源的光栅光谱，若已知某光波的波长，只要测出其某级的衍射角，就可求出光栅的光栅常数，测出某一级待测波长的衍射角，利用光栅公式可算出未知光波波长。

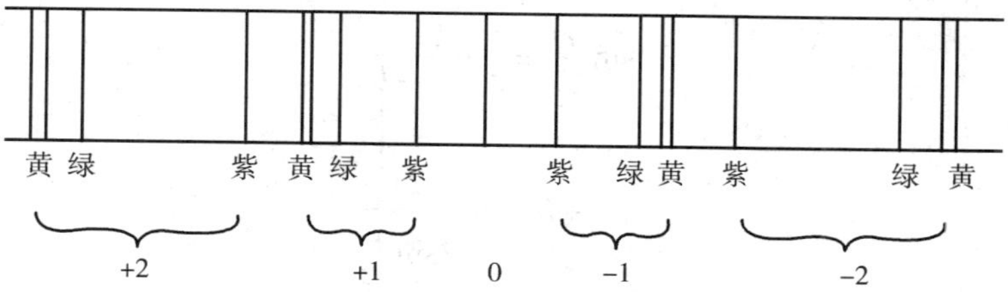

图 16－2　汞灯光栅光谱简图

2. 斜入射时的最小偏向角法

当光线以入射角 i 投射到光栅时，如图 16－3，根据两相邻光束的光程差计算，光栅公式应改为

$$(a+b)(\sin\varphi+\sin i) = k\lambda \qquad k=0,\ \pm1,\ \pm2\cdots \tag{2}$$

当 $k=0$（零级）时，得 $\varphi=-i$，φ 可取正值或负值。由图 16－3 可见，φ 取正值时表明衍射光和入射光在光栅法线的同侧；φ 取负值时，衍射光和入射光在光栅法线的两侧。

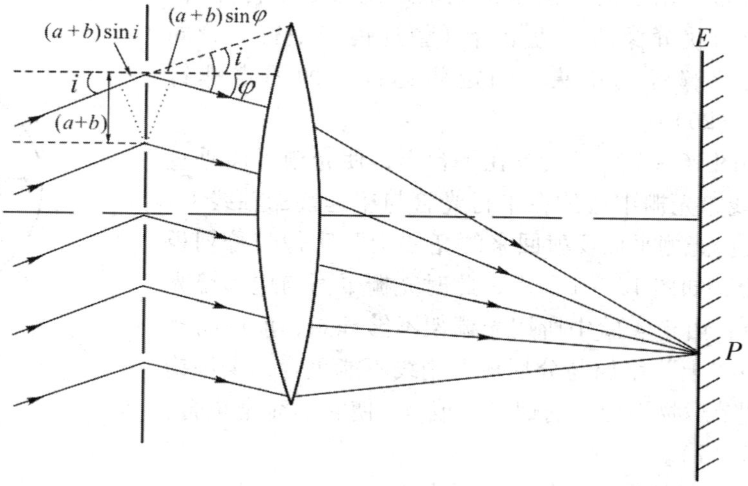

图 16－3　斜入射时光栅衍射

为了避免在实验中既要测 φ 又要测 i 的问题，往往采用最小偏向角法。衍射光线与入射光线的夹角称为偏向角 δ，即 $\delta = i + \varphi$

由式（2）有

$$(a+b)(\sin\varphi + \sin i) = (a+b)\left[2\sin\frac{1}{2}(i+\varphi)\cos\frac{1}{2}(i-\varphi)\right] = k\lambda$$

$$\sin\frac{1}{2}(i+\varphi) = \frac{k\lambda}{2(a+b)\cos\frac{1}{2}(i-\varphi)} \qquad (3)$$

从式（3）可知，当 $\cos\frac{1}{2}(i-\varphi)=1$，亦即当 $i=\varphi$ 时，$\sin\frac{1}{2}(i+\varphi)$ 得极小值。也即 $\delta = i+\varphi$ 为最小，称为最小偏向角，记作

$$\delta_{\min} = D$$

从公式（3）得

$$\sin\frac{D}{2} = \frac{k\lambda}{2(a+b)}$$

即

$$(a+b) = \frac{k\lambda}{2\sin\frac{D}{2}} \qquad (4)$$

由式 (4) 可知，只要测出谱线 λ 第 k 级衍射条纹的最小偏向角 D 即可求出光栅的光栅常数。

【仪器的调节】

1. 分光计的调节

调节的要求同实验十五

2. 垂直入射时光栅的调节

（1）光栅平面（刻痕所在平面）与仪器中心轴平行，且垂直于平行光管。调节的方法是：把望远镜对准平行光管，使分划板竖线与狭缝像的一边重合（望远镜与平行光管同轴），横线在狭缝像的中央。固定望远镜［即固定螺钉 (17)、(18)、(26)］。

将光栅如图 16－4 所示放置在平台上，使光栅平面垂直平分 a、b 连线，光栅中心约在平行光管与望远镜的轴线上，转动平台，使由光栅平面反射回来的亮“十”字像与分划板上叉丝线重合，如图 15－6 所示，此时光栅平面与望远镜光轴垂直。注意：由于实验中所用光栅架不够理想，前后略有俯仰，致使亮“十”字像与分划板上刻线不能重合，少许错位，此时可调节载物平台下的螺钉 a 或 b，使它们完全重合，然后固定螺钉（7）。

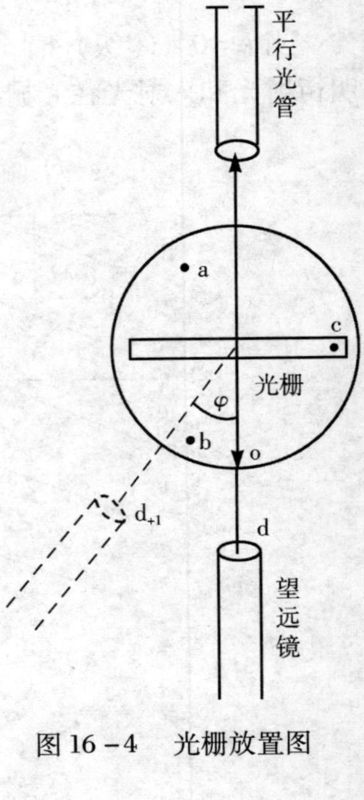

图 16－4　光栅放置图

（2）调节光栅刻痕，使其与仪器中心轴平行，如果两者不平行，谱线会偏斜，一边高一边低。松开望远镜止动螺钉

（18），左右移动望远镜，观察两边各级条纹。注意分划板上叉丝交点是否在各条谱线中央，如果不是则调节 c 予以校正（a、b 保持不动）。

3. 斜入射时最小偏向角的调节方法

将望远镜对准某级已知谱线条纹的一边，松开螺钉（7）并转动平台以带动光栅；螺钉（26）应锁紧，即游标盘不得随之移动。这时从望远镜中可以看到条纹向光线 0 方向移动，光线 1′和光线 0 的夹角就是偏向角 δ（图 16－5）。继续转动平台，条纹继续向 0 方向移动；直到 1″位置时条纹反方向移动，光线 1″和光线 0 的夹角就是最小偏向角 D。将平台固定，转动望远镜，使分划板的竖线与条纹的一边重合，读出 1″的位置。再根据 0 及 1″的位置，即可求得最小偏向角 D。

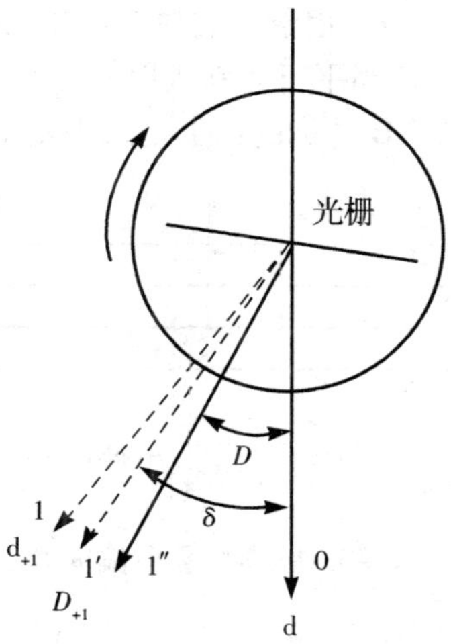

图 16－5　最小偏向角示意图

【实验内容】

1. 按实验十五的要求调节分光计

2. 观察汞灯经光栅后的衍射光谱

以汞灯垂直入射为条件，观察它经过光栅衍射后共能看到几级光谱，除中央明纹外，其余每级光谱由几条谱线组成；画出光谱示意图，并标出每条谱线的颜色。比较光栅光谱与棱镜色散光谱的异同。

3. 测定光栅常数

平行光管垂直入射光栅，转动望远镜，使分划板竖线对准中央亮条纹的一边，记下此时 A、B 游标的读数，转动望远镜（游标盘和平台不动），使分划板竖线分别与 ±1 级绿光（$\lambda = 546.07\text{nm}$）谱线的一边重合，将相应的读数记录到表 16－1，并由此计算光栅的光栅常数。

表 16－1　正入射时　±1 级绿光谱线读数

	$k=+1$	$k=0$	$k=-1$
θ_A			
θ_B			

+1 级谱线偏角：$\varphi_{+1} = \dfrac{1}{2}\left(|\theta_{A+1} - \theta_{A0}| + |\theta_{B0} - \theta_{B+1}|\right) =$

－1 级谱线偏角：$\varphi_{-1} = \dfrac{1}{2}\left(|\theta_{A0} - \theta_{A-1}| + |\theta_{B-1} - \theta_{B0}|\right) =$

则：$(a+b)_{+} = \dfrac{\lambda}{\sin\varphi_{+1}} =$

$(a+b)_{-} = \dfrac{\lambda}{\sin\varphi_{-1}} =$

结果：$\overline{(a+b)} = \dfrac{1}{2}\left[(a+b)_{+1} + (a+b)_{-1}\right] =$

4. 用最小偏向角法测光栅常数

当平行光斜入射光栅时，找出 $\lambda=546.07\text{nm}$ 的绿光在 $k=\pm1$ 级时的最小偏向角，将A、B 游标的读数记入表格 16－2 中，计算光栅常数。

表 16－2　斜入射时 ±1 级绿光谱线读数

	$k=+1$	$k=0$	$k=-1$
θ_A			
θ_B			

+1 级谱线最小偏向角：$D_{+1}=\dfrac{1}{2}\left(\left|\theta_{A+1}-\theta_{A0}\right|+\left|\theta_{B0}-\theta_{B+1}\right|\right)=$

－1 级谱线最小偏向角：$D_{-1}=\dfrac{1}{2}\left(\left|\theta_{A0}-\theta_{A-1}\right|+\left|\theta_{B-1}-\theta_{B0}\right|\right)=$

则：$(a+b)_{+}=\dfrac{\lambda}{2\sin\dfrac{D_{+1}}{2}}=$

$(a+b)_{-}=\dfrac{\lambda}{2\sin\dfrac{D_{-1}}{2}}=$

结果：$\overline{(a+b)}=\dfrac{1}{2}\left[(a+b)_{+1}+(a+b)_{-1}\right]=$

【思考题】

1. 用式（1）测（$a+b$）时应保证什么条件？实验时这些条件是如何保证的？

2. 当用钠光（波长 $\lambda=589.3\text{nm}$）垂直入射到 1mm 内有 500 条刻痕的平面透射光栅上，试问最多能看到第几级光谱？并说明理由。

3. 在光栅常数一定时，光栅的缝宽 a 对衍射条件有何影响？

实　验　指　导

【预习要求】

1. 复习分光计的调节和使用方法。
2. 用衍射光栅测定光栅常数或光波波长的理论依据是什么？
3. 光垂直入射和斜入射时光栅公式在形式上有何变化？为什么？
4. 判断入射光是平行光的依据是什么？
5. 如何使入射光垂直于光栅平面？如何使光栅的刻痕与仪器中心轴平行？
6. 为了便于调节，光栅应如何放置于平台上？这样放置有什么好处？
7. 如何使望远镜的转动平面与衍射角所在平面一致？
8. 斜入射为什么要用最小偏向角法？如何找最小偏向角？
9. 参阅讲义及操作要点写出实验步骤。

【操作要点】

1. 本实验是在分光计上进行，所以实验的第一步仍需熟练地调节好分光计。

2. 光栅如图 16－4 放置的好处是：调节螺钉 a 与 b 就可以有效改变光栅平面的倾斜度，使它垂直于望远镜的光轴。注意此时望远镜已调好，不能再动；然后在调节螺钉 c 时，就能使光栅刻痕与仪器中心轴平行，而不会再影响光栅平面的倾斜度。注意此时不能再动 a 与 b。

3. 在调节望远镜光轴垂直于光栅平面时，从光栅反射回来的亮十字像往往有几个重影。这是因为光栅有好几个反射面（玻璃有两个面），调节时以最清晰的为准。

4. 为了使分划板的竖线精确对齐光谱线，最后必须使用望远镜微调螺钉（16）来对准。

5. 找最小偏向角时，必须松开螺钉（7），使平台自由转动，但不能使游标盘随之转动，因为原 0 级位置即为斜入射时入射光线的位置。

6. 汞的光谱在可见光范围内应有 7 条，依次为紫、紫、蓝、绿蓝、绿、黄、黄。本实验中因光栅的质量及汞灯的亮度等原因，不能看清全部可见谱线。实验中请画出能观察到的所有谱线。

7. 测定衍射角 φ 及最小偏向角 D 时应测出 ±1 级的相应值，然后取其平均。应注意 +1 级和 −1 级的衍射角相差不能超过几分。否则垂直入射时应重新检查入射角 i 是否为零；斜入射时，重新检查入射线的位置是否移动。

【注意事项】

1. 同实验十五。

2. 严禁用手擦拭或触摸光栅。

3. 汞灯的紫外光很强，不可直视，以免灼伤眼睛。

【报告要求】

自拟实验步骤；回答思考题。

【实验后的思考】

1. 调节好望远镜垂直于中心轴后，将光栅放置于平台上，理应有望远镜光轴亦垂直于光栅平面。但实验中往往存在偏差。此偏差是如何产生的？如何消除？

2. 如果直接用光栅代替平面镜进行“望远镜光轴垂直于仪器中心轴”的调节，是否可以？为什么？

3. 垂直入射时，如果光栅平面和转轴平行，但刻痕和转轴不平行，那么整个光谱有什么异常？对测量结果有无影响？

4. 斜入射时，用讲义所述的方法寻找到的偏向角，为什么就是最小偏向角？

5. 仍然用本实验的分光计，但给你换一个刻痕数更多的光栅（光栅常数相同），能否提高分辨本领，从而分辨开精细的谱线？

16 Diffraction Grating

Grating is a kind of splitting elements, and made according to multi – slit diffraction principle. It can produce uniform spectrum with good quality. It has a better resolving power than ordinary prism. It can be applied to split spectrum not only in visible light, but also in infrared band and ultraviolet band. It is usually applied in optical spectrometer.

There are several different structures of gratings, such as: flat grating, concave grating, echelle. And there are transmission grating and reflection grating, according to their diffraction way. The grating used in this experiment is a transmission plane grating.

The research on the diffraction properties of the grating can help us to deeply understand the wave character of light, and further study the experimental technology of modern optics, such as spectrum technology, crystal structure analysis, holography, optical information processing and so on.

Objectives

1. Observation of the phenomenon of grating diffraction, understanding its principle.
2. Be familiar with the adjusting and application of spectrometer.
3. Measure the grating constant and the wavelength of light.

Equipments

Spectrometer; grating; mercury lamp

Principle

There are a lot of parallel lines, with equal interval and equal width, on the surface of a transmission grating. Light can pass through the parts without lines. Thus, a grating is a device with many parallel slits on it. We usually use replica grating in experiment.

Figure 16 – 1 is a diagrammatic sketch of grating, a is the width of a slit and b is the interval between slits, $(a+b)$ is called the grating constant. When a parallel light incident on the grating, the light will be diffracted by the slits and the diffraction light will interfere each other. Thus, the fringes on the focal plane of the converging lens is the total effect of diffraction and interference.

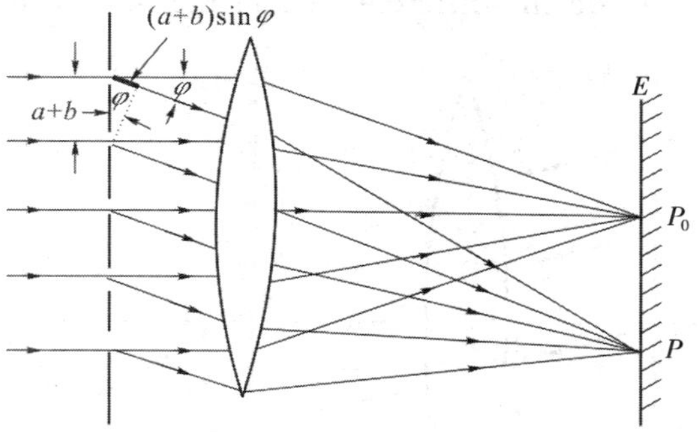

Figure16 – 1 Parallel light is vertically incident on grating

1. The grating formula for vertical incidence.

When a monochromatic parallel light incidents on a grating, if a relationship between the optical path difference in the direction φ and the wavelength λ is:

$$(a+b)\sin\varphi = k\lambda \qquad k=0,\ \pm1,\ \pm2\cdots \qquad (1)$$

Then a bright fringe will be on the screen of focal plane. The k is called series number. The equation (1) is called grating formula.

If the incidence light is not monochromatic, according to equation (1), only the central fringe will be the same color as the incident light, where $\varphi=0$ corresponding to the $k=0$ for all different wavelength. In the other series, the light with different wavelength will form a colorful spectrum with an arranging order of wavelength, because light with different wavelength has different diffraction angle φ. This is the principle of grating spectroscopy. Figure 16 – 2 is a picture of spectrum of mercury – lamp light. From equation (1), we can calculate the grating constant or an unknown wavelength if the relative values are known.

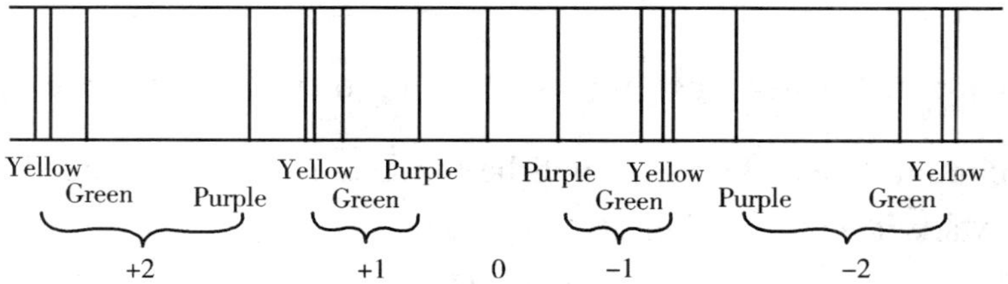

Figure16 – 2 Grating spectrum of mercury lamp

2. The minimum deviation for oblique incidence.

As shown in Figure16 – 3, when the light incidents on the grating with an oblique angle i, the grating formula should be rewritten as:

$$(a+b)(\sin\varphi + \sin i) = k\lambda \qquad k=0,\ \pm1,\ \pm2\cdots \qquad (2)$$

When $k=0$, we get $\varphi = -i$; φ may be positive or negative. If φ is positive, the diffraction light and the incident light will stay on the same side of the normal line; for a negative φ, they will stay on the different side of the normal line.

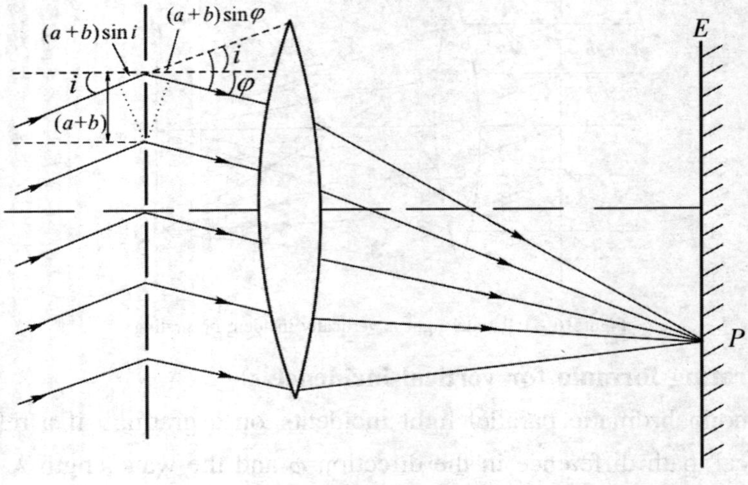

Figure 16 – 3 Grating diffraction for oblique incidence

It is easy to measure the minimum deviation angle in the experiment, and avoid the measurement of the two angles: φ and i. The angle between incidence light and diffraction light is called deviation angle δ, that is:

$$\delta = i + \varphi$$

The equation (2) can be written as:

$$(a+b)(\sin\varphi + \sin i) = (a+b)\left[2\sin\frac{1}{2}(i+\varphi)\cos\frac{1}{2}(i-\varphi)\right] = k\lambda$$

Or:

$$\sin\frac{1}{2}(i+\varphi) = \frac{k\lambda}{2(a+b)\cos\frac{1}{2}(i-\varphi)} \qquad (3)$$

Now we know that when $\cos\frac{1}{2}(i-\varphi) = 1$, or $i = \varphi$, the value of $\sin\frac{1}{2}(i+\varphi)$ will be minimum, i. e. $\delta = i + \varphi$ will be minimum, which is called minimum deviation angle. Mark it with the sign D:

$$\delta_{min} = D$$

The equation (3) will be

$$\sin\frac{D}{2} = \frac{k\lambda}{2(a+b)}$$

Thus:

$$(a+b) = \frac{k\lambda}{2\sin D/2} \qquad (4)$$

According to the equation (4), if the minimum deviation D of light with wave-

length λ in k series is measured, the grating constant $(a+b)$ can be deduced.

Equipment preparation

1. The adjustment of spectrometer.

The adjusting requirement is the same as Exp. 15.

2. The adjustment of grating for vertical incident light.

(1) Make grating plane parallel to central axis of the spectrometer and normal to the collimator.

First, aim the telescope to the collimator, and make the vertical line on the reticule to coincide with the image of the slit (Telescope and collimator has same axis), and the transverse lines is on the centre of the slit, then fix telescope (namely fix screw 17, 18 and 26).

Put the grating on the object stage as shown in figure16 –4, then the plane of grating is the midnormal of the link of a and b, the centre of the grating is about on the axis of telescope and collimator. Now, try to find the bright "+" in the ocular. loosen screw (7) and turn the stage to make the bright "+" coincide with the upper cross of the reticule. Then the plane of grating is vertical to the axis of telescope as shown in figure15 –6. Attention: Since the grating is not vertical to the base frame, so the bright "+" maybe not coincide with the upper cross of the reticule. You can adjust screw a or b, which is under the object stage, to make them coincide and then fix screw coincide and then fix screw (7).

(2) Adjust the grating to make the fringes on the both side of grating at the same level. If the fringes are not parallel with central axis, fringes on one side of grating will be higher or lower than that on other side. In this case, the screw 6 (only c in the figure 16 –4) should be adjusted.

3. Find the minimum deviation angle for oblique incidence.

Aim the telescope at a certain spectrum line. Loosen the stage fixed screw 7, then, turn the stage together with the grating in the direction as shown in Figure 16 – 5. Notice that the screw 25 should be locked in order to keep the vernier plate unchanged. Now, we can find the fringes are moving to the direction of light 0. The angle between light 0 and light 1′ is the deviation angle δ. Keep on turning the

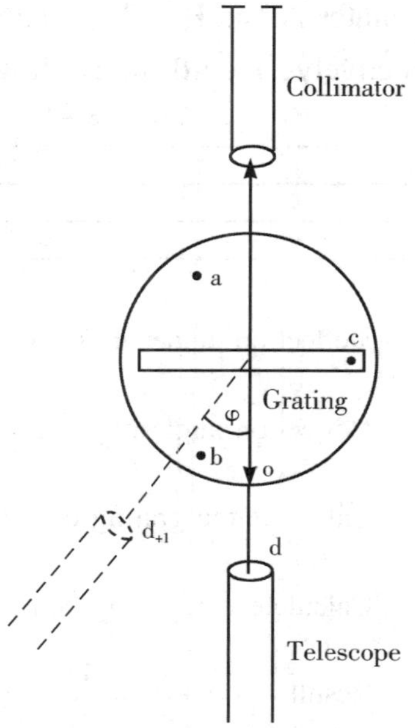

Figure 16 –4 Diagrammatic sketch of grating placement

stage and the fringes moving on the same direction as before, until arrive the place of 1″, where the moving direction of fringes change to the opposite direction. The angle between light 1″and light 0 is the minimum deviation angle *D*. Fix the stage, then, turn the telescope to make the central vertical line of the reticule to coincide with the spectrum line again. Now, the position can be read and the minimum deviation can be calculated.

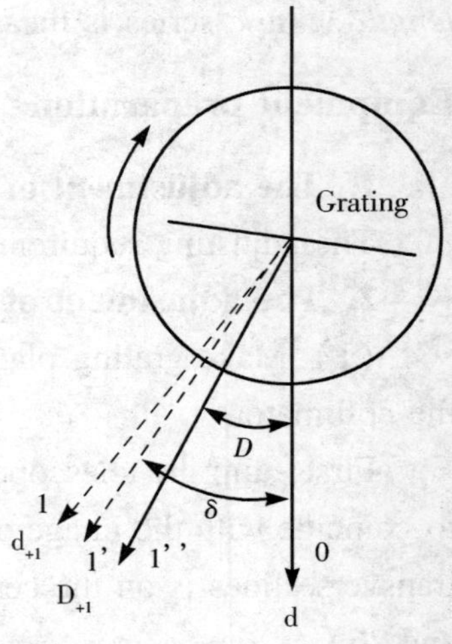

Figure16 – 5 Minimum deviation angle

Experiment

1. Adjust the spectrometer according to the requirement of Exp. 15

2. Observe the grating diffraction fringes of the mercury light

In the case of normal incidence light, how many series of diffraction fringes can be seen? How many spectrum lines are there in one series? What are the colors of them? Record all the details.

3. Determine the grating constant

Make the incidence light from collimator illuminate the grating vertically. Aim the telescope at the central bright line (the image of bright slit) to make the central vertical line of reticule to coincide with it. Record the readings θ_{A0} and θ_{B0} of the two vernier plates A and B. Then, aim the telescope at the ±1st green light ($\lambda = 546.07\text{nm}$) respectively, record the readings in table 16 – 1. Calculate the grating constant.

Table 16 – 1 Readings of ±1 green lines (vertical incidence)

	$k = +1$	$k = 0$	$k = -1$
θ_A			
θ_B			

Deflection angle of +1 green line: $\varphi_{+1} = \dfrac{1}{2}\left(|\theta_{A+1} - \theta_{A0}| + |\theta_{B0} - \theta_{B+1}|\right) =$

Deflection angle of –1 green line: $\varphi_{-1} = \dfrac{1}{2}\left(|\theta_{A0} - \theta_{A-1}| + |\theta_{B-1} - \theta_{B0}|\right) =$

Calculate the grating constant: $(a+b)_{+} = \dfrac{\lambda}{\sin\varphi_{+1}} =$

Calculate the grating constant: $(a+b)_{-} = \dfrac{\lambda}{\sin\varphi_{-1}} =$

Result: $\overline{(a+b)} = \dfrac{1}{2}\left((a+b)_{+1} + (a+b)_{-1}\right) =$

4. Determine the grating constant with minimum deviation

Find the minimum deviation angle of the ±1st green line ($\lambda = 546.07$nm) respectively under the condition of oblique incidence. Then, record the readings of vernier plates A and B into table 16-2 and calculate the grating constant.

Table 16-2 Readings of ±1green lines (oblique incidence)

	$k=+1$	$k=0$	$k=-1$
θ_A			
θ_B			

The minimum deviation of the +1 green line: $D_{+1} = \frac{1}{2}(|\theta_{A+1} - \theta_{A0}| + |\theta_{B0} - \theta_{B+1}|) =$

The minimum deviation of the -1 green line: $D_{-1} = \frac{1}{2}(|\theta_{A0} - \theta_{A-1}| + |\theta_{B-1} - \theta_{B0}|) =$

Calculate the grating constant: $(a+b)_+ = \dfrac{\lambda}{2\sin\dfrac{D_{+1}}{2}} =$

Calculate the grating constant: $(a+b)_- = \dfrac{\lambda}{2\sin\dfrac{D_{-1}}{2}} =$

Result: $\overline{(a+b)} = \frac{1}{2}((a+b)_{+1} + (a+b)_{-1}) =$

Problems

1. What are the conditions should be satisfied for the equation (1)? How are these conditions satisfied in experiment?

2. For a grating with 500 lines in 1mm, how many series of diffraction spectrums can be seen for a vertical incident light with wavelength $\lambda = 589.3$nm (sodium light)?

3. Keeping the grating constant unchanged, what's the effect of the width a variation of grating?

Additional Instruction

Preparation Requirement

1. Refreshing the adjustment and operation method of spectrometer.

2. What is the theoretical basement of characterization of optical wave length or grating constant by diffraction grating?

3. Is there any difference between the grating equation for the normal incidence light and that for oblique incidence light, and why?

4. How do you consider the incidence light is parallel light or not?

5. How to make the incidence light normal to the surface of the grating? And how to make the nick of the grating parallel to the central axis of the instrument?

6. In order to make the adjustment of the instrument easily, How to put the grating on the stage? Which kind of the benefits will be gotten?

7. How to keep the rotation plane of the telescope in the same plane with diffraction plane of grating?

8. Why should the minimum deviation method be used when the light oblique incident into grating? How to find the minimum angle of deflection?

9. Writing out the experimental steps according to the teaching materials and operation points.

Operation Key Points

1. Because the experiment is operate in the spectrometer, the first step in this experiment is familiar with adjustment of spectrometer.

2. The benefits of grating arrangement like Figure16 – 4 are as follows: Adjusting the screw a and b to change inclined degree of slant of grating plane and make it normal to the optical axis of telescope. When the telescope is ready, keep it stable; after then, adjusting the screw c in order to make the nick of grating parallel to the central axis of instrument, and the slant of grating plane is unchanged. Don't adjust the screw a and b in this time.

3. When you adjust the optical axis of telescope normal to the gating plane, there are several fold – over image of bright cross symbol in the field of view. It is because there are several reflective surface of grating; the sharpest image is the adjustment target.

4. In order to make the vertical line of the reticule align with the spectrum lines, the micrometer screw (16) should be adjusted to make the telescope aim to the targets.

5. The screw (7) should be loosed in order to move the stage when you want to find the minimal angle of deflection. But keep vernier circle stable, it is because the 0th position is the position of the incidence line of inclined light.

6. The mercury lamp has 7 spectrum lines in visible range, there are purple, purple, blue, turquoise, green, yellow and yellow, sequentially. But only several spectrum lines can be seen in this experiment because of the quality of the grating and the brightness of the mercury lamp. Please draw all of the spectrum lines, which can be seen in your experiments.

7. Measuring diffraction angle φ and the minimal angle of deflection D of grating in ± 1 order, and then take the average of the detection angles as finial result, respec-

tively. The difference between the diffraction angle of $+1^{st}$ order and -1st order is no more than several divisions. Otherwise, it should be checked that the incidence angle i is equal to zero or not when the incidence light is at normal condition; or the position of the incidence light is moved or not for the oblique incidence light.

Matters Need Attention

1. As description in experiment 15.
2. Don't clean or touch the grating with you hands.
3. Since ultraviolet radiation of mercury is very strong, don't watch the mercury lamp directly to avoid damaging your eyes.

Requirement for report

Determine the experimental steps by yourself and answer the question.

Problems after experiment

1. Putting the grating on the stable after you make the telescope normal to the central axis ready, in this case the optical axis of the telescope is normal to the grating plane. In fact, there still have some deviation, do you know the reasons? How to erase these deviations?
2. If the grating is used to replace the plane mirror directly when you adjust the telescope normal to the central axis of the instrument. Is it possible or not? And why?
3. If the plane of the grating is parallel to the rotation axis at the normal incidence condition, but the nick of the grating is not. Is there any abnormality for the spectrum in this case? Is there any influence on the detection results?
4. Why the finial angle is the minimal deflection angle according to the adjustment method as described as in this teaching materials when the light oblique incident on the grating.
5. If a new grating, which have more nick than old one, used in this spectrometer, is it possible to improve the resolution capability?

实验十七　光栅衍射法测里德伯常数

氢原子结构简单，线光谱规律明显，从而成为实验室研究各种原子光谱线的突破口。从1885年巴尔末公式的提出到1913年玻尔建立半经典的氢原子理论，再到1949年根据量子动力学理论成功解释氢原子波谱的蓝姆移动，对氢原子光谱的种种研究在量子理论的发展过程起到了重要的作用。

【目的要求】

1. 观测氢原子光谱在可见光范围内的谱线。
2. 测定氢原子光谱的谱线波长。
3. 由巴尔末公式求出里德伯常量 R_H。

【仪器器材】

分光计，氢灯，衍射光栅，霓虹灯变压器，调压变压器。

【实验原理】

1885年瑞士科学家巴尔末首先将氢原子光谱可见光区谱线的波长归纳为如下公式：

$$\lambda = B\frac{n^2}{n^2-4} \tag{1}$$

式中 B 是常量，其数值为364.56nm，n 为大于2的正整数。当 $n=3$、4、5、6时，上式分别给出氢光谱中可见光部分四条谱线的波长。这四条谱线的颜色分别为红、深绿、青和紫，用 H_α、H_β、H_γ、H_δ 表示。式（1）称为巴尔末公式，其计算结果与实验结果很吻合。氢原子中的电子从不同的外层轨道跃迁到第二轨道时所发射的谱线，统称为巴尔末系谱线，如图17－1所示。

为了得到光谱学中的通式，1886年瑞典物理学家里德伯用波数 $\tilde{\nu}$（波长的倒数，即单位长度内所含的光波数目）将巴尔末公式改写为

$$\tilde{\nu} = R_H\left[\frac{1}{2^2}-\frac{1}{n^2}\right] \qquad n=3,4,5,6\cdots \tag{2}$$

式中，R_H 为里德伯常数。当时的实验值为 $R_H=1.096776\times10^7\text{m}^{-1}$。

1913年玻尔理论问世，由玻尔理论导出的氢原子光谱的巴尔末公式为：

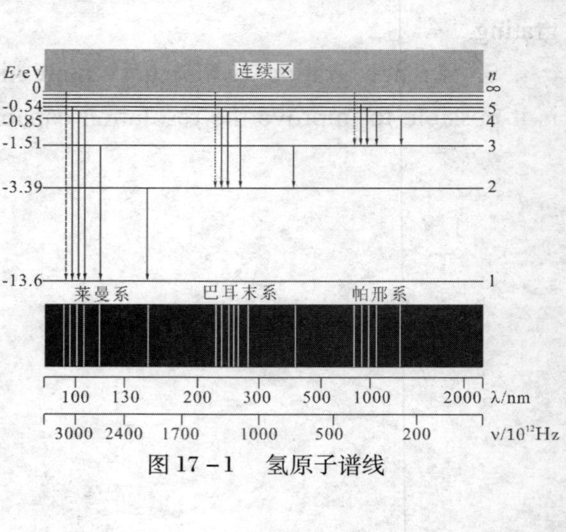

图17－1　氢原子谱线

$$\tilde{\upsilon} = \frac{1}{\lambda} = \frac{me^4}{8\varepsilon_0 h^3 c}\left(\frac{1}{2^2} - \frac{1}{n^2}\right) \tag{3}$$

式中 ε_0 为真空介电常数，h 为普朗克常数，c 为光速，e 为电子电荷，m 为电子质量。这样，玻尔理论不仅对巴尔末的经验公式作出了物理解释，而且把里德伯常数和许多基本物理量联系了起来。由玻尔理论算出的值为

$$R_H = 1.097373 \times 10^7 \mathrm{m}^{-1}$$

近代实验测定值为

$$R_H = 1.096776 \times 10^7 \mathrm{m}^{-1}$$

氢原子的可见光谱能用巴尔末公式表示，而且结果又是如此精确，充分表明这个式子不单纯是个经验公式，而是氢原子内在规律的表现。玻尔正是受巴尔末公式的启发，提出了著名的玻尔理论。

氢灯是观测氢原子谱线的光源。将玻璃管抽成真空后充入氢气，在玻璃管两端装上电极。当两电极间加上几千伏高压电后，管内气体放电，气体原子或离子在导电过程中受到碰撞而被激发，氢原子在各能级间跃迁时发射光谱线。用它作光源，通过衍射光栅和分光计，根据光栅公式

$$(a+b)\ \sin\varphi = k\lambda \qquad k=0,\ \pm 1,\ \pm 2\cdots \tag{4}$$

测出氢原子光谱中可见光部分谱线的波长，从而可算得 R_H 值。

【实验内容】

1. 根据分光计的调节要求调节分光计。

2. 根据光垂直入射到光栅的调节要求调节光栅。

3. 转动望远镜观察氢灯经光栅衍射后 ±1 级可见光部分的衍射光谱。记录能观察到几条谱线，并标明它们的颜色。

4. 根据已知的光栅常数，在入射光线垂直入射时测定氢原子光谱巴尔末系中的 H_α（红，$n=3$）、H_β（深绿，$n=4$）、H_γ（青，$n=5$）、H_δ（紫，$n=6$）的波长。分别测定三次，取其平均值。

【数据处理】

1. 记录各条谱线的位置，计算衍射角并求出三次测量的波长的平均值。

表 17－1　可见光范围内的 ±1 级衍射谱线

观测谱线	$k=+1$		$k=-1$		衍射角	波长 λ	波长 $\bar{\lambda}$（nm）
	θ_{+1}	θ_{+1}'	θ_{-1}	θ_{-1}'	φ	（nm）	
H_α（$n=3$）							
H_β（$n=4$）							

续表

观测谱线	$k=+1$		$k=-1$		衍射角 φ	波长 λ (nm)	波长 $\overline{\lambda}$ (nm)
	θ_{+1}	θ_{+1}'	θ_{-1}	θ_{-1}'			
H_γ ($n=5$)							
H_δ ($n=6$)							

其中：$\varphi=\dfrac{|(\theta_{+1}-\theta_{-1})+(\theta_{+1}'-\theta_{-1}')|}{4}$

$\lambda=(a+b)\sin\varphi$

2. 以 $\dfrac{1}{\lambda}$为 y 轴，$\dfrac{1}{n^2}$为 x 轴作图，可得到一条直线，该直线的斜率的绝对值，即里德伯常数。将求出的里德伯常数和标准里德伯常数相比较，计算其相对误差。

$$E=\frac{|R_H-R_H'|}{R_H'}\times 100\%$$

其中 R_H表示测量值，R_H'表示标准里德伯常数。

3. 用最小二乘法求里德伯常数，然后与作图法相比较。

【思考题】

1. 每一条谱线都是一条单独的线吗?
2. 由玻尔理论推导里德伯常数的值。

实　验　指　导

【预习要求】

1. 熟练掌握分光计和光栅的调节方法。
2. 掌握巴尔末公式，并能用巴尔末公式求出里德伯常数 R_H。
3. 能够运用作图法和最小二乘法对实验数据进行处理和计算。

【操作要点】

1. 实验中使用的光源为氢光源，氢放电管两极间所加电压高达数千伏，使用时务必注意安全。

2. 根据实验十五和实验十六介绍的方法仔细调节分光计及光栅，测量衍射角时注意A、B 两个游标分别读数，消除偏心误差对测量的影响。

【实验后的思考】

1. 用最小二乘法和作图法分别求出里德伯常数，就本实验而言，哪种方法更好?

2. 一共能观察到几级可见光区的谱线，每一级谱线都能看到四条可见光谱吗？能否利用 $k=\pm 2$ 级谱线测量里德伯常数?

17 Determination of Rydberg Constant by Diffraction Grating

Hydrogen atom has simple structure and regular spectrum, which is the breakthrough to research the spectrums of other atoms. From Balmer formula proposed in 1885 to semi – classical Bohr Theory proposed in 1913, and then hydrogen spectrum's Lamb shift explained with quantum dynamics in 1949, the study on hydrogen spectrum played an important role in the development of quantum theory.

Objectives

1. Observe the spectrum lines in the range of visible light.
2. Measure the wavelength of certain lines of hydrogen atom.
3. Calculate the Rydberg constant according to Balmer formula.

Equipments

Spectrometer, hydrogen light, diffraction grating, neon transformer, voltage regulator transformer.

Principle

In 1885, Switzerland scientist Balmer found that some of the spectrum lines could be expressed by the following empirical formula:

$$\lambda = B\frac{n^2}{n^2-4} \tag{1}$$

B is a constant, 364.56nm, and n is positive integer greater than 2. For $n=3$, 4, 5, 6, we can get the wavelength of four lines of the hydrogen spectrum, which can be denoted as H_α, H_β, H_γ, H_δ and the color are red, olive, green, purple respectively. The formula (1) is called Balmer formula. Its calculation results are in the same with experimental results. The spectrum resulting from the electrons in the outer orbit of hydrogen atom transmitting to the second orbit is called Balmer series, which is shown in figure 17 – 1.

In 1886, Sweden scientist Rydberg expressed the Balmer formula by the wave number $\tilde{v}$, (the reciprocal of wavelength), and got the following formula:

$$\tilde{v} = R_H\left[\frac{1}{2^2}-\frac{1}{n^2}\right] \qquad n=3,\ 4,\ 5,\ 6\cdots \tag{2}$$

R_H is the Rydberg constant, $R_H = 1.096776 \times 10^7 \mathrm{m}^{-1}$.

In 1913, Bohr found his atomic theory and deducted the Balmer formula theoretically as the following:

$$\tilde{v} = \frac{1}{\lambda} = \frac{me^4}{8\varepsilon_0 h^3 c}\left(\frac{1}{2^2} - \frac{1}{n^2}\right) \quad (3)$$

Where ε_0 is dielectric constant for vacuum, h is Plank constant, c is the speed of light, e is electron charge, and m is electron mass. In this way, Bohr Theory not only explained the Balmer empirical formula but also made Rydberg constant contact with many basic physical quantities.

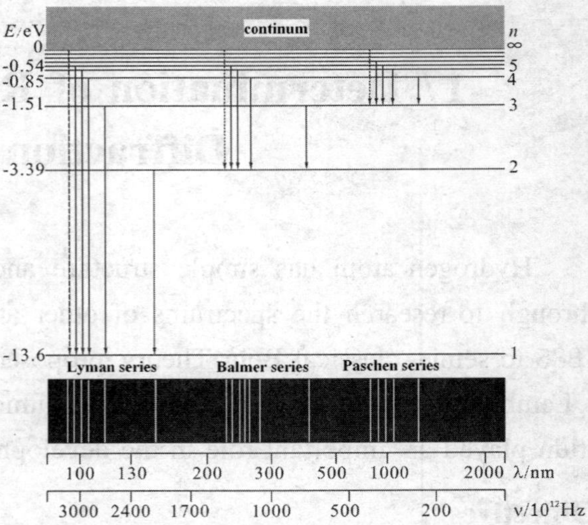

Figure 17 – 1 Hydrogen spectrum

The theoretically calculated R_H value is

$$R_H = 1.097373 \times 10^7 \mathrm{m}^{-1},$$

and the modern physics experimental value is:

$$R_H = 1.096776 \times 10^7 \mathrm{m}^{-1}.$$

The hydrogen spectrum can be expressed by Balmer formula and its highly exactness means that the Balmer formula is not only a empirical formula, but also the presentation of physical rules. Inspired from Balmer formula, Bohr proposed the well – known Bohr Theory.

Hydrogen light is the common light to research hydrogen spectrum. Pump a glass tube into vacuum and fill hydrogen into it, then equip the glass tube with electrodes at both ends. When put a few thousand volts on the electrodes, the gas in tube discharge and the gas atoms or ions are excited by collision in the conduction process. Hydrogen atom emit spectra when transit at different energy levels.

In this experiment, the light source is hydrogen light. The grating formula is:

$$(a+b)\sin\varphi = k\lambda \quad k = 0, \pm 1, \pm 2\cdots \quad (4)$$

We can measure the wavelengths of hydrogen spectrum in the range of visible light by the spectrometer and diffraction grating, and then calculate the value of R_H.

Experiment

1. Adjust the spectrometer for meeting the requirement of measuring.
2. Adjust the spectrometer to make the grating being vertical to the incidence light.
3. Turn the telescope arm to observe the ±1 order diffraction light in the range of

visible light, record their colors.

4. For the known grating constant, measure the wavelength of four lines of Balmer series, respectively they are H_α (red, $n=3$)、H_β (olive, $n=4$)、H_γ (green, $n=5$)、H_δ (purple, $n=6$). Repeat the measurement for three times, calculate their mean values.

Data Processing

1. Record the positions of the hydrogen spectrum lines, calculate the diffraction angles and the mean values of the wavelength.

Table 17 -1 ±1 order diffraction light in the range of visible light

Spectrum lines	$k=+1$		$k=-1$		Diffraction angle	Wavelength λ	Wavelength $\overline{\lambda}$
	θ_{+1}	θ_{+1}'	θ_{-1}	θ_{-1}'	φ	(nm)	(nm)
H_α ($n=3$)							
H_β ($n=4$)							
H_γ ($n=5$)							
H_δ ($n=6$)							

Where, $\varphi = \dfrac{|(\theta_{+1}-\theta_{-1}) + (\theta_{+1}'-\theta_{-1}')|}{4}$

$\lambda = (a+b)\sin\varphi$

2. Calculate the wavelengths of the four lines first, let y axis denote $\dfrac{1}{\lambda}$ and x axis $\dfrac{1}{n^2}$, plot the four experimental dots and draw a straight line. The absolute value of the slope of the line is the Rydberg constant, compare it with the standard value and calculate its relative error.

$$E = \frac{|R_H - R_H'|}{R_H'} \times 100\%$$

Where R_H is measured value and R_H' is standard Rydberg constant。

3. Calculate the Rydberg constant by least square method and compare it with the above result.

Problems

1. Do you think every spectrum line is an ideal line?
2. How to get the Rydberg constant from Bohr Theory?

Additional Instruction

Preparation requirement

1. Master the adjustment method of spectrometer and diffraction grating.

2. Master Balmer formula and derive Rydberg constant R_H from it.

3. Be able to handle experiment data using mapping method and the least square method.

Operation key points

1. The light source used in the experiment is hydrogen source, a few kV voltage will be added up to the two poles of the hydrogen discharging tub, the operator must be careful.

2. Adjust the spectrometer and the grating for meeting the requirement of measuring. The reading of vernier plate A and B must be read respectively, and the average of them is adopted in order to avoid the decentration error.

Problems after experiment

1. Compare the least square method with the mapping method, as far as this experiment is concerned, which is the better method?

2. How many orders of the spectrum line in the range of visible light can be seen? If Rydberg constant can be measured with ± 2 order spectrum line?

实验十八　密立根油滴实验

电子电荷是一个十分重要的基本物理量，对它的准确测定有很大的意义。美国物理学教授密立根（R. A. Milikan）从1907年开始，历经七年时间，对微小油滴上所带的电荷进行测量，直接证实了电荷的不连续性，并测定了电子的电荷量，即 $e=1.602\times10^{-19}$C。这就是著名的密立根油滴实验。本实验就是利用密立根油滴仪验证电荷的不连续性，测定电子所带的电量。从实验结果可以看出，任何油滴从空气中捕获的电量都是电子电荷的整数倍。

【目的要求】

1. 验证电荷的不连续性，并测定电子电荷的绝对值。
2. 培养科学实验应有的严谨态度。

【实验原理】

用喷雾器将油滴喷入两块相距为 d 的水平放置的平行板之间，（图18－1）。油滴在喷射时由于摩擦，一般都是带电的。设油滴的质量为 m，带电量为 q，两极板间加的电压为 U，则油滴在平行板间同时受到两个力的作用，而且可以使这两个力的方向相反：一个是重力 mg，方向为铅直向下；另一个是静电力 $qE=\frac{qU}{d}$，方向向上。若调节极板间的电压，使两个力达到平衡，则油滴静止不动。这时有：

$$mg=\frac{qU}{d} \tag{1}$$

显然，只要设法测出油滴的质量 m，就可以求出所带电量 $|q|$。

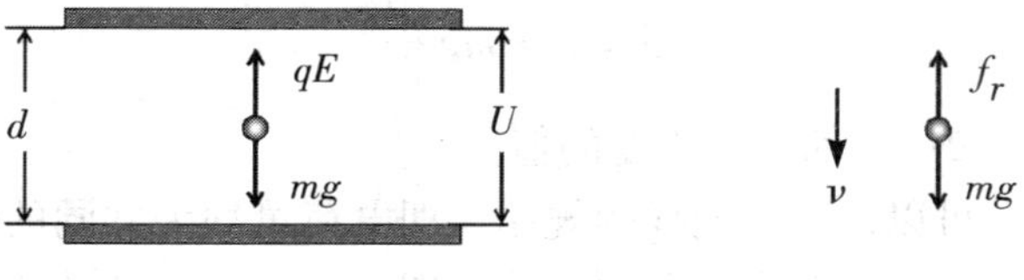

图18－1　静止和匀速运动的油滴

1. 油滴质量的测定

平行板未加电压时，油滴受重力而加速下降。但由于空气阻力的作用，油滴下降一段距离后，将作匀速运动。设匀速运动的速度为 v，这时重力和空气阻力平衡（忽略空气浮力）。根据流体力学的斯托克斯定律，在静止的黏性流体中运动的刚性小球受到的黏力（即空气阻力）为

$$f=6\pi r\eta v \tag{2}$$

式中 η 是空气的黏度，r 是油滴的半径（由于表面张力，油滴呈小球状，半径大小约为 10^{-6}m，质量约为 10^{-15}kg）。设油的密度为 ρ，则油滴的质量 m 为

$$m=\frac{4}{3}\pi r^{3}\rho \tag{3}$$

将式（3）代入式（2）可得

$$r=\sqrt{\frac{9\eta v}{2\rho g}} \tag{4}$$

但是，油滴并非刚性的，而且体积太小，其线度与室温时空气气体分子的平均自由程（约 10^{-7}m）相差不大，故此时斯托克斯定律不能严格成立。公式（2）中的 η 要修正为 η'：

$$\eta'=\frac{\eta}{1+b/pr} \tag{5}$$

式中 b 为一常量，$b=8.21\times10^{-3}\text{m}\cdot\text{Pa}$，$p$ 为大气压强，单位用 Pa。

由（3）、（4）式，并用 η' 替代式中的 η，可得

$$m=\frac{4}{3}\pi\left[\frac{9\eta v}{2\rho g}\cdot\frac{1}{1+b/pr}\right]^{3/2}\rho \tag{6}$$

2. 油滴下降速度的测定

当两极板间的电压 $U=0$ 时，设油滴在两板间匀速下降 l 路程所需时间为 t，则有

$$v=\frac{l}{t} \tag{7}$$

由式（7）、（6）、（1）整理可得

$$q=\frac{18\pi}{\sqrt{2\rho g}}\left[\frac{\eta l}{t\ (1+b/pr)}\right]^{3/2}\frac{d}{U} \tag{8}$$

式中 $r=\sqrt{\frac{9\eta l}{2\rho gt}}$。

实验发现，对于同一油滴，如果设法改变油滴所带电量，则使油滴达到平衡的电压必须是某些特定值 U_n。研究这些电压变化的规律，可发现它们都满足下列方程：

$$q=ne=mg\frac{d}{U_n}$$

式中的 $n=\pm1,\ \pm2\cdots$，而 e 是不变的值。

对于不同的油滴，可以发现有同样的规律：即电荷量均为共同的常数 e 值的整数倍。这就证明了电荷的不连续性，并存在基本电荷，即电子电荷的绝对值 e。

$$ne=\frac{18\pi}{\sqrt{2\rho g}}\left[\frac{\eta l}{t\ (1+b/pr)}\right]^{3/2}\frac{d}{U} \tag{9}$$

式（8）或式（9）就是本实验测量油滴电荷的理论公式。

【实验仪器】

密立根油滴实验仪包括油滴仪、电源、计时器、喷雾器等，电源同时作为油滴仪的座架，以使仪器结构紧凑。

1. 油滴仪结构

MOD－2 型油滴仪简图如图 18－2 所示。

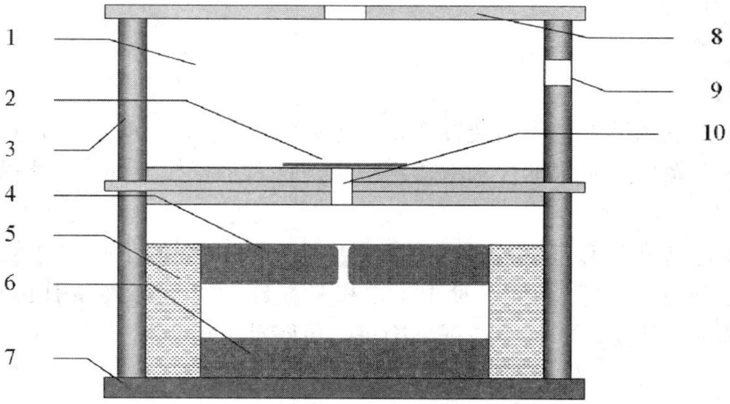

图 18－2　油滴仪结构

1. 油雾室；2. 油雾室开关；3. 防风罩；4. 上电极板；5. 油滴盒；
6. 下电极板；7. 座架；8. 上盖板；9. 喷雾室门；10. 油雾孔；

油滴仪由三只水平调节螺丝和电源箱连接起来。油滴盒（5）由两块经过精磨的平行电极板（4）和（6）组成，间距 $d=0.500\text{cm}$。上电极板（4）中央有一个 $\phi0.4\text{mm}$ 的小孔，以供油滴落入。整个油滴盒装在有机玻璃防风罩（3）中，以防止周围空气流动对油滴的影响。防风罩上面是油雾室（1），油滴用喷雾器由喷雾口喷入，并经油雾孔（10）落入油滴盒，油雾孔底部有油雾室开关（2），关闭后防止油滴不断落入油滴盒，以免干扰对已经选定的油滴的观察测量。照明装置包括灯室和导光玻璃棒，灯室中装 2. 2V 聚光小灯珠。

显微镜是用来观察和测量油滴的运动的。目镜中装有分划板（图 18－3），上下共分六格，每格相当于视场中 0. 050cm，六格共长 0. 300cm。利用分划板可测量油滴运动的距离 l，以测出油滴运动的速度 v。

2. 电源

电源共提供三种电压：

（1）2. 2V 交流电压，供聚光小灯珠用。

（2）500V 直流平衡电压，该电压的大小可以连续调节，数值可以从电压表上读出来。当标有“平衡电压”的开关拨至中间“0”的位置时，上下电极板被短路，并接零电位。当开关拨至“＋”位置时，能达到平衡的油滴带正电荷；反之，带负电荷。

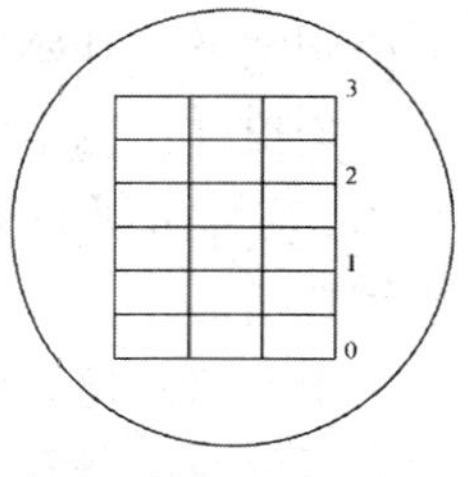

图 18－3　分划板

（3）250V 左右的直流升降电压，该电压大小也可以连续调节，并可通过标有“升降电压”的拨动开关将其叠加在平衡电压上。由于该电压只起移动已经平衡的油滴在两块平行极板间的上下位置的作用，并不需要知道它的大小，因此没有读数。

3. 计时器

计时器是液晶显示的数字秒表，能准确读出0.1秒。

【实验内容】

1. 仪器调节

（1）将油滴仪电源接上，照明装置2.2V电源及平行极板500V直流电源都已接好。

（2）调节水平螺丝，使水准仪水泡移到中央，这时平行极板处于水平位置，电场方向与重力平行。

（3）将调焦针（在油雾室中）插入电极板中央ϕ0.4mm的小孔内（切记：此时“平衡电压”开关必须置于“0”位置，使上下电极板短路，以免打火和引起人身触电）。调节显微镜，对调焦针聚焦。如针不在视场中央，可转动上电极板，使它移到中央。然后即取出调焦针。

（4）将数滴油注入喷雾器，并从喷雾口喷入油雾室（喷一下即可），视场中将出现大量微小油滴，犹如夜空繁星。

2. 测量前的练习

（1）练习控制油滴。加上平衡电压（在300V左右，“+”、“-”均可），驱走不需要的油滴，直到剩下几颗为止。注视其中的一颗，调节平衡电压值，使这颗油滴平衡。然后去掉平衡电压，让这颗油滴匀速下降（在视场中看上去是上升）一段距离后，重新加上平衡电压使其静止（即平衡），再加上升降电压使其上升（在视场中看上去是下降）。反复多次练习，直至掌握控制油滴的方法。

（2）练习选择油滴。要做好本实验，选择被测油滴非常关键。油滴的体积不能太大，也不能太小。选择时，可根据平衡电压的大小（约300V）和油滴匀速下降的时间（约10～30秒）来判断油滴所带电量的多少及油滴的大小。可通过测试进行练习。

3. 正式测量

由式（9）可知：进行本实验真正要测量的只有两个量。一个是平衡电压U_n，另一个是油滴匀速下降一段距离 l 所需的时间 l。

仔细调节平衡电压（此时升降电压为零），将油滴置于分划板上的某条横线上，以便准确判断这颗油滴是否处于平衡状态。

为保证油滴下降时速度均匀，应先让它下降一段距离后再进行测量。选择测量的一段距离l应靠近平行极板的中央。一般取$l=0.200$cm（相当于分划板中间纵向四格的距离）比较合适。

由于有涨落，对于同一油滴必须进行10次左右的测量。同时还应对2～5个不同的油滴进行反复的测量。

（1）数据处理

将下列数据：

油的密度	$\rho=977\text{kg}\cdot\text{m}^{-3}$（30℃时）
重力加速度	取$g=9.80\text{m}\cdot\text{s}^{-2}$
空气黏度	$\eta=1.83\times10^{-5}\text{Pa}\cdot\text{s}$
油滴下降距离	$l=0.200\text{cm}=2.00\times10^{-3}\text{m}$

平行极板间距　　　$d=5.00\times10^{-3}\text{m}$

大气压强　　　　$p=1.013\times10^{5}\text{Pa}$

常数　　　　　　$b=8.21\times10^{-3}\text{m}\cdot\text{Pa}$

代入式（9）整理后得

$$q=ne=\frac{18\pi}{\sqrt{2\rho g}}\left(\frac{\eta l}{t\left[1+b/p\sqrt{\frac{9\eta l}{2\rho gt}}\right]}\right)^{3/2}\frac{d}{U_n}=\frac{1.43\times10^{-14}}{[t\ (1+0.0196\sqrt{t})]^{3/2}}\cdot\frac{1}{U_n}$$

上式中ρ、η都是温度的函数，g和p又随实验地点和条件的变化而变化，因此上式是近似的。在一般条件下，由此引起的误差仅 1% 左右，带来的好处是使计算方便得多，这是可取的。

（2）为了证明电荷的不连续性和所有电荷都是基本电荷 e 的整数倍，并得到基本电荷 e 的值，我们应将实验测得的各个电荷值求最大公约数。这个最大公约数就是基本电荷 e 的值，即电子的电荷值。但由于实验误差，求最大公约数有时比较困难，通常我们用“倒过来验证”的办法进行数据处理，即用公认的电子电荷的绝对值 $e=1.60\times10^{-19}\text{C}$ 去除实验测得的电荷值，得到一个接近于某一整数的数值，然后取其整数，该整数就是油滴所带的电荷数 n。再用这个整数 n 去除实验测得的电荷值，所得结果即为实验测得的电子电荷的绝对值 e。

【注意事项】

1. 用金属细丝调焦针对显微镜进行调焦时，两电极板绝不能加电压，否则会因短路而造成仪器的损坏，甚至还会引起人身触电。

2. 喷雾时，喷雾器应竖着拿，食指堵住出气孔，喷雾器口对准油雾室的喷雾口，轻轻喷入少许油即可。切勿将喷雾器深深插入油雾室，甚至将油倒出来。否则，很容易堵塞落油孔，并将油滴盒周围弄脏。

3. 对油滴进行跟踪测量时，如油滴的像变得模糊不清，可稍稍调节显微镜筒的前后位置，使其聚焦。

4. 测量时，应记住使油滴平衡的开关位置。每次测量结束，应先把平衡开关拨回原来位置。如果忘了拨回平衡开关，或拨错了方向，会使油滴跑掉，而影响多次测量。

5. 如发现油滴速度有明显的变化时，应重新调节平衡电压，再进行重测。

【思考题】

1. 在测量时，油滴的象可能会变得模糊，这是为什么原因？

2. 如何判断油滴处于匀速运动状态？

3. 如何选择合适的油滴进行测量？

4. 在实验过程中，如果未调节水平螺丝（即平行极板未处于水平位置），则会对实验结果有何影响？

实 验 指 导

【预习要求】

简要地说，本实验是通过测量油滴匀速下降 l 距离所用的时间 t 及平衡这颗带电油滴所加的电压 U，来测定该油滴所带电量的。因此，平行极板一定要呈水平，以使电场方向与重力场方向一致。

【操作要点】

1. 平行极板一定要调节到水平状态，这样在长时间的观察测量过程中，方能不让油滴跑出视场范围。

2. 由于油滴非常小（大约 10^{-6}m），因此用显微镜对其进行观察时，调焦工作一定要做好。

3. 对选定的油滴进行测量时，控制油滴所加的平衡电压及升降电压的使用要熟练。并随时注意调整。

4. 本实验时间较长，须对每一微小油滴做反复多次的测量，因此要有严格的实验操作和严谨的科学态度。

5. 通常要求被测油滴带电量在十倍电子电荷值之下。

【实验后思考】

1. 油滴在空气中运动，则除空气阻力、重力外，还有空气浮力，试考虑空气浮力对实验结果的影响。

2. 油滴进入平行极板间时具有各方向速度的油滴均有，而通过显微镜只能看清楚其焦平面上的油滴，为什么在实验过程中见到的油滴基本都是在上下竖直方向上运动的呢？

3. 油滴在空气中要挥发，在长时间的测量过程中，可能挥发掉很多，但对测量结果没有影响，为什么？

18 Millikan Oil–drop Experiment

The value of electric charge of an electron is a basic physics constant. It is very important for the precisely determination of it. From 1907 on, American physics professor R. A. Millikan had been done the experiment to measure the value of electric charge carrying on a tiny oil – drop for 7 years. His experiment confirmed the discontinuity of electric charge, and determined the value of charge of an electron, $e = 1.602 \times 10^{-19}$C. That is the famous Millikan oil – drop experiment. Now, we verify the discontinuity of electric charge and measure the quantity of electric charge of an electron by the help of the device of Millikan oil – drop. The result will show that the electric charge carrying by a tiny oil – drop is multiple of an electron.

Objectives

1. Verify the discontinuity of electric charge, and determine the value of electric charge of an electron.
2. To train rigorous attitude by scientific lab working.

Principle

Spray oil – drops into the space between the two metal parallel plates with a distance of d (figure 18 – 1). Commonly, the oil – drops will carry electric charge because of their friction with air when spouting out. Suppose the mass of a certain oil – drop is m and carrying a charge of q, the electric voltage between the two plates is U. There are two forces act on the oil – drop, and it is possible to make them in opposite directions: one is gravity mg, straightly downwards; the other is electrostatic force $qE = \frac{qU}{d}$ with a direction upwards. Adjust the voltage to balance the two forces, that is:

$$mg = \frac{qU}{d} \qquad (1)$$

So, if we can measure the mass m, we can get the value of charge $|q|$.

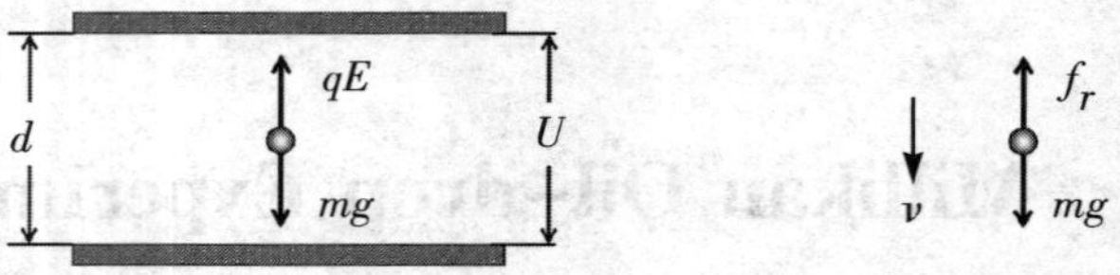

Figure 18 – 1 Oil drop: at rest and in uniform motion

1. Determination of the mass of the oil – drop

If there is no voltage between the two plates, the oil – drop will be accelerated falling down. Because of the resistance of air, the oil – drop will be uniform motion after a falling distance. When the gravity equals the air resistance (neglect the air buoyancy), the speed of uniform motion is v. According to Stokes law of fluid mechanics, the viscous force acting on a moving ball is:

$$f = 6\pi r\eta v \tag{2}$$

where, η is the viscosity, and r is the radius of the oil – drop (about a size of 10^{-6}m and mass of 10^{-15}kg). If the density of oil is ρ, the mass m of oil – drop will be:

$$m = \frac{4}{3}\pi r^3 \rho \tag{3}$$

Combining equation (2) and (3), we get

$$r = \sqrt{\frac{9\eta v}{2\rho g}} \tag{4}$$

But, the oil – drop is not a rigid body and has a small size of about the mean free path of air molecules (about 10^{-7}m), the Stokes law is not valid strictly and the η in equation (2) should be modified as η':

$$\eta' = \frac{\eta}{1 + b/pr} \tag{5}$$

b is a constant, $b = 8.21 \times 10^{-3}$ m · Pa, p is the atmosphere pressure in unit Pa. Substitute η with η', and combine equation (3) and (4), we get:

$$m = \frac{4}{3}\pi \left[\frac{9\eta v}{2\rho g} \cdot \frac{1}{1 + b/pr}\right]^{3/2} \rho \tag{6}$$

2. Determination the falling speed of oil – drop

When the voltage between the two plates is $U = 0$, say the oil – drop falls down a distance of l within time t, the speed of the oil – drop will be:

$$v = \frac{l}{t} \tag{7}$$

Combing (7), (6), and (1) we get:

$$q = \frac{18\pi}{\sqrt{2\rho g}} \left[\frac{\eta l}{t\,(1 + b/pr)}\right]^{3/2} \frac{d}{U} \tag{8}$$

where, $r = \sqrt{\frac{9\eta l}{2\rho g t}}$.

It is found in this experiment that for the same oil – drop, if its quantity of electric charge is changed, then the voltage keeping it stable will change into some certain value U_n. All these values can be expressed in the following equation:

$$q = ne = mg\frac{d}{U_n}$$

$n = \pm 1, \pm 2 \cdots$, and the e keeps unchanged.

There are the same fact for different oil – drops: the quantity of electric charge will be multiple times of a common constant e. This is the proof of discontinuity of charge and the existence of basic charge with a quantity of e.

$$ne = \frac{18\pi}{\sqrt{2\rho g}}\left[\frac{\eta l}{t(1 + b/pr)}\right]^{3/2}\frac{d}{U} \qquad (9)$$

Equation (8) or (9) is the key point for determining charge quantity in this experiment.

Equipments

The device of Millikan oil – drop experiment includes oil – drop device, electrical source, chronograph, mist atomizer.

1. Oil – drop device

Oil – drop device, type MOD – 2, is shown in figure 18 – 2. It is linked with the electrical source box by three adjustable screws.

The oil – drop box (5) is made of two parallel plates (4) and (6) with fine surface, space between them is $d = 0.500$cm. There is a small hole of $\phi 0.4$mm in the central of upper plate (4), which is the entrance of oil – drops. The oil – drop box is placed in the Perspex cover (3) to lessen the air streaming. There is an oil – drop fog chamber (1) on the top of Perspex cover, at the bottom a switch (2) can open or close the entrance of oil – drop (10) to oil – drop box.

Lighting set includes a light room and a light pipe glass rod; there is a bulb of 2.2V in the light room.

The moving oil – drops can be observed and measured with the microscope. There is a graticule in the ocular piece (figure 18 – 3), on which there is 6 divisions, every division equals a actual distance of 0.050cm, totally 0.300cm. Checking the moving distance l of oil – drop contrasting by graticule, we can measure the speed v.

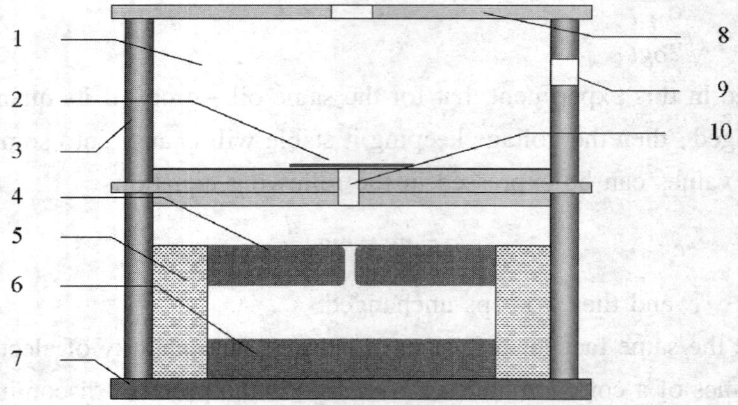

Figure 18 -2 Oil - drop device sketch

1. Oil - drop fog chamber; 2. Chamber hole cover; 3. Perspex cover;
4. Upper plate; 5. Oil - drop box; 6. Lower plate; 7. Frame;
8. Cover; 9. Chamber door; 10. Chamber hole

2. Electrical source

There are three kinds of electrical source:

(1) 2. 2V alternating current, supply for the bulb.

(2) 500V direct current voltage, which can be adjusted continuously and indicated by a voltage meter. Turn the switch with symbol "Balance Voltage" to the middle position "0", the two plates will be short circuit. When the switch is at the position "+", the voltage can balance an oil - drop carrying positive charge; and the "-" can balance negative.

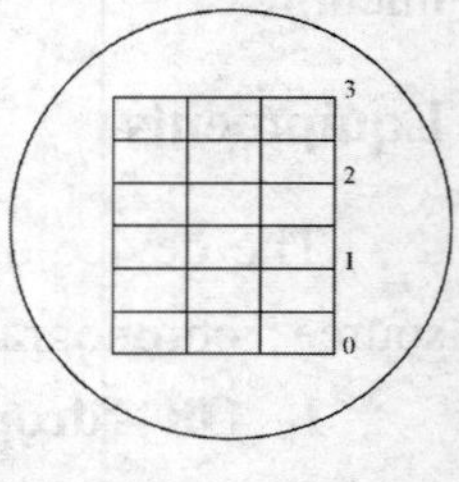

Figure 18 -3 Graticule

(3) A direct current voltage is about 250V, which can also be adjusted continuously. It is added up to the balance voltage to move the oil - drops. It is used to drive the charged oil - drops only, thus its value is unnecessary to be known.

3. Chronograph

A stop - watch with liquid crystal display can show 0. 1 second exactly.

Experiment

1. Adjust equipment to meet the requirement according to instruction.

(1) After power on the device, the electric power sources of 2. 2V for the lightening set and 500V for the parallel plates have all connected.

(2) Adjust horizontal screw to make the air bulb at the center of the level gauge. Now the parallel plates are horizontal and the electric field parallel with gravity.

(3) Check the "Balance Voltage" switch at the place of "0", then put the focus

needle into the central hole on the up parallel plate. Adjust the microscope to focus on the needle. Don't forget take out the needle.

(4) Spray oil – drops just one time into the oil – drop fog chamber. Observe from the microscope. The tiny oil – drops just like stars in a night sky.

2. Practice of controlling oil – drop before measuring.

(1) Practice of controlling oil – drop. Put on balance voltage (about 300V, "+" or "–"), drive out drops until there are only several drops on the scene of view. Focus on one drop and adjust balance voltage on it to make it stop. Switch off balance voltage to let the selected drop fall down with uniform speed for a distance. And put on balance voltage again to stop it. Practice the above process several times.

(2) Select proper oil – drop for measurement. It is very important to select proper drop in this experiment: the drop should not be too big or small. For a suitable drop, the balance voltage maybe about 300V and its falling time for a 0.300cm distance maybe about 10 ~ 30s.

3. Measuring

According to equation (9) there are only two quantities should be measured: one is balance voltage U_n; the other is falling time t of a distance l.

By carefully adjusting the balance voltage, let an oil – drop stay on a graticule line for estimating it is still or not.

To make sure that the oil – drop falling uniformly, let the drop fall a certain distance before measuring. The distance l for measurement should be near the central of plates. It is suitable for a distance of $l = 0.200$cm, that is four divisions of graticule.

Because of fluctuation, it's better to measure one drop for 10 times, and take 2 ~ 5 drops for measuring.

4. Data handling

(1) The constants in equation (9) is listed below:

Oil density	$\rho = 977\text{kg} \cdot \text{m}^{-3}$ (30℃)
Gravity acceleration	$g = 9.80\text{m} \cdot \text{s}^{-2}$
Air viscosity	$\eta = 1.83 \times 10^{-5}\text{Pa} \cdot \text{s}$
Falling distant	$l = 0.200\text{cm} = 2.00 \times 10^{-3}\text{m}$
Space between the two plates	$d = 5.00 \times 10^{-3}\text{m}$
Atmosphere pressure	$p = 1.013 \times 10^{5}\text{Pa}$
Constant	$b = 8.21 \times 10^{-3}\text{m} \cdot \text{Pa}$

Then, the equation (9) will be:

$$q = ne = \frac{18\pi}{\sqrt{2\rho g}}\left[\frac{\eta l}{t\left[1 + b/p\sqrt{\frac{9\eta l}{2\rho g t}}\right]}\right]^{3/2}\frac{d}{U_n} = \frac{1.43 \times 10^{-14}}{[t\ (1 + 0.0196\sqrt{t})\ ]^{3/2}} \cdot \frac{1}{U_n}$$

Because ρ and η are all functions of temperature, and g, P will change as experiment place and surrounding changes, the above equation is approximately correct. Usually the relative error is about 1%, but it's worthy and convenient for calculation.

(2) To prove the discontinuity of charge, and all the quantities of charge is multiple times of basic charge e, and to get the quantity of e, we have to get the greatest common divisor of all the values of electrical charge we have gotten in this experiment. But it is difficult because of experimental errors. Here, we handle it by "Inverse Process", that is, to divide all the values by the known basic constant of electron charge $e = 1.60 \times 10^{-19}$C, and round the result to a integer, this integer n is the number of redundant electrons carrying by that oil – drop. Then, by dividing the measurements by the number n we get the final result of electron charge e of this experiment.

Notice

1. There should have no voltage between the two plates when focus the microscope by a fine needle, or there will cause short circuit to damage device.
2. Keep the mist atomizer on end when spray oil – fog.
3. When following an oil – drop, if the image becomes faint, focus again.
4. If find the speed of oil – drop change obviously, it is necessary to adjust the balance voltage again.

Problems

1. What is the reason for the image becoming faint?
2. How to judge an oil – drop being uniform motion?
3. What kind of effective will it cause if the plates are not horizontal?

Additional Instruction

Preparation Requirements

In brief, the value of charge carried by an oil – drop is determined by measuring the falling time t for a distance of l of uniform motion and its balance voltage U. For this reason, the parallel plates must be horizontal to make the electric field force parallel with gravitational force.

Operation key points

1. Adjust the parallel plates being horizontal, or the oil – drop may move out of the view under the condition of long time of observation.

2. Because the oil – drop is very small (about a size of 10^{-6}m), it must be carefully focused.

3. Operate the balance voltage and controlling voltage carefully for a selected drop and it maybe take some slight change.

4. It maybe takes quite a long time to measure a tiny drop through observation on the microscopy for repeated measurement. One must be strict with and precise on the experimental operation.

5. Generally, it is a requirement that the value of charge for a drop must under 10 times of an electron.

Problems after experiment

1. An oil – drop moving in the air is experienced air friction, gravity and air buoyancy. Consider the effect of air buoyancy on the final result.

2. Oil – drops have moving directions not only in the vertical direction at the beginning of entering the parallel plates, why the drops being measured are all moving in the vertical direction?

3. Oil – drops may volatilize a lot during measuring process. How can the volatilization of oil – drops affect the final result?

实验十九　用光电效应测普朗克常量

普朗克常量是物理学中一个非常重要和基本的常量。1905 年爱因斯坦在普朗克量子假说的基础上圆满地解释了光电效应，约十年后密立根以精确的光电效应实验证实了爱因斯坦的光电效应方程，并测定了普朗克常数。光电效应实验及其光量子理论的解释在量子理论的确立与发展上，在解释光的波粒二象性等方面都具有深刻的、决定性的意义。如今光电效应已经广泛地应用于各个科技领域，利用光电效应制成的光电器件（如：光电管、光电池、光电倍增管等）已成为生产和科研中不可缺少的器件，并且还在不断开辟新的应用领域，具有广阔的应用前景。

【实验目的】

1. 测定光电效应的伏安特性曲线，加深对光的量子性的理解。
2. 通过爱因斯坦光电效应方程，测定不同频率下的截止电压，并由此测定普朗克常数。
3. 测定实验中光电管的截止频率（红限频率）。

【实验器材】

GP－1 型普朗克常数测定仪，包括高压汞灯、光电管、滤色片和微电流放大器等四部分。见图 19－1。

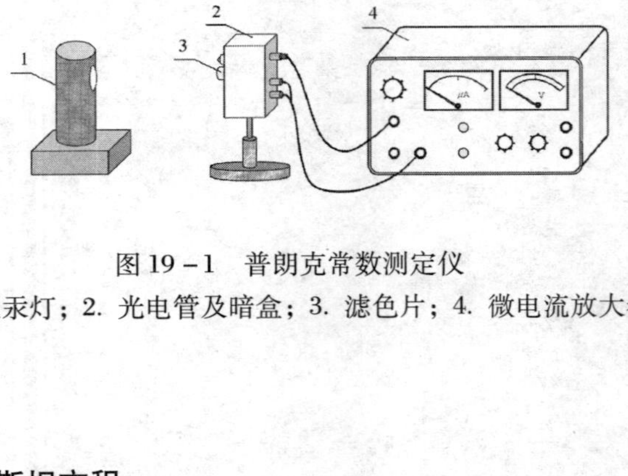

图 19－1　普朗克常数测定仪

1. 高压汞灯；2. 光电管及暗盒；3. 滤色片；4. 微电流放大器。

【实验原理】

1. 光电效应与爱因斯坦方程

光电效应现象是指用合适频率的光照射在某些金属表面上时，会有电子从金属表面逸出。从金属表面逸出的电子叫光电子。实验原理如图 19－2，图中的双刀双掷开关可以起到在阴极 A 和阳极 K 之间的电压换向的作用。

为了解释光电效应现象，爱因斯坦提出了“光量子”的概念，认为对于频率为 f 的光波，每个光子的能量为：

$$E = hf$$

式中，h 为普朗克常数，它的公认值是 $h = 6.626 \times 10^{-34}$ J · s。

按照爱因斯坦的理论，光电效应的实质是当光子和电子相碰撞时，光子把全部能量传递给电子，电子所获得的能量，一部分用来克服金属表面对它的约束，其余的能量则成为该光电子逸出金属表面后的动能。爱因斯坦提出了著名的光电方程：

$$hf = \frac{1}{2}mv^2 + A \qquad (1)$$

式中，f 为入射光的频率，m 为电子的质量，v 为光电子逸出金属表面的初速度，A 为被光线照射的金属材料的逸出功，$\frac{1}{2}mv^2$ 为从金属逸出的光电子的最大初动能。

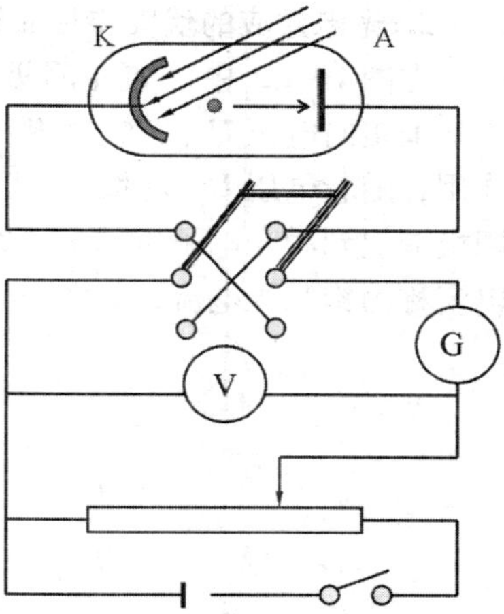

图 19－2　实验原理图

由式（1）可见，入射到金属表面的光频率 f 越高，逸出的电子动能必然也越大，所以即使阳极 K 不加电压也会有光电子到达阳极而形成光电流，甚至阳极电位比阴极电位低时也会有光电子到达阳极。只有当反向的阳极电位增大到某一数值时，所有光电子都不能到达阳极，光电流才为零。这个相对于阴极为负值的阳极电位被称为光电效应的截止电压 U_0。

显然，有

$$eU_0 = \frac{1}{2}mv^2 \qquad (2)$$

代入式（1），即有

$$hf = eU_0 + A \qquad (3)$$

由光电效应能量关系可知，若光电子能量 $hf < A$，则不能产生光电子。产生光电效应的最低频率是 $f_0 = A/h$，称为光电效应的截止频率（红限频率）。不同材料有不同的逸出功，因而 f_0 也不同。由于光的强弱决定于光量子的数量，所以光电流与入射光的强度成正比。又因为一个电子只能吸收一个光子的能量，所以光电子获得的能量与光强无关，只与光子的频率 f 成正比，将式（3）改写为：

$$U_0 = \frac{hf}{e} - \frac{A}{e} = \frac{h}{e}(f - f_0) \qquad (4)$$

上式表明，截止电压 U_0 是入射光频率 f 的线性函数，如图 19－3，当入射光的频率 $f = f_0$ 时，截止电压 $U_0 = 0$，没有光电子逸出。图中的直线的斜率 $k = \frac{h}{e}$ 是一个正的常数。由此可见，实验中作出不同频率下的 $U_0 - f$ 曲线，并求出此曲线的斜率 k，就可以通过下式求出普朗克常数 h。

$$h = ek \qquad (5)$$

其中 $e = 1.60 \times 10^{-19}$ C，是电子的电量。

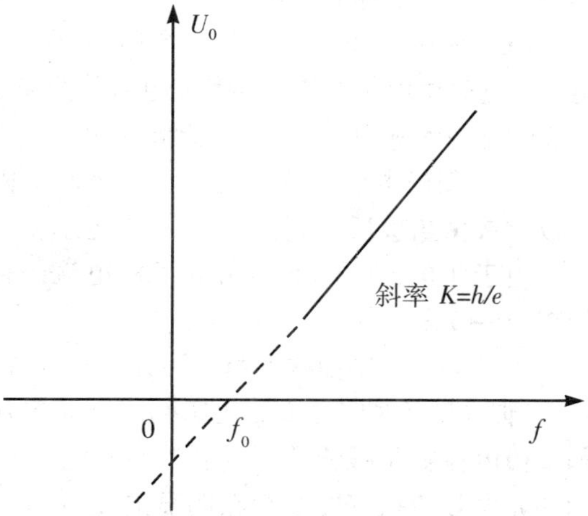

图 19－3　$U_0 - f$ 图

2. 光电效应的伏安特性曲线

在图 19－2 中，如在阴极 K 和阳极 A 之间加正向电压 U_{AK}，它使 K、A 之间建立起的电场对从光电管阴极逸出的光电子起加速作用，随着电压 U_{AK}的增加，到达阳极的光电子将逐渐增多。当正向电压 U_{AK}继续增加，阳极捕捉到的光电子也越多，光电流增加但不能一直增加，而是达到饱和状态，对应的光电流称为饱和光电流，如图 19－4。

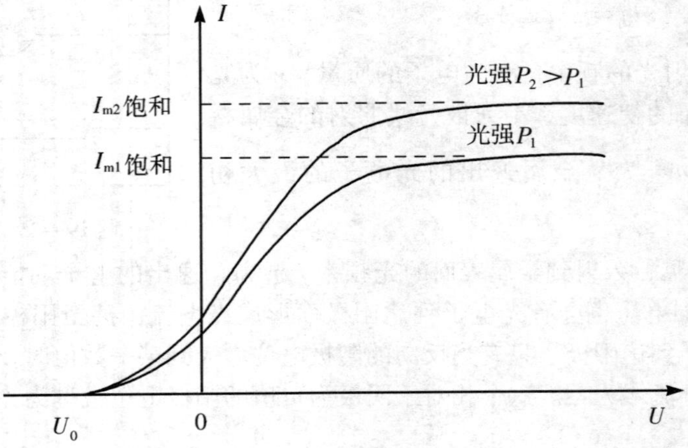

图 19－4　伏安特性曲线

由于光电子从阴极表面逸出时具有一定的初速度，所以当两极间电位差 $U_{AK}=0$ 时，仍有光电流 I 存在。若在两极间施加一反向电压，光电流随之减少；当反向电压达到截止电压 U_0时，光电流 I 才为零。并且，根据光量子理论，增加光强只能增加了单位时间内入射到阴极金属表面的光子数，而由于频率没有改变，每个光子的能量没有变化，因此饱和光电流的增大了，但截止电压没有改变。见图 19－4 中的曲线 P_2。

请注意爱因斯坦光电效应方程的导出是基于同种金属做阴极和阳极，且阳极无电子发射的理想状态。实际上只是阴极的金属逸出功比作阳极的金属逸出功小很多，也就是说，阳极金属在实验中也会有电子逸出。所以实验中存在着如下问题。

（1）暗电流和本底电流。当光电管阴极没有受到光线照射时也会产生电子流，称为暗电流。它是由电子的热运动和光电管管壳漏电等原因造成的。室内各种漫反射光射入光电管造成的光电流称为本底电流。暗电流和本底电流随着 K、A 之间电压大小变化而变化。

（2）阳极电流。制作光电管阴极时，阳极上也会被溅射有阴极材料，所以光入射到阳极上或由阴极反射到阳极上，阳极上也有光电子发射，就形成阳极电流。

由于暗电流、本底电流和阳极电流的存在，使得实际的 $I-U$ 曲线较理论曲线下移，如图 19－5 所示。

由图 19－5 中曲线的特点可以看出，由于上述问题的存在，使得实际的伏安特性曲线与 U 轴的交点对应的 U'_0 比实际的截止电压数值上小一点。但注意到暗电流、本底电流和阳极电流等是很快达到饱和的，光电管的伏安曲线在截止电压附近又较为陡峭，因此实际的光电效应伏安曲线存在明显的拐点 A，对应截止电压 U_0。通过伏安曲线上的拐点即可测出截止电压。

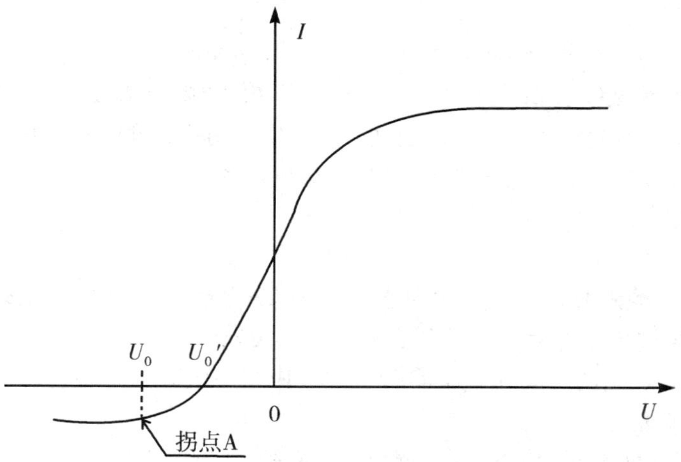

图 19－5　伏安特性曲线的拐点

3. 高压汞灯强谱线列表

表 19－1　汞灯谱线

波长/nm	频率/10^{14}Hz	颜色
577.0	5.198	黄
546.1	5.492	绿
435.8	6.882	蓝
404.7	7.410	紫
365.0	8.216	近紫外

注：各谱线可通过加相应的滤色片获得。

【实验内容】

1. 调整仪器

对照仪器和仪器说明书熟悉各部件的作用和调节方法。连线，并按照说明书要求预置各开关位置。接好电源，打开电源开关，充分预热（不少于20分钟），然后按照要求进行调零。

在暗盒通光窗上盖有遮光罩的情况下，将为电流放大器"倍率"旋至"$\times 10^{-7}$"或"$\times 10^{-6}$"挡，测量相应的电压、电流值。这时的光电流即为光电管的暗电流。

取下暗盒光窗口遮光罩，换上365.0nm滤光片，取下汞灯出光窗口的遮光罩，装好遮光筒，调节好暗盒与汞灯距离，进行粗测，注意电流变化大的电压位置。

2. 测量光电管的伏安特性曲线

预先设计记录数据的表格。

（1）暗盒光窗口装365.0nm滤光片和2mm光阑，缓慢调节电压旋钮，令电压输出值缓慢由－2V增加到＋20V，－2～0V之间每隔0.2V记一个电流值，0～20V之间每隔3V记一个电流值。但注意在电流值为零处记下截止电压值。

（2）在暗盒光窗口上换上404.7nm滤光片，仍用2mm的光阑，重复步骤（1）。

（3）换上4mm的光阑重复步骤（1）、（2）。

（4）选择合适的坐标，分别作出两种光阑下上述两个波长（365.0nm、404.7nm）的光电管伏安特性曲线 $U-I$。

3. 测量普朗克常数 h

将直径为4mm的光阑和365.0nm的滤色片装在光电管电暗盒输入口上。从高到低调节电压，用“零电流法”测量该波长对应的 U_0，记录数据。依次换上404.7nm、435.8nm、546.1nm、577.0nm的滤色片，重复测量。对这5个波长的光电效应，分别找到截止电压 U_0，作 U_0-f 曲线。求出曲线斜率，并由此求出普朗克常数和光电管的截止频率。

【注意事项】

1. 微电流放大器和汞灯的预热时间必须长于20分钟。实验中，汞灯不要随意关闭。如果关闭，必须经过5分钟后才可重新启动，且须重新预热。

2. 汞灯与暗盒之间的距离在整个实验过程中应当一致。

3. 注意保护滤光片，防止污染。

4. 每次换滤光片时要先遮住汞灯的出光口，以免汞灯的光直射光电管，而致光电管损坏。

5. 微电流放大器每改变一次量程，必须重新调零。

【思考题】

1. 光电管为什么要装在暗盒中？为什么在非测量时，用遮光罩罩住光电管窗口？

2. 为什么当反向电压加到一定值后，光电流会出现负值？

3. 入射光的强度对光电流的大小有无影响？实验中你是如何进行控制光强的？

实　验　指　导

【预习要求】

1. 对光电效应原理有比较好的理解

爱因斯坦光电效应方程证明了在微观领域能量守恒定律依然成立。由方程得出的截止频率与阴极材料有关，但是在实验中所用的光的频率是不能连续调节的，而是通过求 U_0-f 曲线的 x 轴交点得到的。对于能出现光电效应的每一波长的光，都要找到其对应的截止电压 U_0。

2. 实验仪器的认识

GP-1型普朗克常数测定仪的面板如图19-6。

该仪器的主要功能是进行微电流放大。由仪器上直接读出的是对应电压下的光电流值。预习时应先熟悉仪器面板。

【操作要点】

1. 根据实验仪器说明书，调整、操作仪器。实验开始时应检查面板上个开关是否置于实验室要求的位置。

2. 本实验的关键所在是准确地找出各个频率入射光所对应的外加截止电压。这就需要对五个波长的光分别做出伏安特性曲线，再找到曲线的拐点对应的电压，就是截止电压。反向电压段曲线上拐点之后的阳极反向电流很快饱和。这是确定拐点的主要依据。本实验中是将零电流对应的电压作为截止电压的。

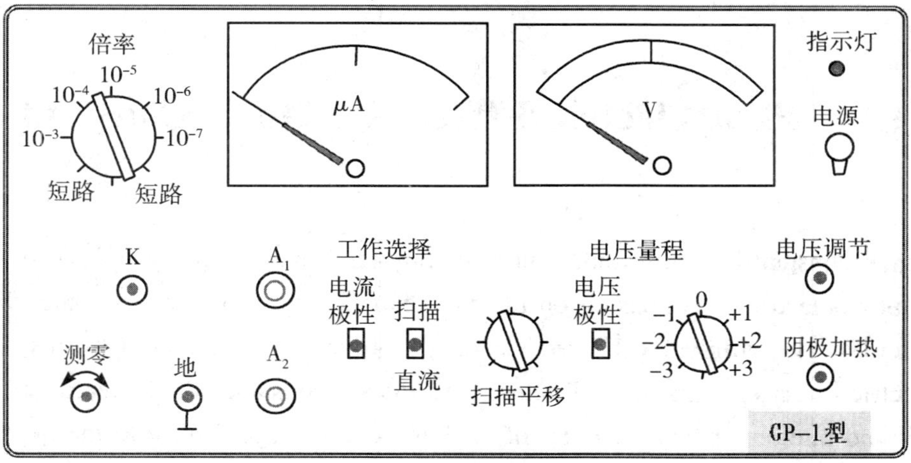

图 19－6　GP－1 型普朗克常数测定仪面板

【报告要求】

作出合适的数据记录表格。并记录下每一波长实验数据表对应的光电管和汞灯间距离、光阑孔径等。作图要求见实验一。

对本实验数据尝试用 Origin 软件处理和作图。

【实验后的思考】

1. 从实验获得的数据和处理结果，请你再次说明光电效应规律。
2. 分析实验中影响截止电压测量的主要因素。

19 Photoelectric Effect & Planck Constant

Planck constant is a very important basic physics constant. In 1905 A. Einstein explained photoelectric effect basing on the M. Planck's quantum theory, and about 10 years later R. A. Millikan validated Einstein's photoelectric equation by his accurate photoelectric effect experiment and determined Planck constant. Photoelectric effect experiment and photons theory have a profound and crucial significance to the explanation of wave – particle duality nature of light as well as the development of quantum theory of physics. Nowadays photoelectric effect has a widely application in many fields of science and technology. Phototube such as photoelectric cell and photo multiplier tube, which are made upon the photoelectric principle, not only has been an indispensable and basic electric apparatus in the fields of manufacture and scientific research, but also having been applied to more and more areas.

Objectives

1. Plotting the graph of photoelectric effect and deepen understanding of photon nature of light.

2. According to the Einstein's photoelectric effect equation measure the cutoff voltages for lights of different frequency, and thus get the Planck constant.

3. Measure the cutoff frequency of the phototube.

Equipments

Planck constant measurement instrument, including four parts as: high voltage mercury lamp, phototube, color filter and weak current amplifier. It is showed in figure 19 – 1.

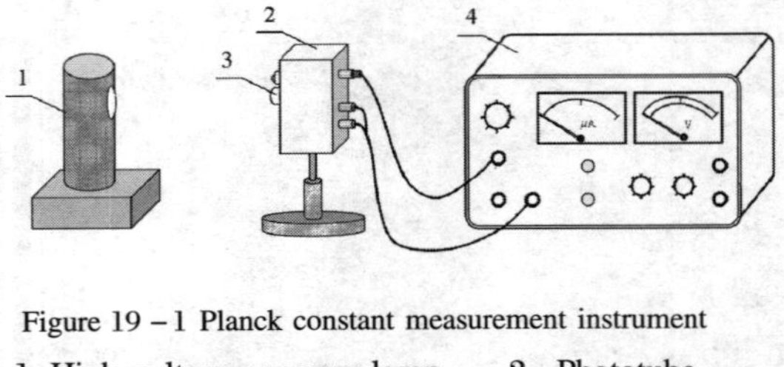

Figure 19 – 1 Planck constant measurement instrument

1. High voltage mercury lamp; 2. Phototube;

3. Color filter; 4. Weak current amplifier.

Principle

1. Photoelectric effect and Einstein' s equation

Incidence light with suitable frequency upon the surface of some kinds of metal will cause electrons emitting from the surface, this phenomenon is called photoelectric effect. The electrons emitting from the surface are called photoelectron. Its principle circuit is showed in figure 19 –2. In the Figure the double pole double throw switch can reverse the voltage between cathode A and anode K.

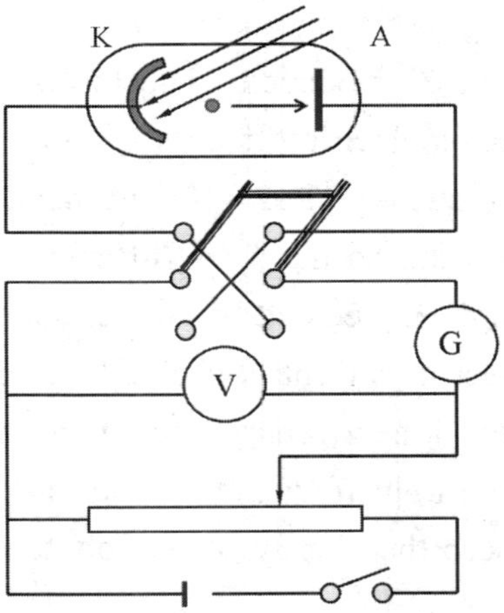

Figure 19 –2 Principle circuit

To explain the photoelectric effect, Einstein put forward the concept of "photon": in his point of view every photon of light of frequency f has the energy:

$$E = hf$$

where, the h is Planck constant, of which the acknowledged value is $h = 6.626 \times 10^{-34}$ J · s.

According to Einstein's theory, the essence of photoelectric effect is when a photon collides with an electron the electron gets all the energy of the photon. A portion of the energy overcomes the bound of the metal surface and another portion of energy becomes the initial kinetic energy of the electron which escapes from the metal surface. Einstein's photoelectric effect equation is about the conservation of energy:

$$hf = \frac{1}{2}mv^2 + A \qquad (1)$$

Where, f is the frequency of the incident light, m is the mass of electron; v is the initial speed of the electron escaping from the metal; A is the escape work or work function of the metal; $\frac{1}{2}mv^2$ is the maximum initial kinetic energy of the escaping electron.

It can been seen from equation (1) that the higher of the frequency f the higher of the initial kinetic energy of the escaping electron; thus, the photoelectron can arrive the anode with it initial speed with an anode applying no positive voltage or even an inverse voltage. Only the inverse voltage reaches a certain value the photoelectric current can become zero. This voltage with a negative value comparing the cathode is called the cutoff voltage U_0 of photoelectric effect.

Obviously that:

$$eU_0 = \frac{1}{2}mv^2 \qquad (2)$$

And the equation (1) can be:

$$hf = eU_0 + A \tag{3}$$

According to the photoelectric equation if the energy of a photoelectron $hf < A$ the photoelectron can not be produced. The lowest frequency for generating photoelectric effect is $f_0 = A/h$, which is called cutoff frequency. For different metal the work function is different, so the f_0 is different. Because the intensity of light is decided by the amount of photons, the photoelectric current is decided by that amount too. The initial kinetic energy of photoelectron is proportional to the light frequency f and independent of the light intensity for one electron can only absorb the energy of one photon. Rewrite the equation (3):

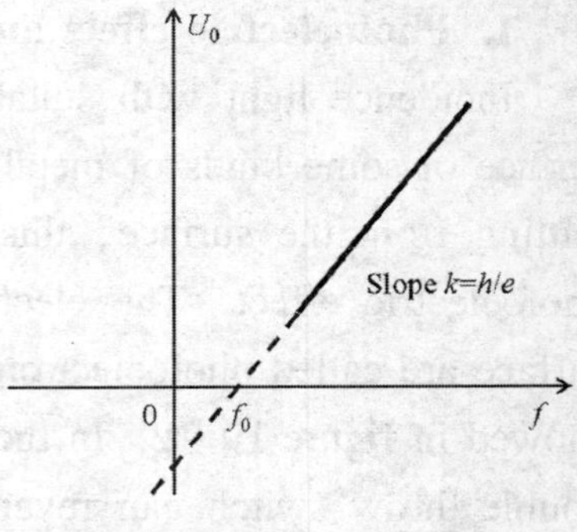

Figure 19 – 3 Relationship of $U_0 - f$

$$U_0 = \frac{hf}{e} - \frac{A}{e} = \frac{h}{e}(f - f_0) \tag{4}$$

The above equation indicated that the cutoff voltage U_0 is a linear function of the frequency f of the incident light, as showing in Figure 19 – 3. When the frequency $f = f_0$ the cutoff voltage $U_0 = 0$, that means there is no photoelectron escaped. The slope of the line in the Figure is $k = \frac{h}{e}$, which is a positive constant. In this experiment the graph of $U_0 - f$ can be made and then the slope k. Finally, the Planck constant h:

$$h = ek \tag{5}$$

where, $e = 1.60 \times 10^{-19}$ C is the charge of an electron.

2. The volt – ampere characteristic graph of photoelectric effect

In the figure 19 – 2 the positive voltage U_{AK} between cathode A and anode K will accelerate the photoelectron from the cathode. More and more photoelectrons will reach anode under an increasing U_{AK}. Thus the photoelectric current will increase. But the current cannot increase on as it will reach a saturation state, which is called saturate photoelectric current, as figure 19 – 4.

Because the photoelectrons have initial velocities, they can reach the anode even if the $U_{AK} = 0$ to form a photoelectric current. If put on an inverse voltage between the two poles, the current will cut down; when the inverse voltage reaches the cutoff value U_0, the current will be zero. According to quantum theory, in-

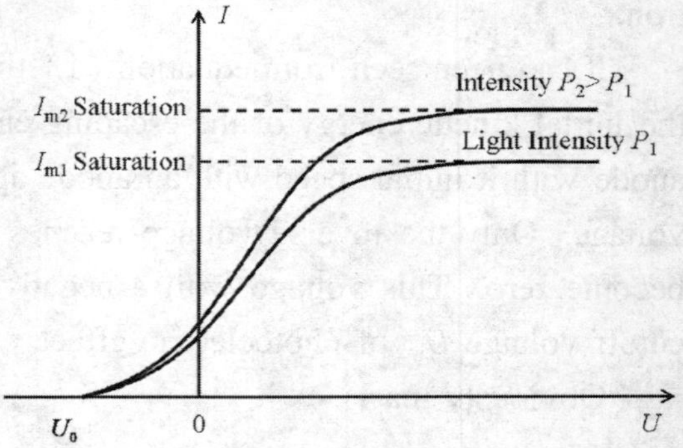

Figure 19 – 4 $U - I$ Graph of photoelectric effect

creasing the light intensity can only make the number of photons incident on cathode in per unit time big, and their frequency are not changed. For a more intensive light the saturation current become bigger, but a same value of cutoff voltage, as showed in the curve P_2 in figure 19 – 4.

It must be paying more attention that Einstein's equation is based on the same kind of metal of cathode and anode as well as there is no electron emitting from anode. In practice, the work function of cathode is much smaller than anode and there is also some electrons can escape from anode. Thus, the following problems should be taken into account.

(1) Dark current and background current. There is a photoelectric current when no light incident on the cathode, which is called dark current. Dark current is caused by the electronic thermal motion and the electric leakage of the photoelectric tube shell. Any type of diffuse reflection goes into the tube can cause photoelectric current too, which is called background current. The dark current and background current can change as the voltage between K and A changes.

(2) Anode current. When making a photoelectric tube, the cathode material can contaminate the anode. Thus, light incident on or reflecting on the anode can also electron emitting, causing an anode current.

Because of the dark current, background current and anode current, the actual I (U curve will shift down a little, as showed in figure 19 – 5.

In figure 19 – 5, the actual curve intersects the U axis at U_0', which is small than the cutoff voltage U_0. Because the dark current, background current and anode current can saturated easily under a small inverse voltage, the volt – ampere curve has an obvious inflexion point A, corresponding to the cutoff voltage. Therefore, the cutoff voltage can be read out by the inflexion point.

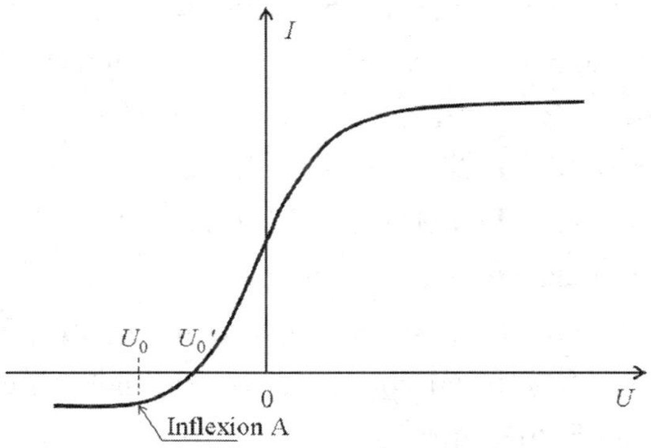

Figure 19 – 5 Inflexion point on the $U-I$ Graph

3. List of spectral lines of high voltage mercury lamp

Table 19 – 1 Spectrum of mercury lamp

Wavelength /nm	Frequency /10^{14} Hz	Color
577.0	5.198	Yellow
546.1	5.492	Green
435.8	6.882	Blue
404.7	7.410	Purple
365.0	8.216	Near Ultraviolet

All the spectral line can be gotten by adding color filter.

Experiment

1. Adjust instrument

Consult the instruction booklet to be familiar with the main parts of instrument. According to instruction book, joint wires and set the switches on proper positions. Put on the instrument power to warm up it (more than 20 minutes). Then do the preparing work of zeroing.

Cover the window of cassette with lens hood, and set the current amplifier to the rate of "$\times 10^{-7}$" or "$\times 10^{-6}$". Measure the corresponding voltage and current, the current is dark current.

Take off the lens hood and put on the color filter of 365. 0nm. Take off lens hood from the window of mercury lamp, and then adjust the distance between cassette and lamp. Now you can begin to measure the voltage and current roughly, and notice the places of current strongly changing.

2. Measure the volt – ampere curve

Design a data table beforehand.

(1) Covering the window of cassette with a color filter of 365. 0nm and a diaphragm of 2mm, record the current value by points of voltage output from (2V to + 20V. Record a current value by step of 0. 2V between (2V and 0V, and step of 3V from 0V to 20V. Notice that cutoff voltage is at the point of zero current.

(2) Switch the color filter with 404. 7nm, yet the diaphragm is also 2mm. Repeat the measurement of above.

(3) Switch the diaphragm of 4mm, repeat the measurement above.

(4) In an appropriate coordinate draw the volt – ampere curves of the two wavelength 365. 0nm and 404. 7nm.

3. Measure the Planck constant *h*.

Fix the diaphragm of 4mm and color filter of 365. 0nm on the cassette window. Scan voltage from high to low to get the cutoff voltage U_0 corresponding to zero current, record the result. Switch the color filter of 404. 7nm, 435. 8nm, 546. 1nm and 577. 0nm to find their cutoff voltage. Draw the straight line of $U_0 - f$. Get the slope from the graph, and then Planck constant and cutoff frequency of the photoelectric tube.

Notice

1. Weak current amplifier and mercury lamp must have been warmed up for more than 20 minutes. During experiment the mercury lamp should not be switch on and off frequently. If the lamp have to be power on again it should wait more than 5 minutes af-

ter it is power off.

2. Keep the distance between mercury lamp and cassette be unchanged during the experiment.

3. Keep the color filter clean.

4. Avoid the lamp light directly incident on the photoelectric tube, especially when switch the color filter, or the photoelectric tube maybe damaged.

5. Before measuring with a new range of the weak current amplifier adjust zero point first.

Problems

1. Why the photoelectric tube should be installed in a cassette? The window of photoelectric tube should be covered by a lens hood when it is not used, why?

2. Why the photoelectric current will be negative for a certain inverse voltage?

3. How can the incident light intensity affect the photoelectric current? What method did you take to control the incident light intensity?

Additional Instruction

Preparation requirement

1. It is required that a good understanding of photoelectric effect principle

Einstein's photoelectric equation shows that the law of conservation of energy is also correct in the microcosmic field. The cutoff frequency getting from the equation is decided by the cathode material. The light frequency in this experiment cannot be changed continuously, and the cutoff frequency is gotten according to the point of intersection of the line $U_0 - f$ with x axis. In this experiment, we should have to find the corresponding cutoff voltage U_0 for every frequency of light causing photoelectric effect.

2. Knowledge about the weak current amplifier

The panel of Planck constant measurement instrument type GP – 1 is showed in figure 19 – 6.

The key function of this instrument is to amplify weak electric current. The reading of current is corresponding to a certain voltage U_{AK}. Be familiar with its panel in preparation.

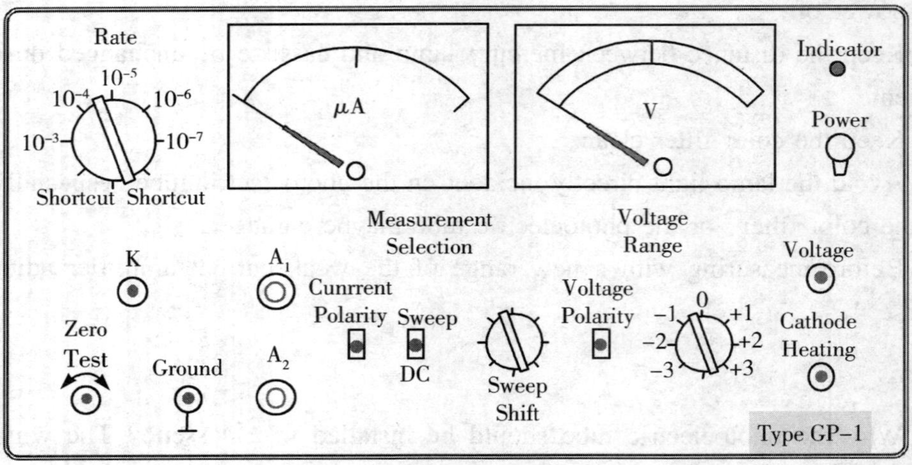

Figure 19 − 6 The Panel of Plank Constant measurement instrument type GP − 1

Operation key points

1. Adjust and operate instrument according to the instruction book. Before start, check all the switches on the required position.

2. The key point of this experiment is to find the cutoff voltages corresponding to every frequency. It's better to draw five volt − ampere curves for five frequencies, and then find the voltage corresponding to the inflexion point on every curve. That is the cutoff voltage. The inverse current will easily be saturated after the inflexion point on the curve segment of inverse voltage. This is the basis for determine the inflexion point. In this experiment the voltage corresponding to zero current is taken as cutoff voltage.

Report requirement

Write down suitable data tables, data such as the wavelength, distance between lamp and tube, diaphragm aperture should also be recorded. See detail requirement of graph drawing in experiment 1.

Try to draw graph and process data by software ORIGIN in computer.

Problems after experiment

1. Explain the photoelectric effect according to your own data.
2. Analyze the main factors affect the cutoff voltage.

实验二十　用阿贝折射仪测定液体的折射率

折射率是透明物质的重要光学参数之一，它反映了物质的光学性质、纯度等方面的信息。阿贝折射仪是根据全反射原理制成的测量透明或半透明固体、液体折射率的仪器，在药品检验等方面有着广泛应用。通过本实验可了解阿贝折射仪的结构原理，掌握其使用方法，并用阿贝折射仪测量不同浓度 NaCl 溶液折射率，熟悉用图示法和最小二乘法由折射率求液体浓度的方法。

【实验目的】

1. 熟悉阿贝折射仪的结构原理，学会其正确的使用方法。
2. 学会用阿贝折射仪测定液体折射率，并由此确定液体浓度的方法（折光分析法）。
3. 熟悉实验数据的图示法和一元线性回归法。

【实验仪器】

阿贝折射仪、蒸馏水、无水乙醇、几种不同浓度的氯化钠溶液、滴管、擦镜纸。

1. 阿贝折射仪的光学系统

阿贝折射仪光学系统主要由两部分组成：望远镜系统和读数系统（图 20－1）。

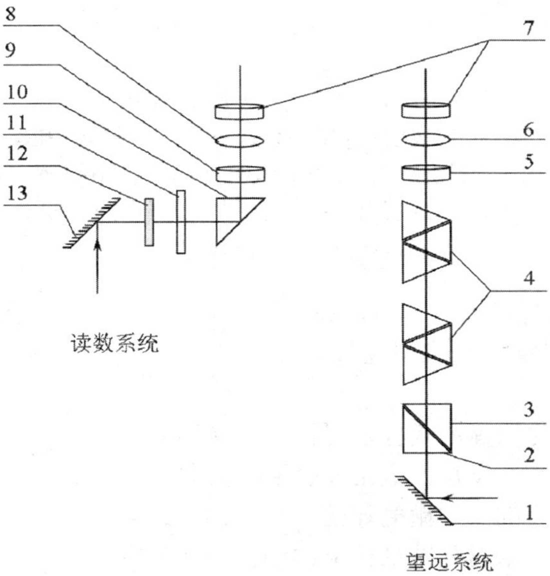

图 20－1　阿贝折射仪的光学系统示意图

1. 反光镜；2、10. 棱镜；3. 折射棱镜；4. 阿西米棱镜；5、9. 物镜；6、8. 场镜；7. 目镜；11. 照明度盘；12. 毛玻璃；13. 小反光镜望远镜系统：光线由反光镜（1）进入进光棱镜（2）及折射棱镜（3），被测定液体放在（2）、（3）之间，经阿米西棱镜（4）以抵消由于折射棱镜及被测物体所产生的色散。由物镜（5）将明暗分界线成像在场镜（6）的平面上，经场镜（6）、目镜（7）放大后成像于观察者眼中。

读数系统：光线由小反光镜（13）经过毛玻璃（12）和照明度盘（11），经转向棱镜（10）及物镜（9）将刻度成像于场镜（8）的平面上，经场镜（8）和目镜（7）放大后成像于观察者眼中。

2. 阿贝折射仪的机械结构

图 20－2 为阿贝折射仪的机械结构示意图。

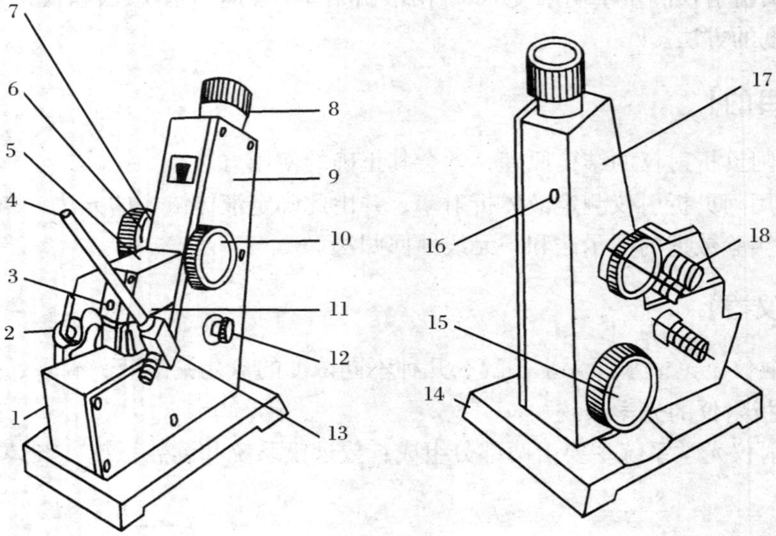

图 20－2 阿贝折射仪机械结构图

1. 反射镜；2. 转轴；3. 遮光板；4. 温度计；5. 进光棱镜座；6. 色散调节手轮；7. 色散值刻度圈；8. 目镜；9. 盖板；10. 锁紧手轮；11. 折射棱镜座；12. 照明刻度聚光镜；13. 温度计座；14. 支撑座；15. 折射率刻度调节手轮；16. 校正螺钉孔；17. 机壳；18. 恒温器接头

因测量时的温度对折射率有影响，在温度变化大时，为了保证测量精度在必要时可开启恒温器。

3. 准备工作

在开始测定前必须先用标准式样校对读数。在标准试样的抛光面上加 1～2 滴溴代萘，贴在折射棱镜的抛光面上，如图 20－3 所示，标准试样抛光面的一端应向上，以接受光线。当读数镜内指示于标准试样上的刻值时，观察望远镜内明暗分界线是否是在十字线中间，若有偏差则用附件校正扳手转动示值调节螺钉，调整明暗分界线至中央（图 20－7）。在以后测定过程中不允许再动该螺钉。

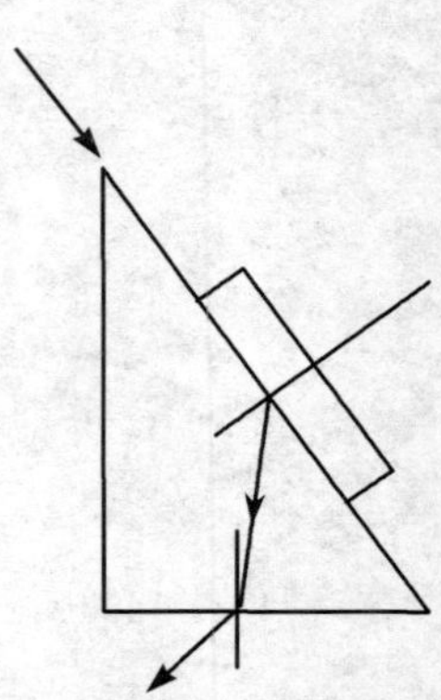

图 20－3 标准式样校对示意图

【实验原理】

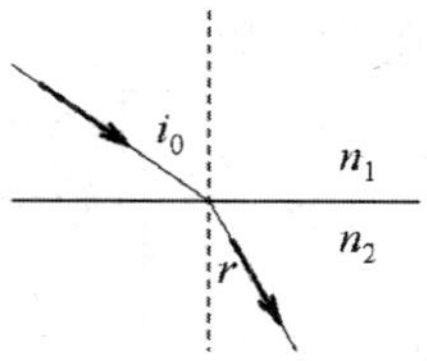

图 20－4 折射定律示意图

光线从一种介质斜入射到另一种介质时光路会发生偏折，这种现象称为光的折射现象。

当光线由折射率为 n_1 的介质射入折射率为 n_2 的介质时，如图 20－4 所示，偏折的角度可由折射定律得知。

$$n_1 \sin i = n_2 \sin r \qquad (1)$$

式中，i 表示入射角，r 表示折射角。

若 $n_1 > n_2$，即光由光密介质射向光疏介质时有 $i < r$，当 i 增大到某一值 i_0，折射角为 90°，入射角再增大，光线将全部反射回光密介质中，这种现象称为全反射现象，而 i_0 即为发生全反射时的特定入射角，称为临界角。

根据光路的可逆性，若光由光疏介质 n_2 射向光密介质 n_1 时，当光在光疏介质中入射角在 0°～90°范围内，即 0°～90°范围均有光线时，在光密介质中只有在临界角 i_0 内有光线，而在大于 i_0 范围为暗区，因而形成明暗分界线（用望远镜观察）如图 20－5 所示。

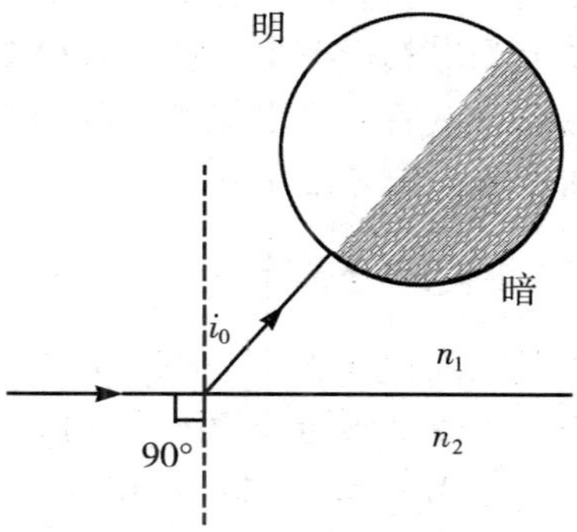

图 20－5 明暗分界线示意图

阿贝折射仪就是根据上述完全反射原理制成的，它是药物检验中常用的分析仪器，其外型如图 20－2 所示。它的主要部分是由两个同样的直角棱镜 ABC 和 DEF 所组成（图20－6），两棱镜面夹待测液体薄层（设折射率为 n'）。棱镜 DEF 的 DF 面为磨砂表面，由普通光源发来的光由镜面 M 反射通过透光孔 P 进入棱镜 DEF，经折射后射到磨砂表面 DF，这使磨砂表面被照亮而成为发光面。由于磨砂面使光向各方向漫射，因此由 DF 面发出的漫射光线通过液层入射到棱镜 ABC 的 AC 面。因液层很薄，总可以有入射角非常接近 90°的入射光，如图 20－6 所画出的光线 SO，称为掠射光线。设棱镜的折射率为 n，如果 $n > n'$，则 SO 射入 AC 面的折射角为棱镜对液体的临界角 i_0。光线 OR 在 BC 面的入射角 i 及折射角 r 都由 i_0 而定。即出射线的位置也就由待测液体的折射率 n' 所决定。由于 SO 是所有入射 AC 面的光线中入射角最大的，故所有射入 AC 面的光线经两次折射后其出射线的方向只能在 Rl 的左边。若射入光线为单色光，则对准 Rl 方向的光线所成的像就是这明暗的分界线，我们只要测定这分界线 Rl 的出射角 r 就可以求出待测液体的折射率 n'。

图 20－6 中，设棱镜的棱角∠ACB 为 φ，由三角形 ORC 可知：

$$i_0 + 90° = i + 90° + \varphi$$

即：

$$i_0 = i + \varphi \qquad (2)$$

对光线 SO 和 OR，由折射定律得：

$$n' \sin 90° = n \sin i_0 \qquad (3)$$

所以：

$$n' = n \sin i_0 = n \sin(i + \varphi) = n \sin i \cdot \cos\varphi + n \cos i \cdot \sin\varphi \qquad (4)$$

在 BC 界面 R 点有：

$$n\sin i = n_a \cdot \sin r = \sin r \qquad (5)$$

其中 $n_a=1$，为空气的折射率。故：

$$\sin i = \frac{\sin r}{n}$$

$$\cos i = \sqrt{1-\sin^2 i} = \frac{1}{n}\sqrt{n^2-\sin^2 r} \qquad (6)$$

将式（6）代入式（4），得：

$$n' = \sin r \cdot \cos\varphi + \sin\varphi\sqrt{n^2-\sin^2 r} \qquad (7)$$

上式中，棱镜的棱角 φ 和折射率 n 均为定值，因此 n' 由 r 决定，从折射计测得角 r，即可确定液体的折射率 n'。

实际上，阿贝折射仪是经过换算校正的，可以从仪器刻度上直接读出待测液体的折射率。通常在测量溶液浓度时，我们可用已知浓度的若干标准溶液在阿贝折射仪上测出其折射率，从而求得该种溶液的折射率（浓度曲线，然后，测出待测溶液的折射率 n_x，再根据此标准曲线求出未知浓度 c_x。

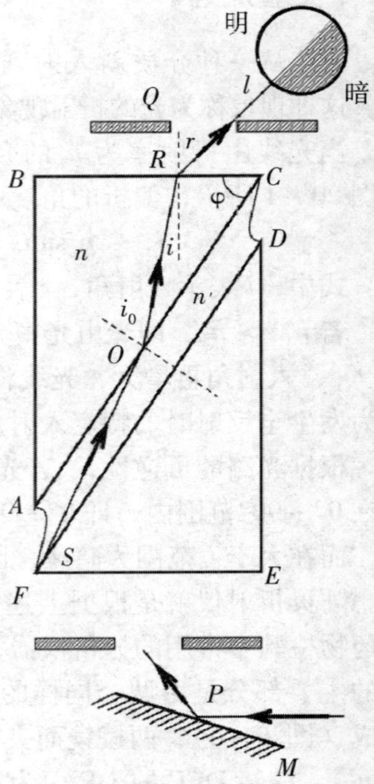

图 20－6　阿贝折射仪原理图

【实验步骤】

如图 20－2 所示，打开反光镜，旋转目镜，使视野中的十字叉线成像清晰，然后合上反射镜。

1. 转动锁紧手轮（10），打开棱镜，用滴管将被测液体加在折射镜表面，合上棱镜并锁紧。打开遮光板观察视场，要求液层无气泡。

2. 旋转手轮（15），在目镜视场中找到明暗分界线的位置，再旋转手轮（6），使分界线不带任何色彩（且不呈某一单色，如红色或蓝色的边线），再微调手轮（15），使分界线处于十字线中心，如图 20－7 所示。此时目镜视场下方显示出的与竖线对齐的刻度值即为该液体的折射率（上方标度为糖的百分含量），将实验数据记录在表 20－1 内。

3. 将棱镜表面用擦镜纸擦干净，重复步骤 2，测定乙醇和给定的 7 种氯化钠标准溶液的折射率，以浓度为横坐标，折射率为纵坐标，用坐标纸画出折射率（浓度关系曲线。

4. 测未知浓度氯化钠溶液的折射率，再由画出的折射率（浓度关系曲线确定未知浓度值。

5. 设折射率（浓度的函数关系为 $y=a+bx$，用最小二乘法由实验数据求出 a、b 值（与由曲线得到的值比较）。并由方程确定未知浓度，求出相关系数。

表 20－1　乙醇和不同浓度氯化钠的折射率表

	乙醇	氯化钠溶液							
		0%	2%	4%	6%	8%	10%	12%	未知浓度
第一次									
第二次									
第三次									
平　均									

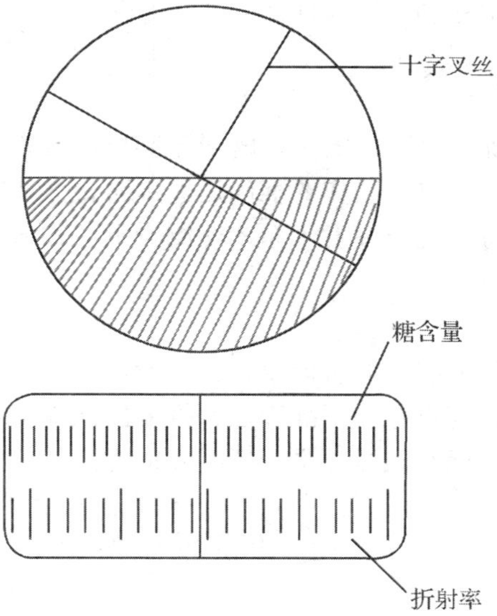

图 20－7　目镜视场示意图

【注意事项】

1. 往棱镜上加待测液体时，不得使滴管与棱镜表面接触。

2. 测完某液体，再测其它液体时，必须将棱镜擦洗干净。棱镜表面只能用擦镜纸擦洗，常用的清洗液有乙醇、乙醚、二甲苯等。糖类和易溶于水的盐类溶液应先用蒸馏水洗擦干净，再用有机溶剂洗涤，擦净并晾干再使用。

3. 测量腐蚀性液体时，应尽量避免被测液体与仪器金属部分接触。

4. 液体的折射率与温度有关，如测量同一种液体在不同温度下的折射率，可将温度计插入孔（13）内，通入恒温水，待温度稳定 10 分钟后方可测量。

5. 实验后，要用清洗液清洗，擦净棱镜，并晾干（15 分钟）后方可合上棱镜。

【思考题】

1. 阿贝折射仪测定折射率的理论依据是什么？
2. 若待测物质的折射率大于折射棱镜的折射率，能否用阿贝折射仪测定？为什么？
3. 阿贝折射仪中的进光棱镜起什么作用？

实　验　指　导

【预习要求】

1. 熟悉阿贝折射仪的结构原理。

2. 简要写出实验步骤。
3. 复习实验数据的图示法和一元线性回归法。

【操作要点】

1. 实验中要掌握好所加待测液体的量，既要足够形成液膜又不能溢出太多。
2. 对乙醇等液体进行测量时，动作要准确迅速以防止液体挥发影响测量。
3. 擦洗棱镜后一定要待擦洗液挥发干净才能测量下一种溶液。

【实验后思考】

1. 结合本实验思考如何测固体的折射率，并请设计具体的实验步骤。
2. 若阿贝折射仪在实验前没有校正准确可否推测出未知溶液浓度。
3. 除了使用前面介绍的标准试样外是否还有其他方法能够对阿贝折射仪进行校准。

20 Determination Index of Refraction of Liquid by Abbe Refractometer

Refraction index is one of the most important optical parameters of transparent materials. From it we can know the information of optical properties, purity of materials, etc. Abbe refractometer is build upon the principle of total internal reflection, and can measure liquid refraction index of transparent or translucent solid and liquid. It is widely used in the area of medicines examining. Through the experiment students should be familiar with the principle and structure of Abbe refractometer, learn the correct operation of it. Measure the index of refraction of several kinds of standard solution of NaCl with the Abbe refractometer. Then confirm liquid concentration with graphic method of experiment data as well as the least square method.

Objectives

1. Familiar with the structure of Abbe refractometer, and learn the correct operation of it.
2. Learn to determine the liquid refraction index with the Abbe refractometer, and confirm liquid concentration (refraction analytical method).
3. Familiar with graphic method of experiment data and the least square method.

Equipment

Abbe refractometer; distilled water; absolute alcohol; NaCl solution of different concentration; dropper; lens paper.

1. The optical system of Abbe refractometer

The optical system of Abbe refractometer basically consists of two parts: telescope system and reading system (as show in figure 20 – 1).

Telescope system: light from reflector (1) enters the incidence prism (2) and refraction prism (3), the liquid to be measured is placed between (2) and (3), the Amici compensator (4) is used to eliminate the dispersion, which is produced by the refraction prism and the object being measured. By the objective lens (5) light and shade boundary will image in the plane of the mirror (6), and enlarged by mirror (6) and eyepiece (7), and finally imaged in the eyes of the observer.

Readings system: light from small mirror (13) passes through ground glass (12), illuminating dial (11), then through prism (10) and objective lens (9), and will image scale in the plane of the field mirror (8), which is enlarged by mirror (8) and eyepiece (7). It is imaged in the eyes of the observer finally.

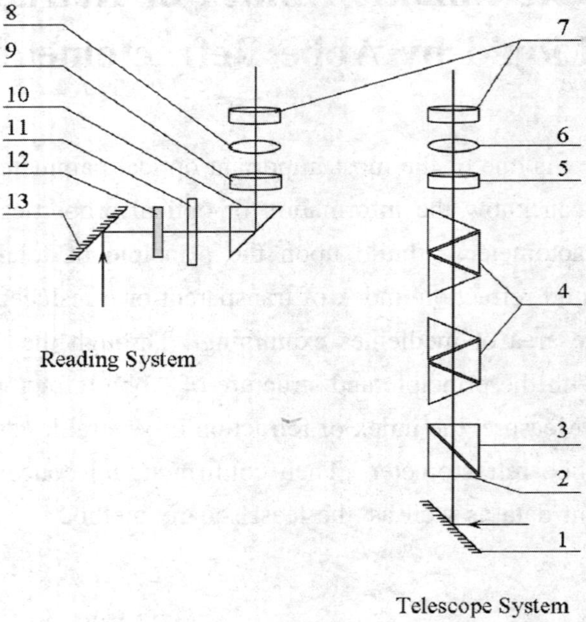

Figure 20 – 1 optical system of Abbe refractometer

1 – mirror; 2, 10 – prism; 3 – refraction prism; 4 – Amici prism; 5, 9 – objective lens; 6, 8 – field mirror; 7 – eyepiece; 12 – groundglass; 13 – small mirror.

2. Mechanical structure of Abbe refractometer

Figure 20 – 2 is the schematic diagram of mechanical structure of Abbe refractometer.

For temperature has influence on the index of refraction, thermostat should be power on to ensure measurement accuracy if necessary.

3. Preparation

Before measuring, reading must be calibrated with standard sample. Add 1 ~ 2 drops naphthalene bromide on the polishing surface of the standard sample, as shown in figure 20 – 3, the polishing surface should be upward in order to receive light. When the indicating in the reading mirror is just on the mark of the standard sample, check - whether the light and shade boundary in the telescope is at the middle of the cross hair. Or, the reading adjustment screw should be adjusted to make boundary appear in the central (as shown in figure 20 – 7). This screw can not be turned any more during the subsequent measurement.

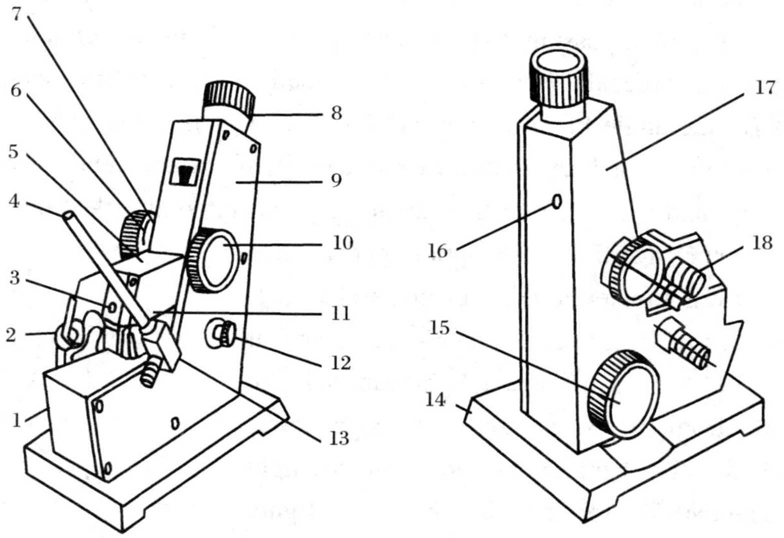

Figure 20 – 2 Mechanical structure of Abbe refractometer

1. mirror; 2. rotation axis; 3. shadow sheet; 4. thermometer; 5. incoming prism seat; 6. dispersion handwheel; 7. dispersion scale loop; 8. eyepiece; 9. cover board; 10. lockup handwheel; 11. refraction prism seat; 12. lighting lens; 13. thermometer seat; 14. supporting seat; 15. refration index scale handwheel; 16. correcting screw hole; 17. housing case; 18. thermostat joint

Principle

When a ray of light is transmitted obliquely through the boundary between two materials of unlike index of refraction, the ray bends, and this phenomenon is called refraction. The way in which a ray refracts at the interface two materials with indices of refraction n_1 and n_2 , as shown in figure 20 – 4, is given by Snell′ law.

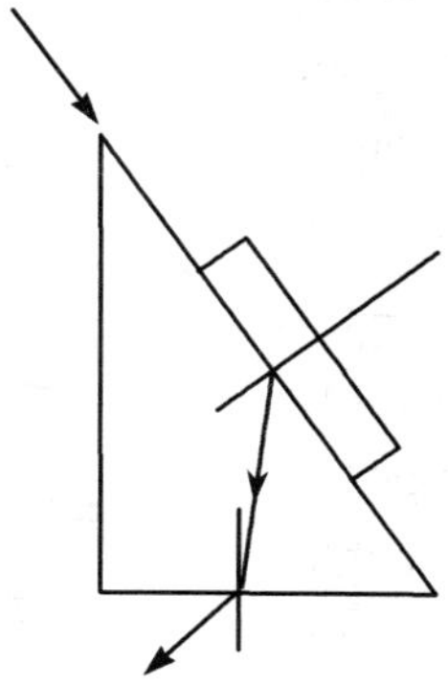

Figure 20 – 3 Correcting standard sample

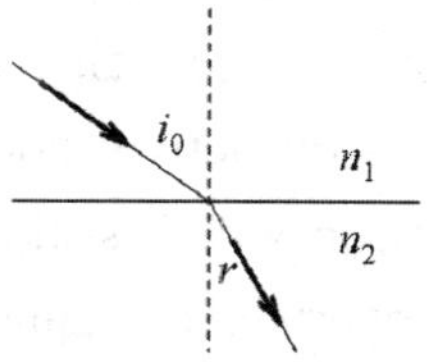

Figure 20 – 4 Snell′ law

$$n_1 \sin i = n_2 \sin r \tag{1}$$

Where, i is the angle of incidence and r is the angle of refraction.

When $n_1 > n_2$, light passing from a material of high index of refraction to lower index must have a r larger than i (according to Snell' law). When increase i to a certain value of i_0, the angle of refraction can be 90. And if angle of incidence is lager than i_0, the entire refracted ray should be reflected. This phenomenon is called total internal reflection, and i_0 is called critical angle for total internal reflection.

Because of the invertibility of optical path, when a ray incident from a material of lower index of refraction to higher index, the incident angle can vary of from 0 to 90, but the refracted light can only be found at angle from 0 to i_0, and the scope corresponding the angle lager than i_0 is dark. Thus the light and shade boundary is formed, as shown in figure 20-5.

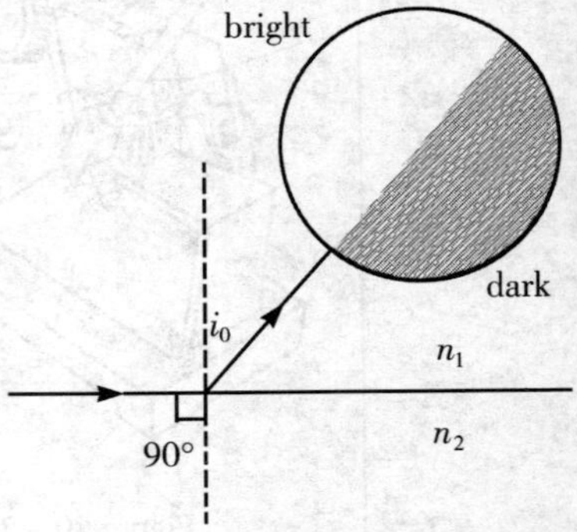

Figure 20-5 Light and shade boundary

Abbe refractometer is build according to the principle of total internal reflection. It is widely applied in the area of medicines examining. Its structure is shown in figure 20-2. The main part of it consists two same right angle prisms ABC and DEF (as shown in figure 20-6). The liquid being measured (refraction index n') is placed between the two prisms' surface. Surface DF of the prism DEF is frosted glass. The light from light source reflected by mirror M enters the prism DEF by passing through hole P, strikes the frosted surface DF, which is illuminated and becomes a radiant surface. The light beam diffuses in different directions because of the frosted surface, therefore the diffused light emitting from DF is incident to the AC surface of the prism ABC through the liquid layer. Because the liquid layer is very thin, there will have light of incident angle about 90°, as the light SO shown in figure 20-6, which is called glancing incidence light. Suppose the prism's refractive index is n, if $n > n'$, the angle of refraction which strikes surface AC is the critical angle i_0 from prism to liquid. The angle of incidence i and the angle of refraction r of the ray OR in the surface BC are both decided by i_0. That is the location of the emerging ray and it is also obtained by the refractive

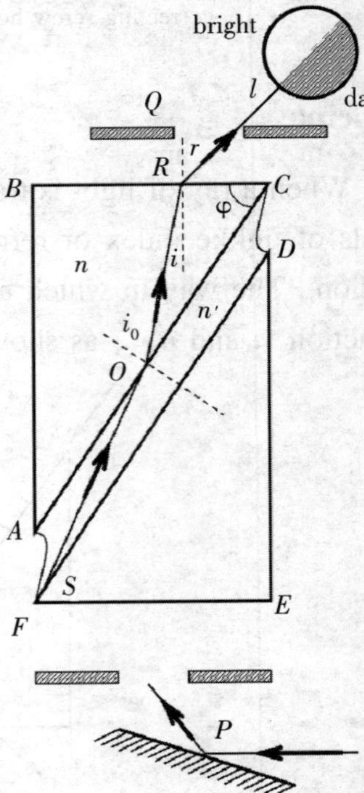

Figure 20-6 Principle of Abbe refractometer

index n' of the measured liquid. Because the angle of incidence of SO is the largest one in the surface AC, all the emergent light will be on the left side of *Rl*. For a monochromatic incident light, a parting line *Rl* or boundary of bright and dark appears. And then we can work out the refractive index n' of liquid by the refraction angle *r*.

As shown in figure 20 –6, suppose $\angle ACB$ is φ, for the triangle *ORC*:

$$i_0 + 90° = i + 90° + \varphi$$

Or:

$$i_0 = i + \varphi \tag{2}$$

For the rays SO and OR, according to Snell' law, we can know:

$$n' \sin 90° = n \sin i_0 \tag{3}$$

And we can get:

$$n' = n \sin i_0 = n \sin (i + \varphi) = n \sin i \cdot \cos\varphi + n \cos i \cdot \sin\varphi \tag{4}$$

To point *R* in the BC interface:

$$n \sin i = n_a \cdot \sin r = \sin r \tag{5}$$

The $n_a = 1$, is the refractivity of air.

$$\sin i = \frac{\sin r}{n}$$

$$\cos i = \sqrt{1 - \sin^2 i} = \frac{1}{n}\sqrt{n^2 - \sin^2 r} \tag{6}$$

Substitute formula (6) into formula (4), we get

$$n' = \sin r \cdot \cos\varphi + \sin\varphi \sqrt{n^2 - \sin^2 r} \tag{7}$$

The value of angle φ of prism and refractive index n are all known, so n' can be calculated according to r, which is measured with refractometer.

Actually, Abbe refractometer has been corrected for conversion, so we can directly read the refractive index of the liquid from the instrument scale. Usually, we can measure the concentration of some certain solution under the help of refractive index – concentration curve: we can find out some standard solution with known concentrations, measure their refractive index with Abbe refractometer first, get a refractive index – concentration curve of this kind of solution after that. Then, find out refractive index of the certain solution; calculate the concentration c_x according to the standard curve.

Experiment

The Abbe refractometer is shown in figure 20 – 2. Open reflector, turn the eyepiece and make the cross hair's imaging clear in field of vision, then close the reflector.

1. Turn the lock wheel (10), open the prism, drop measuring liquid on surface of refraction prism with dropper, close prisms and lock them. Open cover board and observe the viewing field. Make sure there is no air bubbles in liquid layer.

2. Turn hand wheel (15), find the position of light and shade parting line in viewing field of the eyepiece, turn hand wheel (6) again to make the color of parting line eliminated (no chromatic aberration). Adjust handle (15) to align the intersection of the cross hairs with the parting line accurately, just as shown in figure 20 – 7. With the lower scale in viewing field aligning the upright line, the value is the refractive index of the measured solution (the upper scale value is the concentration). Write down the data in table 20 – 1.

3. Wipe up the prism with lens paper, repeat step 2, measure the refractive indexes of distilled water, and seven kinds of standard solution of NaCl. Taking concentration as the horizontal axis, the refractive index as the vertical axis, draw a refractive index – concentration graph in the coordinate paper.

4. Measure the refractive index of the NaCl solution that the concentration is unknown, then confirm the unknown concentration from the refractive index – concentration relationship curve.

5. Suppose the function relation of the refractive index – concentration is $y = a + bx$, and get the value of a and b by least square method (compare with the value gotten by curve). Then work out unknown concentration by the equation. Calculate the related coefficient.

Table 20 – 1 Refractive Index of Alcohol and NaCl Solution of Different Concentration

	Alcohol	NaCl solution							
		0%	2%	4%	6%	8%	10%	12%	Unknown concentration
First time									
Second time									
Third time									
Average									

Points for attention

1. While adding the measured liquid on the prism, do not make dropper touch with the surface of prism, .

2. Scrub the prism before measure another liquid. The surface of prism can be scrubbed by lens paper only. The common cleaning liquid may be alcohol, diethyl ether, the xylene etc. After measuring carbohydrates or solution of the salt the prism should be scrub by stilled water first, then wiped by organic solvent. Then dry in the air.

3. Avoid the liquid touching the metal parts of the instrument as far as possible, while measuring the amyctic liquid.

4. The refractive index of the liquid dependent on temperature. If it is required to measure the refractive index of certain kind of liquid under the different temperature, the thermometer should be inserted into the bore (13) and connect constant temperature water. After ten minutes the measurement can be done.

5. After the experiment, clean the prism. Before close the prism dry it in the air about fifteen minutes.

Problems

1. What is the theoretical principle for determination of the index of refraction with Abbe refractometer?

2. If the index of refraction of a substance is larger than that of prism, can we determine the index of refraction by Abbe refractometer? Why?

3. What is the function of the incidence prism in the Abbe refractometer?

Additional Instruction

Prepare requirements

1. Be familiar with the structure of Abbe refractometer.
2. Write down the experiment steps briefly.
3. Review graphic method of experiment data and the least square method.

Operation key points

1. Pay attention to the used amount of measuring liquid, be sure the liquid layer can form and the liquid would not outflow too much.

2. It should be quickly to measure the absolute alcohol, or the alcohol can volatilize easily.

3. Be sure the cleaning liquid has been volatilized completely before measure.

Problems after experiment

1. According to this experiment, how can we determine the index of refraction of a solid? Try to figure out experimental steps.

2. If the Abbe refractometer isn't calibrated with standard sample before the experiment, can we determine the liquid refraction index?

3. Are there any other methods to calibrate the Abbe Refractometer beside the standard sample?

